OXFORD MEDICAL PUBLICATIONS

DISEASES OF INFECTION

Dose schedules are being continually revised and new side effects recognized. Oxford University Press makes no representation, express or implied, that the drug dosages in this book are correct. For these reasons the reader is strongly urged to consult the drug company's printed instructions before administering any of the drugs recommended in this book.

Diseases of Infection

Second Edition

An Illustrated Textbook

NORMAN R. GRIST
Emeritus Professor of Infectious Diseases, University of Glasgow

DARREL O. HO-YEN
Consultant Microbiologist, Raigmore Hospital, Inverness

ERIC WALKER
Consultant Physician/Epidemiologist, Ruchill Hospital, Glasgow

GLYN R. WILLIAMS
Consultant Physician, Ayrshire Central Hospital

OXFORD NEW YORK TOKYO
OXFORD UNIVERSITY PRESS
1993

Oxford University Press, Walton Street, Oxford OX2 6DP
Oxford New York Toronto
Delhi Bombay Calcutta Madras Karachi
Kuala Lumpar Singapore Hong Kong Tokyo
Nairobi Dar es Salaam Cape Town
Melbourne Auckland Madrid
and associated companies in
Berlin Ibadan

Oxford is a trade mark of Oxford University Press

Published in the United States
by Oxford University Press Inc., New York

First edition 1987

A catalogue record for this book is available from the British Library

Library of Congress Cataloging in Publication Data
Diseases of infection: an illustrated textbook/Norman R. Grist . . .
[et al.].—2nd ed.
(Oxford medical publications)
Includes bibliographical references and index.
1. Communicable diseases. 2. Infection. I. Grist, N. R. (Norman Roy) II. Series.
[DNLM: 1. Communicable Diseases. WC 100 D611]
RC111.D54 1993 616.9—dc20 92-16185
ISBN 0-19-262308-7 (hbk)
ISBN 0-19-262307-9 (pbk)

Photoset by Cotswold Typesetting Ltd, Gloucester
Printed in Hong Kong

Preface to the second edition

In preparing this second edition, we have tried to take into account important and continuing advances in the evolving subject of infection and to improve the presentation in the light of the response to the first edition. For this we have been guided not only by our own experience with the book but also many helpful comments and suggestions by other teachers and colleagues using it in clinical practice. We also made some rearrangements of the text which we hope will improve its convenience, particularly for users in developing countries.

We remain grateful to those whose help in preparing the first edition was acknowledged in the previous preface. From many of them we have received further help and comments, for which we thank them. We also thank Dr S. I. A. L. Mathieson of Raigmore Hospital, Inverness.

Once more we are indebted to Miss Edith Simpson for her dedicated secretarial assistance in revising and redrafting the text. For additional secretarial help we thank Mrs Diane Moran, and also Mr Andrew Millar for producing new and amended figures.

Glasgow
September 1992

N. R. GRIST
D. O. HO-YEN
E. WALKER
G. R. WILLIAMS

Preface to the first edition

This book is intended primarily for medical students throughout the developed world. We hope that it may be useful to postgraduates and others, particularly medical and nursing staff involved with the increasingly complex and often unfamiliar problems of infection in today's changing world. We have not limited consideration to classical 'fevers' and communicable diseases but have also mentioned more recently recognized infections and effects of infection in other branches of medical and nursing practice. Illustrative data reflect the first-hand experience of the authors, being taken mainly from United Kingdom sources.

The pattern of infections constantly changes. Microbes and other potential parasites continually adapt to exploit our changing world and way of life. New and unfamiliar problems repeatedly require management at both individual patient and community levels. Many exotic and 'tropical' infections are not only important in developing countries; because of the increased scale and speed of travel they appear more often nowadays in Britain and other developed countries outside the tropics.

Our material is presented primarily on the basis of major disease syndromes encountered in clinical practice. We hope that this will make the book easier to use. Since most infectious agents can cause different syndromes or combinations of syndromes they often appear in several chapters but with major consideration in one. This arrangement can be somewhat arbitrary in some cases, such as the highly 'multipotent' viruses of the herpes group, but we hope that the cross-references and index will simplify the reader's task. Particularly for the beginner, the book is intended to be read and used as an integrated whole. This is true also of individual chapters in which introductory sections usually provide background context and other matter common to the various diseases described in the chapter. Dosages correspond to the British National Formulary and are not usually specified.

The book reflects our practical, academic, and teaching experience but it is not a mirror of past or present teaching programmes. It should not be seen as a last effort by a retired professor but as a new contribution by three of his younger colleagues, closer to the immediacies of student teaching, of patients, and of the laboratory bench, and who have been involved in the teaching programme of the Department of Infectious Diseases of the University of Glasgow. My own role has primarily been editorial and advisory, co-ordinating the ideas and contributions of my co-authors. We have been able to draw on our continuing experience of this developing subject and have had the benefit of critical comments by senior colleagues in the Departments at Ruchill Hospital and others whom it was impracticable to saddle with additional responsibilities and effort as co-authors. We have also been helped by the availability of illustrations from Departmental sources, from the Communicable Diseases, Scotland, Unit and from former colleagues elsewhere.

For comments and suggestions we are grateful to the following colleagues: Drs B. Datta, D. H. Kennedy, W. C. Love, and I. W. Pinkerton (Clinical Department of Infectious Diseases), Dr J. F. Boyd (Brownlee Laboratory), Dr R. J. Fallon (Department of Laboratory Medicine), Professor M. C. Timbury, and Dr D. Carrington (Regional Virus Laboratory), Drs J. A. N. Emslie, D. Reid, and J. C. M. Sharp (Communicable Diseases, Scotland, Unit), and Dr J. H. Cossar (General Practitioner and Research Assistant), all of Ruchill Hospital, to Dr A. McMillan of the Department of Genitourinary Medicine, Royal Infirmary, Edinburgh and to Professor S. Phillips, Department of Zoology, University of Glasgow.

For illustrative material we are grateful to these colleagues and also to Drs E. J. Bell and Dr E. A. C. Follett of the Regional Virus Laboratory, and Miss M. Riding, M.Sc. of the Scottish Serum Bank, all at Ruchill Hospital, Dr A. P. Ball, Cameron Hospital, Fife, Dr J. M. Blair, Ayrshire and Arran Health Board, Dr P. L. Chiodini, Hospital for Tropical Diseases, London, Dr S. Cameron, Professor R. S. Patrick, and Dr J. R. Donaldson, Glasgow Royal Infirmary, Dr D. M. Denning, Northwick Park Hospital, Dr W. Roberts formerly of Ayrshire Central Hospital, Dr R. A. Sharpe, Ninewells Hospital, Dundee, and also to the Audiovisual Department of Stobhill General Hospital, Glasgow.

We particularly thank Miss E. H. Simpson for typing the drafts and final text and Miss S. McDonald for producing most of the diagrams.

Glasgow N.R.G.
March 1987

Contents

1

Introduction to infections

Man shares his world with many other life forms. We interact with many of them as predator but rarely as prey except to some microorganisms and other parasites. Like other species, our population tends to multiply up to the natural limits set by available habitat and food supply. Unless limited voluntarily, it then becomes subject to the natural checks of famine, habitat degradation and pollution, war (aggressive intraspecies competition)—and infection. Infection, the main form of interspecies competition in which our species is the victim, was until a century ago the major factor limiting the population in the more fertile areas of the world. Even today it is usually responsible for most of the deaths during famine and war. Under normal conditions in developed countries the impact of infection is felt more in terms of morbidity, suffering, temporary incapacity, and lasting disability than in mortality—except perhaps for the marked excess of both mortality and morbidity which is still caused by influenza epidemics.

General relationships between host and parasite

Man's body, immediate environment, and general habitat provide favourable niches for microorganisms. Most of these, such as normal skin flora, have a harmless *commensal* relationship with their host, though they may sometimes provide beneficial competition with would-be invaders. Others, such as some normal gut flora, may be *symbiotic*, acting to the mutual benefit of host and microorganism. *Infection*, as distinct from mere passive contamination, implies active colonization of the host's cells, tissues, or body cavities, to the benefit of the invader.

Infection does not always cause illness. Depending upon whether the infecting microbe causes significant damage or provokes marked defence reactions by the host, infection may be *silent* (inapparent, asymptomatic, subclinical) or *overt* (symptomatic, clinical). Overt infection constitutes a 'disease of infection' (*infectious disease*) which is often, but not always, *communicable* to others. Infecting organisms which cause disease are termed *pathogens*, especially those which do so regularly. Many others are considered *potential pathogens*, opportunists which can exploit defects and breakdowns in the physical, chemical, or immunological defences of the body.

Most infections causing disease are *acute* and transient, being arrested by immunological defence mechanisms which may be reinforced by antimicrobial therapy. Some are *persistent*, generally becoming *chronic* infections held in check to various degrees by immune reactions which fail to eradicate them efficiently—for example, many cases of tuberculosis. There may be continuous or intermittent excretion of the persisting infectious agent by symptomless *carrier* hosts (for example, many typhoid carriers), who may not even have a history of preceding acute illness (most carriers of hepatitis B). Some infections, especially with viruses of the herpes group, may be *latent*, the parasite persisting in non-infectious form for long periods or even for life but often with periodic reactivation and release into the outside world to infect others. Persistent virus infections, with varying degrees of integration of viral genomes with those of the host cells, can impair specialized functions of cells and may explain some neoplastic and other 'diseases of infection' not yet recognized or understood. Unconventional agents ('slow viruses': prions) are transmissible agents causing chronic, progressive diseases such as Creutzfeldt–Jakob disease (Chapter 6) after prolonged delay, usually of years.

ROUTES OF INFECTION

The main portals of entry and exit of infectious agents are shown in Fig. 1.1, with examples in Tables 1.1 and 1.2.

Airborne infection

Microorganisms in respiratory secretions, the mouth or throat can be expelled particularly by coughing, sneezing, or speaking. At short range expelled droplets may impinge directly upon the eye, oro-respiratory mucosae, or skin of a close contact—a form of direct transmission, not truly 'airborne'. More usually they evaporate to leave tiny particles termed *droplet nuclei*. The smallest of these can remain airborne for long periods. Larger particles settle as dust to contaminate the floor, textiles, and other objects in the vicinity: from these they can be re-dispersed as *secondary aerosols* and once more be inhaled or settle on skin or other surfaces. Survival of organisms in droplets, droplet nuclei, and dust depends on their innate stability, the temperature and humidity, and exposure to light and other radiations and to atmospheric chemicals. Many respiratory infections (Chapter 2) stimulate increased secretions, coughing, sneezing, and nose-blowing which favour their spread. Most *exanthems* (febrile systemic infections with characteristic skin rash—Chapter 8) involve the respiratory tract and spread similarly, although some of the viruses can also be liberated from the rash; for instance, chickenpox (and formerly smallpox). Many organisms causing infections of the nervous system (Chapter 6) can also spread from the throat and upper respiratory tract by airborne routes.

Particles of 10 μm or greater in diameter are filtered off by nasal hairs and

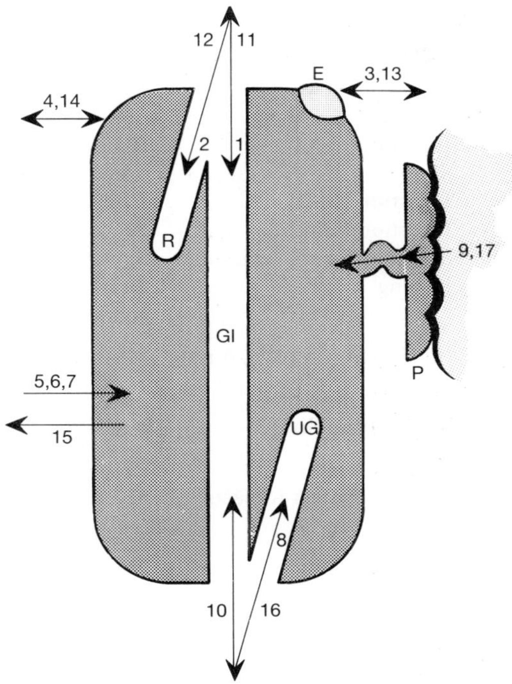

Fig. 1.1 Entry and exit routes for infection. E = eye; GI = gastrointestinal tract; P = placenta; R = respiratory tract; UG = urogenital tract. Numbers refer to Tables 1.1 and 1.2.

turbinates, trapped by the muco-ciliary blanket, and thus removed from the respiratory tract, often by swallowing. Those of 5 μm or less can penetrate deeply and reach the alveoli, from which their removal depends on phagocytosis by alveolar macrophages.

Alimentary infection

Infectious agents in faeces can contaminate the environment, especially under conditions of crowding, poor hygiene and sanitation, and may then be ingested by fresh hosts. They may be transmitted by direct personal contact, especially by contaminated fingers, or indirectly via eating utensils, toilets, and most importantly through food or water in which infection may be amplified by further growth. Numerous viruses, bacteria, protozoa, and other parasites have exploited this faecal–oral route, and many of them enhance their dissemination by provoking diarrhoea (Chapter 4). Several *zoonoses* (natural infections of non-human vertebrates, transmissible to man) can be transmitted by ingestion of infected meat or milk—for instance, salmonellosis, trichinosis, and brucellosis.

Table 1.1. *Portals of entry of infectious agents*

Entry portal*	Mode of entry	Examples
1. Mouth	Ingestion (food)	Bacillary dysentery; salmonellosis
	Ingestion (water)	Typhoid fever; hepatitis A
	Ingestion (milk)	Brucellosis; tuberculosis
	Sucking, licking (objects)	Streptococcal infection
	Kissing	Oral herpes; EBV†; gonococcus
2. Respiratory tract	Inhalation (droplets, dust)	Measles; influenza; Q fever
3. Eye	Inoculation (fingers, instruments, flies)	Epidemic keratoconjunctivitis; trachoma
4. Skin	Direct contact	Herpes; molluscum
	Indirect contact (objects)	Mycosis (athlete's foot); warts
5. Percutaneous	Abrasion	Erysipelas; hepatitis B; tetanus
	Surgical wound	Wound infection; hepatitis B
	Arthropod bite	Malaria; dengue; tick-borne encephalitis
	Animal bite	Rabies
6. Bloodstream	Transfusion	Hepatitis B; cytomegalovirus
	Intravenous drug abuse	Hepatitis B, D, E; HIV†
7. Tissues (direct)	Transplantation	Hepatitis B; cytomegalovirus
	Operative surgery	Tetanus; sepsis
8. Urogenital tract	Catheterization	*E. coli* cystitis/pyelitis
	Sexual intercourse	Syphilis; herpes; *Chlamydia*; HIV†; gonococcus
9. Placenta	Transplacental	Rubella; syphilis; toxoplasmosis

*Numbers refer to Fig. 1.1.
†EBV = Epstein–Barr virus (human herpesvirus 4); HIV = Human immunodeficiency virus.

Contact infection

Infection may be transmitted by *direct contact* from skin to skin, mucosa to mucosa, skin to mucosa, or mucosa to skin of the same or another person. This is important for spreading not only many cutaneous, muco-cutaneous, and sexually transmitted infections and infestations (Chapter 10) but also many alimentary and respiratory infections. Infection may also spread by *indirect*

Table 1.2. *Portals of exit of infectious agents*

Exit portal*	Infectious material	Examples
10. Anus	Faeces	*Shigella*; *Salmonella*; polio virus; hepatitis A virus
11. Mouth	Saliva	Mumps virus; cytomegalovirus
	Secretions	Measles virus; *Streptococcus*
	Sputum	*M. tuberculosis*; *Pneumococcus*
12. Respiratory tract	Secretions	Influenza; RS† virus; *M. pneumoniae*
13. Eyes	Tears; exudates	Adenovirus; herpesvirus; trachoma
14. Body surface	Skin	Tinea pedis
	Hair	Tinea capitis
	Crusts	*Staphylococci* (impetigo)
	Exudates	*Staphylococci* (furunculosis); HIV
15. Skin puncture	Blood	Hepatitis B; arbovirus (via vector); HIV
16. Urogenital tract	Urine	Typhoid
	Secretions/exudates	Gonorrhoea, herpes, syphilis, HIV, neonatal hepatitis B
	Semen	HIV; Marburg virus infection
17. Placenta	Transplacental transmission	Syphilis; rubella; cytomegalovirus; toxoplasma

*Numbers refer to Fig. 1.1.
†RS = respiratory syncytial virus.

contact via water and surfaces ('fomites') as in communal bathing, especially in whirlpool spa baths and hot tubs.

Tissue penetration

Infection can be implanted directly into tissues by penetrating injuries, surgical procedures, and non-sterile injections as in intravenous drug abuse. The common but often inapparent breaches of skin, especially of fingers, and mucous membranes, especially in the mouth, provide access for agents such as hepatitis B virus (Chapter 5) which would not penetrate intact skin. The bites of rabid animals can transmit rabies virus in contaminated saliva (Chapter 10). Bites of mosquito, mite, or tick vectors can transmit many infections such as malaria, typhus, and dengue. Bite wounds or scratches can be infected by the faeces or squashed bodies of fleas (carrying plague) or lice (carrying typhus). Some organisms such as leptospirae or larval cercariae of schistosomes in contaminated waters can penetrate intact skin and mucous membranes. Hookworm larvae from moist ground can enter bare feet.

Transfusion of blood and blood products and *transplantation* of cells, tissues, or organs, unless efficiently screened, provide possible mechanisms for direct transfer of infections from person to person, even from apparently healthy donors in whose cells the infection is latent—examples include hepatitis B, cytomegalovirus, malaria, and the human immunodeficiency virus (HIV).

Congenital transmission

Several infections can be transmitted directly from mother to child either across the placenta before birth (rubella, cytomegalovirus, varicella, parvovirus, HIV, toxoplasma, syphilis) or via the birth canal during childbirth (herpes simplex virus, hepatitis B, group B streptococci). The intimacy of mother and child also facilitates infection after birth by other routes such as via milk during suckling—a potential risk with hepatitis B, HIV, and mumps.

Vertical transmission through the germ line of viruses latent in sperm or ovum is recognized in some leukaemic and other neoplastic diseases of animals, and has been reported for the human T-cell lymphotropic virus type 1, a retrovirus.

DEFENCE MECHANISMS

Anatomical and non-specific defences

The structural and functional integrity of the body provides powerful first-line defences against infection. The skin and mucous membranes form barriers at surfaces exposed to microbes. Microbes which penetrate these meet various antagonistic factors within the tissues and body fluids.

Skin

The continuously desquamating dead surface layers of the skin form an effective barrier to many noxious substances and microbes. Sweat and sebum contain antibacterial substances, and invaders must compete with established normal flora which tolerate these inhibitors.

Mucous membranes

Living surface cells of mucosae are sheltered from direct contact with foreign particles and microorganisms by mucin-containing secretions. These contain inhibitory or lethal substances such as lysozyme in tears, acid in the stomach, bile, and enzymes in the alimentary tract, and inhibitors of attachment of some viruses (such as antineuraminidase in respiratory secretions). Invaders are largely removed by the flushing action of tears and urine, the churning and onward flow of bowel contents, and by efficient ciliary propulsion of the 'muco-ciliary blanket' which clears the respiratory tract of trapped particles. Normal flora compete with invaders of the respiratory and alimentary tracts. Wandering macrophages scavenge, notably in non-ciliated areas such as lung alveoli and the throat.

Tissues and organs

Microbes which succeed in penetrating surface defences meet further structural barriers within organized tissues. These are supplemented by defending phagocytic cells and lymphatic drainage to the reticulo-endothelial system. Local conditions may be unfavourable to the invader—thus essential substrates may be absent or deficient (for example, iron for *Corynebacterium diphtheriae*), oxygen tension may be intolerable for anaerobes, and the temperature may be raised by fever or by local inflammation. Lysozyme and other antimicrobial substances are present in tissues. Even before specific immune responses are activated the inflammatory reaction to tissue damage mobilizes phagocytes, dilutes toxins by fluid exudate and creates fibrin barriers against spread of infection. Fibrin threads assist phagocytes to ingest those bacteria which, by possessing capsules, would otherwise elude phagocytosis. Another aid to phagocytosis is coating of invaders by pre-existing macroglobulin 'natural antibodies' of broad reactivity though low avidity. The interferon response to one virus can increase resistance of uninfected cells to the same or other viruses.

Specific immunological defences

Acquired immunity to substances recognized as 'foreign' (not-self) comes into play quickly through a complex series of interrelated cellular and humoral reactions (Fig. 1.2). The initial cellular reactions lead rapidly to production of antibodies, initially of IgM class, and to release of cytokines such as interferons which stimulate and regulate other cells of the immune system. Specific antibodies may act in various ways; for example, by combination with and neutralization of toxins, inactivation of virus, opsonization of bacteria to facilitate phagocytosis, co-operation with complement to destroy microbes or infected cells which have viral antigen on their surface, and agglutination of microorganisms to facilitate their trapping and destruction.

IgA antibody persists for some time and is secreted into fluids covering mucous membranes to provide local protection. IgG antibody gradually replaces IgM immunoglobulin, persists after recovery, and may provide long-lasting immunity. Cell-mediated immunity clears the battlefield of acute infection, removing both invaders and also those host cells infected by certain viruses (for example, measles) which express viral antigen on the host-cell surface.

Long-lived 'memory cells' confer the ability, for many years after recovery, to respond vigorously and quickly to re-exposure to the same antigens by a rapid

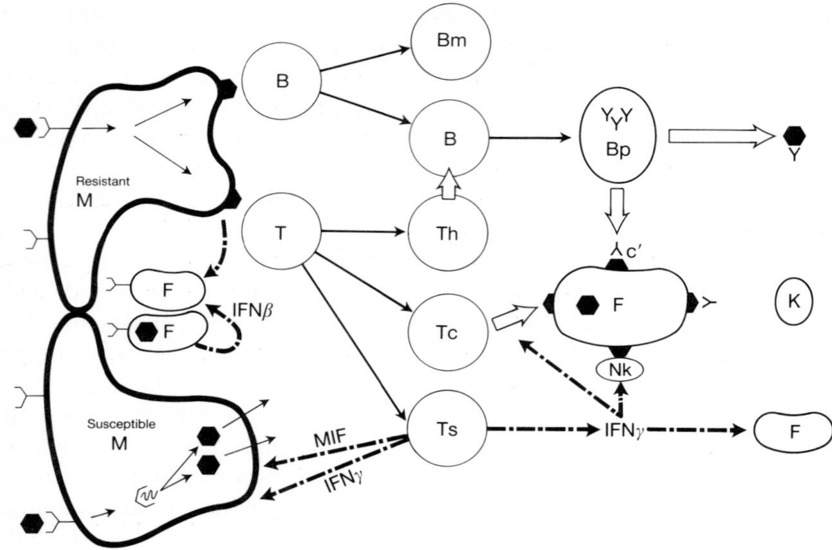

Fig. 1.2 Main immunological host-defence mechanisms against virus infections.

Key:

⬢ = Virus.

M = Macrophage: if susceptible, may enhance viral replication; if resistant, restricts viral growth but can present antigens to stimulate B and T cells and also produce interferon (IFN).

B = B-lymphocyte (bone-marrow derived): after presentation of antigen by macrophage, can proliferate and differentiate into: Bm = memory cell (long-lived) and Bp = plasma cell, secreting antibody (Y).

T = T-lymphocyte (thymus-derived): after presentation of antigen by macrophage, can differentiate into: Th = helper cell (helps B-cells to produce antibody); Tc = cytotoxic cell (destroys fixed cells of body which have viral antigen on surface); Ts = secretory cell producing immunological mediators, e.g. MIF (migration inhibiting factor), IFN.

K = Antibody-dependent cytotoxic cell (recognizes viral antigens via specific antibody).

NK = 'Natural killer' cell (destroys non-specific virus-infected cells).

C′ = Complement (cooperates with antibody to destroy target).

F = Fixed cells of body (when infected, can liberate IFN which increases the resistance of other cells to viral infection).

IFN = Interferon: IFNγ = gamma interferon ('immune interferon'); IFNβ = beta ('type I') interferon.

'secondary response' of IgG immunoglobulin production and by delayed hypersensitivity and other manifestations of cell-mediated immunity.

Immune reactions to infections can often contribute to the disease process and damage cells and tissues during acute and chronic inflammatory processes. Immunopathological mechanisms contribute to many skin rashes such as those of measles and of chronic meningococcal infection. They are also involved in the fulminating haemorrhagic types of infection occasionally seen in chickenpox, meningococcal infection, dengue, and Lassa fever. Complex immunological processes are involved in some chronic infections, for instance tuberculosis and hepatitis B in which the liver is damaged more by the immune response than by the virus itself.

Further information with particular reference to impairments of the immunological system is given in Chapter 12.

HOST–PARASITE INTERACTIONS—INTERACTIONS WITHIN THE INDIVIDUAL HOST: PATHOGENIC PROCESSES

In the battle between invading parasites and the population of host cells organized to repel invaders, the former may fail to establish a foothold, to consolidate or break out. They may be contained locally, with or without severe casualties among defenders with tissue necrosis and pus formation. If not contained, the invaders are likely to be carried by lymphatics to local lymph nodes to be trapped and attacked there by phagocytes of the reticulo-endothelial system and thus also stimulate immunological counteractions. This can be accompanied by lymphangitis and lymphadenitis even to the point of suppuration of involved lymph nodes.

Some viruses and other microorganisms specialize in parasitizing lymphocytes and/or macrophages and, thus sheltered from antibodies in the surrounding fluids, can be carried within these cells into and beyond the lymph flow. Either in this way or by having overwhelmed and broken through the lymph node defences, the invaders can reach the blood circulation and be carried throughout the body. Phagocytes in the liver, spleen, and bone marrow exert a powerful scavenging action to clear these circulating parasites and this phase of the battle is reinforced by the growing humoral and cell-mediated immune response.

Viraemia is usually silent, virus circulating sometimes extracellularly (for example, yellow fever) but more often within mononuclear cells (for example, measles). Bacteria and fungi, if circulating extracellularly and multiplying in the bloodstream, usually produce toxins and severe septicaemic disease (Chapter 8). Mycobacteria, typhoid, and brucella bacteria can circulate within mononuclear cells, as can pyogenic bacteria within polymorphs. Trypanosomes and microfilariae circulate extracellularly, toxoplasmas within mononuclear cells, and malaria plasmodia mainly within red blood cells (Chapter 14).

Toxins

Substances which damage the host are produced by many microorganisms. These include *exotoxins* (specific proteins released by some organisms) and *endotoxins* (lipopolysaccharide–protein complexes in the outer membrane of Gram-negative bacteria).

Exotoxins react with specific receptors and types of cells. Their toxicity to the host may provide no obvious benefit to the microorganism (for example, neurotoxins of *Clostridium botulinum* and *Cl. tetani*). Often they assist the microorganism—thus hyaluronidase and streptokinase help *Streptococcus pyogenes* to spread within tissues. Enterotoxins of some *Escherichia* and *Vibrio* species cause hypersecretion of fluid and electrolytes provoking diarrhoea. This increases dissemination of the organisms into the outside world and increases chances of transmission to other susceptible hosts. Exotoxins are good antigens against which useful antisera and vaccines (for example, tetanus and diphtheria) can often be prepared (Chapter 16).

Endotoxins cause many of the pathogenic effects of infections with Gram-negative bacteria. Their complex effects include stimulation or release of various mediators from host cells, particularly pyrogenic cytokines which cause fever, and also vascular changes with progressive hypotension. This can culminate in 'Gram-negative septic shock', a severe, life-threatening feature of many septicaemias (Chapter 8). In future, specific antibodies to endotoxins may contribute to therapy.

Adaptations of the parasites

The conditions and defence reactions of the host impose severe selection pressure on the population of invaders, just as these do on the host population. Successful parasites evolve methods to elude or subvert the defences. Some, like measles, rubella, cytomegalovirus, and Epstein–Barr virus may even exploit defence systems and cause various degrees of immune depression—most markedly seen with HIV (Chapter 12). Ideally for the parasite, the infected host survives as a continuing or intermittent disseminator of parasites as in typhoid carriage or herpes simplex infection. Long periods of latency alternate with reactivations of herpes virus in nerve-ganglion cells. New-formed herpes virus, protected from extracellular antibody, travels down nerve fibres to infect skin cells which are then destroyed by combined virus action and immune counteraction to viral antigens on cell surfaces. This inflammatory response, together with death of infected cells, produces the familiar 'cold sore' which normally heals rapidly as interferon, antibody, and cell-mediated immunity collaborate to overcome the flare-up. Nerve pathways, both peripheral and central, provide sheltered pathways for the spread of rabies and some other neurotropic viruses, also by leprosy bacilli, and tetanus toxin.

HOST–PARASITE INTERACTIONS—INTERACTIONS WITH THE HOST POPULATION: EPIDEMIOLOGY

Infectious agents exert profound effects and selection pressure on the human population as a whole as well as on its individual members, whose experience of infection is conditioned by their context in the community. The subject of epidemiology, indeed, originally developed from the study of epidemics of communicable diseases, although it now extends to cover all health-related states and events in populations. Its important practical objective is the control of health problems. For infections, this requires study not only of human populations in their environment but also of the distribution and behaviour of the parasite population in the shared environment, of the reservoirs and possible vectors of infection and their interrelationships. In effect, it is a branch of ecology.

Sources of infections

Most infections originate from exogenous sources—human, animal, or sometimes environmental (Table 1.3). When normal resistance to infection is lowered by malnutrition, intercurrent disease, immune depression, trauma, surgical intervention, or certain medical procedures, '*opportunist infection*' may arise from invasion by normally harmless '*potential pathogens*' already in or on the patient (endogenous), or in the environment (such as *Pseudomonas* and other Gram-negative bacteria which may contaminate sinks, wash-cloths, foods, or medicaments). Latent infections such as herpes zoster may also be reactivated. Endogenous and opportunist infections are increasingly important in industrialized countries where technically advanced medical (for example, immunosuppression) and surgical (for example, transplantation) procedures are practised. Infections arising from medical or surgical procedures are termed '*iatrogenic*'; those which result from being in hospital as '*nosocomial*'.

Occurrence of infections

In addition to the inherent virulence and infectivity of the parasite, and the genetic characteristics of the host population and its constituent individuals, the occurrence of infections is determined by factors such as the following:

(a) *The proportion of the population exposed to infection which is not susceptible.* This '*herd immunity*', mainly reflects the cumulated specific immunity acquired from previous infection or immunization. Thus with many common and highly communicable infections, most adults are immune from earlier infection in

Table 1.3. *Exogenous sources of infection*

Reservoir–source	Examples
Human: Case	Measles Streptococcal tonsillitis Bacillary dysentery
Convalescent excreter	Typhoid fever Hepatitis B Diphtheria
Symptomless carrier	Typhoid fever Infectious mononucleosis Hepatitis B
Animal: Case	Rabies *Campylobacter enteritis* Psittacosis
Symptomless carrier	Salmonellosis Lymphocytic choriomeningitis Leptospirosis
Arthropod vector	Malaria Dengue fever Filariasis
Environment: Soil	Tetanus Gas gangrene Botulism
Water	Leptospirosis *Pseudomonas* (opportunist infections) Hepatitis A (water as vehicle)
Food (contaminated)	Salmonellosis Brucellosis Amoebic dysentery
Airborne	Legionellosis Histoplasmosis Q fever

childhood, and the disease is then mainly found in children; for example, measles and chickenpox (Fig. 1.3).

(b) *The ease of transmission of the infectious agent.* This reflects the extent, duration, and method of its dissemination from human or animal sources, the degree of exposure to environmental sources of infection, the ability of the

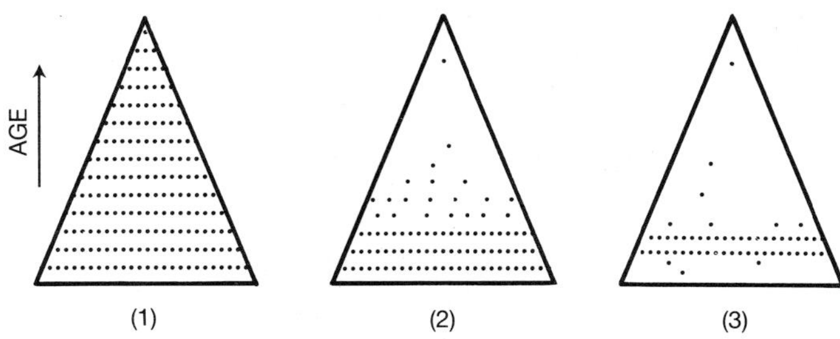

Fig. 1.3 Patterns of infection in populations.
(1) Age distribution of an infection new to the population: no immunity—epidemic affects all age groups. *Example:* Measles in Fiji, 1875. (2) Age distribution of an endemic infection: immunity increases with age—infection mainly confined to young children. *Example:* Respiratory syncytial virus infection. (3) Age distribution of an infection endemic in a community where high standards of hygiene retard transmission: immunity incomplete, mainly in older persons—infection mainly affects children when brought together at school age. *Example:* hepatitis A in Britain.

microbe to survive during transmission, and the infecting dose necessary to initiate a new infection.

(c) *The size, density, location, age, sex, socio-economic status, and pattern of behaviour of the population and its constituent members.* These determine the frequency and effectiveness of contacts between infectious and susceptible persons or of susceptibles with non-human reservoirs and vectors. Thus measles dies out from small, isolated communities when most persons have been infected and become immune so that insufficient susceptibles remain to maintain the chain of transmission. By contrast, measles continues to circulate in large populations and cities where enough new, susceptible children are continually added by birth or immigration, unless the chain is broken by intensive use of vaccine. Measles mainly affects children in these large, inadequately immunized populations. If measles has been absent from an isolated community for many years, unless vaccination has been practised, a pool of non-immunes will have accumulated and re-importation of measles can cause a large epidemic affecting most or all age-groups as, for instance, happened in the Shetland Islands in 1977–78 and the Orkneys in 1980.

(d) *The season of the year.* This often has a profound influence on the timing of epidemics, often by its effects on population behaviour—for example, many respiratory infections are most prevalent in winter when people spend more time together indoors, and common colds break out particularly after schools re-open after the summer when susceptible children spend more time crowded in classrooms. This also affects transmission by vectors such as mosquitos; for example, malaria and arbovirus infections are commoner during and after the rainy season in the tropics.

Distribution of infections

Some infections are *endemic*, i.e. constantly present in the population, although the number of cases may fluctuate with periodic epidemics (for example, measles in mainland Britain). An *epidemic* represents a number of cases in excess of that expected for the population concerned, based on previous experience—that is, an abnormally high incidence of infection. (*Incidence* is the number of new cases occurring in unit time, usually specified as the *incidence rate*, i.e. the ratio of new cases per week, month, year etc. to the population at risk; for example, 40 cases per 100 000 per week.) An *outbreak* is the common term used for a small, usually localized epidemic. A *pandemic* is a world-wide epidemic such as is periodically caused by a new variant of influenza virus.

The term 'incidence' should be clearly distinguished from *prevalence*, a static concept which refers to the total number (or rate) of persons in a population at a particular time (or defined period) with a particular disease or other condition. Thus an incidence of 34 new cases of tuberculosis per 100 000 persons per year can coexist with a prevalence of 1700 cases of tuberculous infection, old and new, per 100 000 in the same population. In Ghana where most unvaccinated children acquire poliovirus infection, an annual incidence of paralytic poliomyelitis of 28 per 100 000 (mostly children) was found to çoexist with a prevalence of lameness attributable to past poliomyelitis of 700 per 100 000 schoolchildren.

An epidemic may last for a year or more, more often for some weeks or months. The epidemic curve can give a clue to the nature and cause of the outbreak, rising and falling abruptly in a *point source outbreak* where cases result from a single exposure to a common source of infection such as infected food at a banquet (Fig. 1.4: A)—all cases then arise within the *incubation period* (interval between acquiring infection and developing first symptoms—Table 1.4) of the infection following the exposure. If these patients are infectious to others, as with salmonellosis, the pattern may be complicated by secondary cases arising from person to person spread with the group (Fig. 1.5). When spread is essentially from person to person as with measles, the epidemic curve is more protracted (Fig. 1.4: B). It is useful to know the different incubation periods of infections when studying an outbreak, trying to establish its cause and assessing the outlook for others who shared the exposure or are contacts of the primary group of cases. The incubation period should not be confused with the shorter *prodromal period* which is the interval between appearance of first symptoms and characteristic illness such as the typical skin rash of an *exanthem* (the characteristic skin rash of measles, scarlet fever, and other infections with fever and eruption—Chapter 7). An *enanthem* is an equivalent eruption on mucous membranes, usually seen in the mouth.

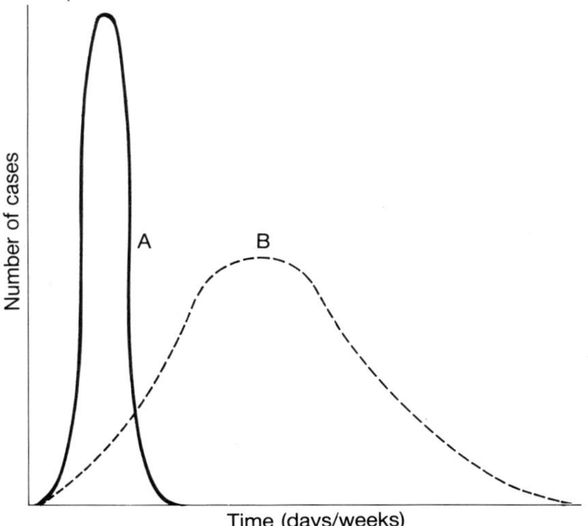

Fig. 1.4 Epidemic curves: A = abrupt rise, peak and fall of cases in outbreak from common point source; B = gradual rise and prolonged epidemic with slow decline due to case-to-case transmission of infection until insufficient susceptibles remain to sustain epidemic.

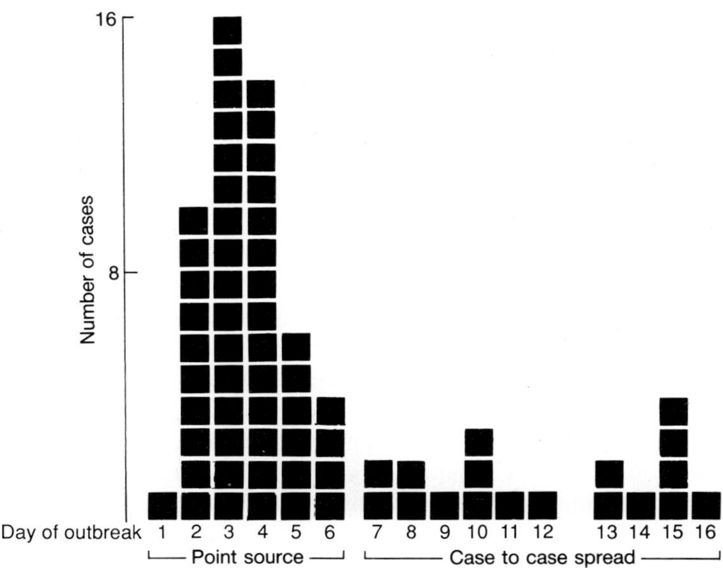

Fig. 1.5 Epidemic curve of an outbreak of infectious diarrhoea in a Nursing Home.

Table 1.4. *Some common incubation periods of infectious diseases*

Incubation period	Disease
Less than 1 week:	Anthrax
	Bacillary dysentery
	Campylobacter enteritis
	Common cold
	Diphtheria
	Infant gastroenteritis
	Viral gastroenteritis
	Hand, foot, and mouth disease
	Herpangina
	Influenza
	Legionnaires' disease
	Meningococcal disease
	Scarlet fever
1–2 weeks:	Measles
	Poliomyelitis
	Psittacosis
	Typhoid fever
	Typhus fever
	Whooping cough
2–3 weeks:	Chickenpox
	Mumps
	Q fever
	Rubella
2–6 weeks:	Hepatitis A
2–6 months:	Hepatitis B
	HIV
1–12+months:	Rabies

PROSPECTS FOR DISEASES OF INFECTION

Many major, classical and epidemic infections have declined during this century in response to improved living conditions, better hygiene and sanitation, and to some extent in response to curative and preventive medical measures. However, natural selection and the adaptation of infectious agents to exploit new opportunities provided by the changing modern world have maintained the challenge of infection. A recent review of the situation in Europe published in 1984 by the World Health Organization listed 21 infections as declining or disappearing, 12 as static, 6 as uncertain, 15 increasing and 17 more as recently recognized. Because of HIV infection, not only is AIDS increasing world-wide but various related infections, some previously rare such as pneumocystis pneumonia and severe cryptosporidial diarrhoea, are no longer uncommon; others are even resurgent, such as tuberculosis, previously declining but first noted to be increasing in the USA (Fig. 1.6) and now rampant through much of Africa. Other infections noted as increasing in Europe included salmonellosis and other food-borne infections, sexually transmitted diseases, hepatitis B, streptococcal infections, pneumococcal pneumonia, toxoplasmosis, malaria and other imported infections, pediculosis, and scabies. Other infections recently recognized include campylobacter and yersinnia infections, viral diarrhoeas, 'non-A non-B' hepatitis (hepatitis C, D, E), Lyme disease (borreliosis), hantavirus infections, viral haemorrhagic fevers, infection by low-grade opportunists in hospitals, by cytomegalovirus, mycoplasmas, chlamydias, helicobacters, human herpesvirus 6, and retroviruses (HIV1, HIV2, HTLV1, HTLV2).

Factors responsible for these changes include alterations in life-style (including increased sexual promiscuity, the drug cult, and foreign travel), altered and intensive methods of food production and catering, and various high-technology advances in industry and medicine, all providing new openings for transmission of infectious agents. Meanwhile, urban decay and the deterioration of hygiene and sanitation services are recreating in many developed countries the conditions in which infections flourished in earlier years, similar to the even worse conditions which have developed in the burgeoning urban concentrations of many developing and tropical countries. Even the decline in some infections in response to improved conditions may be a mixed blessing when, as with poliomyelitis (unless vaccination is widely used) and hepatitis A, postponement of the average age of becoming infected and incomplete immunization coverage

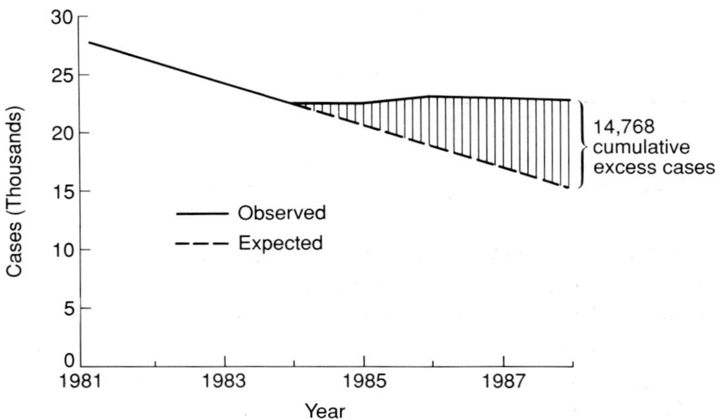

Fig. 1.6 Observed and tuberculosis cases in the USA, 1981–88: the AIDS era. (Based on WHO *Epidemiological Reports*, 1990, **39**, 153–5.)

build up a growing population of non-immune adults. These may then sustain periodic epidemics, often with illnesses more severe than with infection in earlier childhood. Also, the narrowed range of immunity then possessed by women of child-bearing age means that they and their unborn or newborn children become more susceptible to infections with such agents as rubella (unless vaccine has been utilized), cytomegalovirus, and enteroviruses (Chapter 11).

PROSPECTS FOR CONTROL OF INFECTIONS

Much depends upon effective measures to interrupt the chain of transmission of infectious agents by personal and community hygiene and sanitation, by techniques to contain sources of infection and minimize exposure to these, and by avoiding hazardous environments and promiscuous, intimate contact with other persons. Clearly these precautions will be imperfect.

Specific prevention of selected infections by active and occasionally by passive immunization can be effective. Good progress is being made to suppress major infections of children in the world by the Expanded Programme of Immunization of the World Health Organization. This approach should benefit greatly from application of the new methods of biotechnology, monoclonal antibody techniques and the genetic manipulation of infecting agents to provide improved vaccines. However, many persons, especially in developed countries, are disinclined to accept and make use of available vaccines and health education. Others, by contrast, demand more health promotion for example to avoid toxoplasma and cytomegalovirus infection in pregnancy.

During the present decade immunization control programmes are threatened by the AIDS epidemic. This is estimated currently (1992) to infect about 10 million adults and is expected to infect 30–40 million persons by the year 2000, a third of them children of whom most will have died from AIDS.

Production of new chemotherapeutic and antibiotic compounds is a race to keep ahead of the development and spread of microbial resistance. Several antiviral drugs are now available and others will follow. Here also the application of biochemical and modern biotechnological methods should help greatly to define targets more accurately and provide the most appropriate 'magic bullets', interferons, and other modulators of immunity.

Where serious infections become infrequent, it will become more and more important to determine just why any particularly severe case of infectious disease is so severe, in terms of host characteristics. Research will be required to identify which individual genetic and immunological factors are involved, and to find how to modulate the patient's response to infection and compensate for deficiencies in ability to handle infection. HIV infection already provides one important model for the study of these problems.

IMPLICATIONS FOR MEDICAL PRACTICE

Every doctor's work is affected by problems of infection—irrespective of his specialty or otherwise, whether qualified, or in training. Infections are common in all countries—even in developed countries they cause most illnesses of childhood and most of the new illnesses encountered in general practice. While these diseases are often mild and transient, they include many which are serious, even life-threatening, or which cause acute or lasting disability. Infections also cause a high proportion of the complications, failures, and deaths in modern high-technology medicine and surgery. More and more of every doctor's patients will have been infected by HIV, raising difficult ethical issues in relation to testing, confidentiality, and the relative rights of patients and the community.

Unlike most chronic and degenerative diseases of developed countries, many infections can be cured if accurately diagnosed, and many can be prevented by measures of personal or community hygiene and sanitation or by specific immunoprophylaxis which offers the most efficient prevention in the field of medicine (Chapter 16).

Management of infections requires an integrated understanding of both host and parasite at individual and population levels. In addition to clinical expertise, one needs working familiarity with microbiology, pathology, immunology, therapeutics, epidemiology, and community medicine. Although the infection specialist normally practises in just one of these disciplines, conditions particularly in many developing countries require a broader approach and ability to collaborate with colleagues in public health services, veterinarians, parasitologists, entomologists, ecologists, and others.

Every doctor and microbiologist who troubles to notify an infection, formally or otherwise, is contributing his mite to the body of information which provides the data for 'epidemiological surveillance' (Chapter 17) and enables appropriate control measures to be decided upon, initiated, sustained, modified, or abandoned. This activity is organized at national and international levels since infections recognize no boundaries.

2

Infections of the respiratory tract, middle ear, and eyes

Respiratory infections are major causes of illness and death throughout the world. They are the commonest illnesses in industrialized countries including Britain, the commonest cause of consultation with family practitioners and absence from work, and make a major contribution to hospital admissions. All ages are affected, with maximal impact on young children, who mix together, often for the first time in play groups or at school, and commonly suffer six or more colds per annum. Mortality is considerable in infancy, minimal in later childhood and early adult life, rising progressively thereafter to a peak in old age. Pneumonia remains one of the leading causes of death, causing respiratory failure, septicaemia, heart failure, or cerebral ischaemia. This chapter also includes infections of the middle ear and eyes which are connected through the eustachian tube and nasolacrimal duct to the respiratory tract and are affected by similar organisms, often as part of the same illness.

CHARACTERISTICS OF RESPIRATORY TRACT INFECTIONS

Aetiology

Most (over 90 per cent) acute respiratory infections are caused by viruses or occasionally other non-bacterial agents for which treatment, other than symptomatic and supportive, is usually unhelpful. The most severe and potentially lethal infections are due to bacteria, either as primary infections or as secondary invaders, for which antimicrobial therapy is appropriate. Tables 2.1 and 2.2 list the commonest agents and their usual clinical effects.

Epidemiology

Against a constant background of respiratory infections by numerous viruses and other agents there are marked seasonal and annual fluctuations in incidence (Fig. 2.1). In both tropical and temperate regions they are commonest during the cooler months. The seasonal incidence in most temperate climates is dominated by influenza which causes epidemics of lower respiratory tract disease and deaths,

Table 2.1. *Typical features of virus infections of the respiratory tract (including ear and conjunctiva)**

	Coryza	Conjunctivitis	Sore throat	Otitis	Croup Laryngitis Tracheitis	Bronchitis	Pneumonia	Bronchiolitis
Influenza viruses	+	+	+	+	++	+	+	
Parainfluenza viruses	+	+		+	++	+	+	+
Respiratory syncytial virus	+	++		++	+	+	+	+++
Measles virus	+++	+++		++	+	++	++†	
Rhinoviruses	+++	+		+		+		
Coronaviruses	++							
Echo and coxsackie viruses	+	+	+					
Adenoviruses	+	++	++		+	+	+	
Rotavirus	+							
Varicella-zoster virus	+	+ (due to vesicles)					++†	
Herpes simplex		+ (keratitis)						
Epstein–Barr virus			++					

*Presentations: +++ = characteristic; ++ = common; + = occasional.
†In immunocompromised patients.

Table 2.2. *Typical features of bacterial and other infections of the respiratory tract (including ear and conjunctiva)**

	Pharyngitis	Conjunctivitis	Sinusitis	Otitis	Epiglottitis laryngitis	Bronchitis	Pneumonia
Streptococcus pyogenes	++	+	++	++			+
Corynebacterium diphtheriae	++				+		
Haemophilus influenzae	+	++	++	++	++	++	+
Streptococcus pneumoniae		++	++	++		++	++
Staphylococcus aureus		+	+	+			++ (post-influenza or intravenous drug misuse)
Legionella pneumophila							++
Mycobacterium tuberculosis					+		++
Mycoplasma pneumoniae	+						++
Klebsiella species							+
Branhamella catarrhalis			+	+		+	+
Anaerobic organisms (e.g. bacteroides)			++	+			+
Chlamydia trachomatis		++ (neonates)					+ (infants)
Chlamydia psittaci							++
Chlamydia pneumoniae (TWAR)							++
Coxiella burnetii							++

*Presentations: ++ = common; + = occasional.

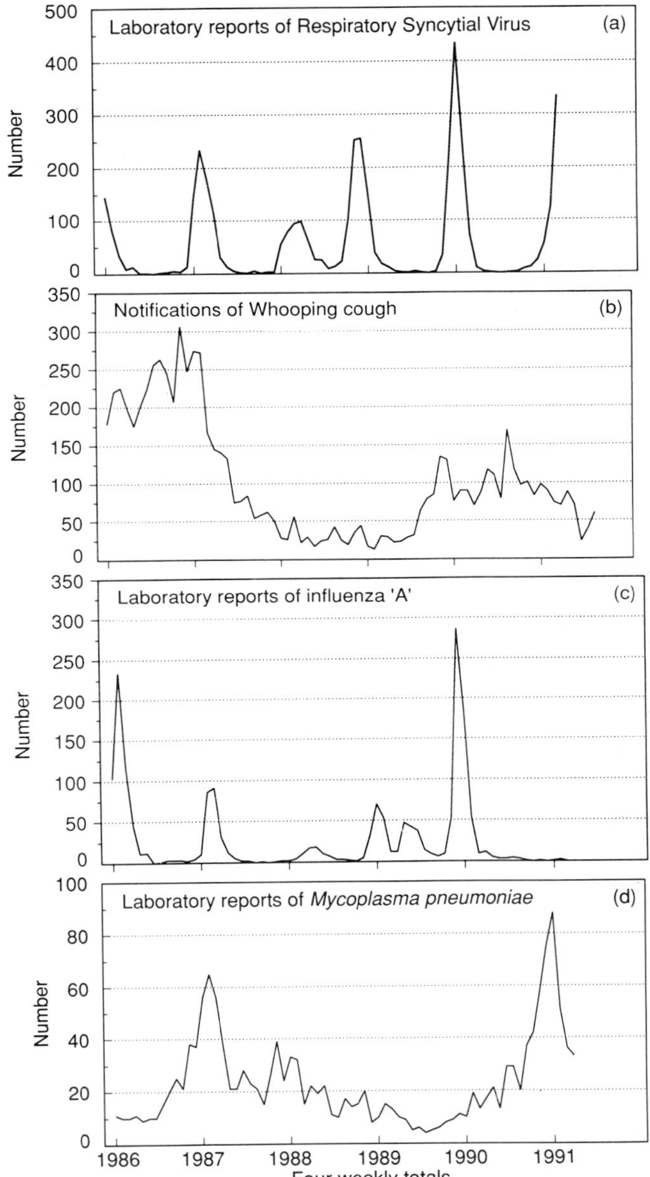

Fig. 2.1 Incidence patterns (Scottish reports) of some common infections of the respiratory tract: **(a)** annual winter epidemics of respiratory syncytial virus infection; **(b)** four-yearly epidemics of pertussis; **(c)** annual winter epidemics of influenza A; **(d)** endemic *Mycoplasma pneumoniae* infection with four/five-yearly epidemics.

especially in the elderly, and by respiratory syncytial virus which annually causes epidemics of bronchiolitis in young children. Pertussis causes epidemics every third or fourth year as does *Mycoplasma pneumoniae*. At intervals of usually more than 10 years influenza causes major epidemics or pandemics.

The main sources of these infections are other infected human beings, especially when coughing and sneezing or crowded in schools, hospital wards, and work-places, or spending more time indoors during the winter. In addition to airborne infection, fingers readily transfer organisms to the eye and nose, and contagion from contaminated surfaces helps to spread agents such as rhinoviruses. Animals are sources of infections, such as ornithosis/psittacosis from ducks, pigeons, and budgerigars, or Q fever from sheep and cattle. Environmental sources may be created by modern technology as when mechanical ventilation systems disseminate organisms causing legionellosis which flourish in warm wet conditions such as in air-conditioning cooling plants (Fig. 2.2). Humidifiers and mechanical respirators readily become contaminated with Gram-negative bacteria which grow within them and infect patients.

Acquired post-infection immunity has limited efficiency in preventing superficial reinfection of the respiratory mucous membranes but usually prevents deeper invasion and significant symptoms. A widening range of immunity to disease caused by numerous agents is built up during life—but ageing results in a higher proportion of infections causing serious disease. Immunological sensitization can modify response to reinfection—thus the first infection by influenza virus usually causes only a feverish cold, but subsequent infections by related strains of the same type can cause classical influenza.

Fig. 2.2 Infected moisture (simulated with smoke) from hospital cooling system when sucked into air conditioning vents has caused outbreaks of legionellosis.

Age-related anatomical factors also modify the response to infection. The narrower airways of young children may be obstructed by infections with viruses which cause only hoarseness and cough in adults.

Smoking damages the respiratory tract and inhibits ciliary action, thus increasing the severity and frequency of respiratory infections—even in 'passive smokers' such as the young children of parents who smoke. Impaired mucociliary clearance also enhances the probability of secondary bacterial infections after damage by influenza or other respiratory viruses. Commensal organisms already at the site are usually involved; for example, staphylococci, pneumococci, or *Haemophilus influenzae*. These may also cause disease when clearance of airways has been obstructed (for example, by tumour growth or inhalation of a foreign body) leading to local pneumonia or lung abscess, or in shallow-breathing elderly or postoperative patients who cough inadequately.

Immunosuppression by disease, drugs, or radio-therapy can lead to unusual and severe infections such as ulcerative pharyngitis or pneumonia due to varicella, measles, or *Pneumocystis carinii*.

Clinical features

Tables 2.1 and 2.2 show the usual types of respiratory disease syndromes. The pattern of illness in a patient usually changes as the disease evolves—for example, from an initial coryza to nasopharyngitis, possibly with fever, which may progress to hoarseness; or a febrile cough and tracheitis may progress to pneumonia. Features which indicate that the use of antibiotics or admission to hospital should be considered include fever, rapid or obstructed breathing, cyanosis, tachycardia, pleuritic pain, and sputum which has become purulent or blood-stained.

Investigations

Chapter 19 describes laboratory investigations for respiratory and other infections.

Microbiology (Table 2.3)

Identification of the infecting organisms can be attempted by (a) *culture* of respiratory secretions or exudates, of laryngeal swabs or gastric washings (for *Mycobacterium tuberculosis*), or of blood (important in pneumococcal and some other bacterial pneumonias); (b) *immunofluorescence* staining of nasopharyngeal aspirates for rapid detection of some microorganisms (Fig. 2.3); and (c) *serological tests* for antibody responses to viruses, chlamydias, *Mycoplasma pneumoniae*, legionellas, and some other bacteria. The results can confirm a provisional diagnosis, guide the use of antimicrobial therapy or give warning of an outbreak requiring specific local or community control measures (Chapter 17).

Table 2.3. *Investigations for infections of the respiratory tract**

	Culture	Immunofluorescence	Serology
Viruses			
Influenza	+ (respiratory secretions or throat swab)	+	+
Parainfluenza	+ (respiratory secretions or throat swab)	+	+
Respiratory syncytial virus	+ (respiratory secretions or throat swab)	+	+
Measles	— (respiratory secretions or throat swab)	+	+
Rhinovirus	+ (respiratory secretions or throat swab)	—	—
Coronavirus	— (respiratory secretions or throat swab)	—	—
Adenovirus	+ (respiratory secretions or throat swab)	+	+
Coxsackievirus	+ (faeces and throat swab)	—	+
Echovirus	+ (faeces and throat swab)	—	—
Epstein–Barr virus	—	—	+
Herpes simplex	+	+	+
Bacteria, etc.			
Strep. pneumoniae	+ (sputum or blood)	—	—
Staph. aureus	+ (sputum or blood)	—	—
L. pneumophila	—	+	+
Mycoplasma pneumoniae	—	—	+
Chlamydia	+	+	+
Cox. burnetii	—	—	+
Mycobacteria	+	—	—

* + = useful; — = not often useful; direct microscopy of stained sputum useful in tuberculosis and staphylococcal pneumonia.

Radiology

Changes in radiolucency of lungs or sinuses can aid diagnosis, monitor recovery and perhaps unmask underlying lung pathology such as tumour. Computerized axial tomography (CAT) scans can provide further definition of pulmonary, sinus, or mastoid lesions.

Other investigations

The blood film and differential white cell count may help to distinguish a bacterial (usually polymorphonuclear leucocytosis) from viral infection (often lymphocytosis). Biochemical tests may reveal disturbance of non-pulmonary organs (helpful in legionellosis and some 'atypical pneumonias'—see below). In respiratory failure, monitoring of blood gases can give warning of the need for assisted respiration.

Specific treatment

Table 2.4 summarizes the antimicrobial drugs used in upper respiratory tract

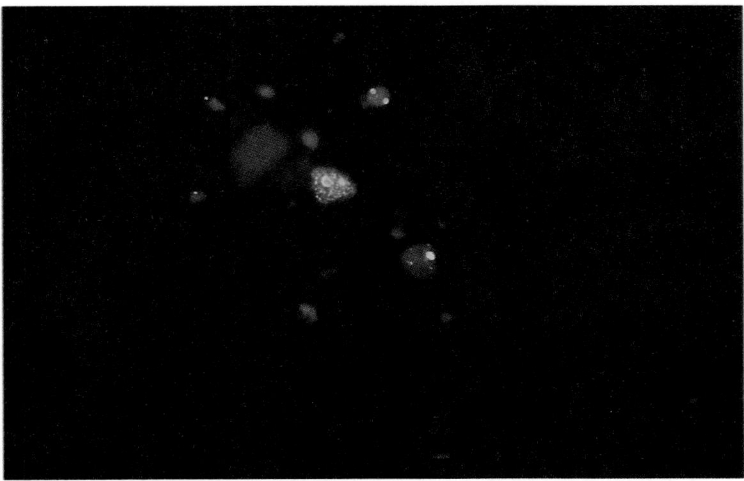

Fig. 2.3 Respiratory syncytial virus in nasopharyngeal aspirates (fluorescent antibody technique).

infections and Table 2.6 those used in pneumonia. General management is discussed below under the separate diseases and tuberculosis is covered separately in Chapter 9.

Prevention and control

The multiplicity of agents, their antigenic variations, the limited efficacy of immunity against superficial mucosal infections, and the difficulty of controlling spread by airborne routes limit severely our ability to prevent these infections. In the long term, improvements in housing, nutrition, reduction of overcrowding, and better working conditions may have more impact than specific medical measures.

Immunization. Vaccines are effective to various degrees for protection against diphtheria, pertussis, measles, influenza, pneumococcal infection, and tuberculosis. Vaccine has been introduced for *H. influenzae* and may be for several other agents. Immunization is discussed in Chapter 16.

Reducing exposure to infection. Containment of infection within controlled environments with sterile-filtered ventilation is an expensive facility available for a few circumstances of special danger. It is also appropriate for 'protective isolation' of vulnerable patients. The ordinary fabric face-mask reduces the amount of potentially infectious aerosol reaching the environment from the wearer but does not significantly protect the user. At least one should try to avoid exposing infants to family or other contacts with streaming colds or coughs; debilitated patients should not, if possible, remain in an open ward with a case of influenza; a patient with Hodgkin's disease or leukaemia should not occupy the next bed to one with shingles or chickenpox; and spread of infectious secretions can be minimized by simple measures such as the use of disposable handkerchiefs, hygienic disposal of sputum, good ventilation, and hand washing.

Other measures. Iatrogenic immunosuppression can render a patient vulnerable to severe respiratory infection. In such circumstances, prophylactic antibiotics have little value except in AIDS where inhaled pentamidine or oral cotrimoxazole can reduce *P. carinii* infection, and when penicillin or erythromycin is given after splenectomy to prevent overwhelming pneumococcal infection.

Prophylactic physiotherapy is helpful for the bed-bound, comatose, or postoperative patient.

UPPER RESPIRATORY TRACT SYNDROMES

This section deals with illnesses which mainly affect areas from the larynx upwards.

The common cold (coryza) and related infections of sinuses and ear

The familiar common cold is normally brief and mild. Affecting the nasal passages and pharynx, it is highly prevalent and causes much inconvenience, interruption of work, and consumption of medicaments. Catarrhal blockage of normal drainage and forceful nose-blowing can encourage secondary bacterial infection of paranasal sinuses and middle ear. Rhinoviruses, the commonest cause, can affect the lower respiratory tract in infants and those with 'chronic bronchitis' (see below).

Aetiology

Rhinoviruses, viruses of the picornavirus, myxovirus, adenovirus, and paramyxovirus groups and many others cause 'colds'. Various bacteria, notably *Streptococcus pyogenes*, *Strep. pneumoniae*, anaerobes, *H. influenzae* and *B. catarrhalis* can cause sinus and middle-ear complications.

Epidemiology

Colds occur throughout the year, more often in winter in temperate zones and after the rainy season in the tropics. Because numerous viruses are involved (over 100 rhinoviruses alone), one is constantly exposed to viruses not met before and immunity is only slowly and incompletely accumulated. Up to 6 colds per year may be experienced by a child, reducing to two or three yearly on average for adults. The epidemiology of acute sinusitis is similar, usually following colds but sometimes complicating allergic rhinitis, dental or other local sepsis. Otitis media

Table 2.4. *Specific antimicrobial therapy for infections of the upper respiratory tract*

Organism	Suitable antimicrobials
Strep. pyogenes	Penicillin or erythromycin
C. diphtheriae	Benzylpenicillin or erythromycin (+antitoxin)
H. influenzae	Amoxycillin + clavulanic acid (Augmentin); chloramphenicol; cefotaxime
Strep. pneumoniae	Penicillin, amoxycillin, or erythromycin
Anaerobic organisms (e.g. *Bacteroides*)	Amoxycillin and/or metronidazole
Vincent's organisms	Benzylpenicillin
Candida spp.	Fluconazole
Herpes simplex	Acyclovir

follows a similar pattern but becomes less common after the age of about five years as the eustachian tube, the usual portal of entry, lengthens and narrows. In younger children otitis media is common, and perforation of the eardrum may allow entry of unusual organisms. The eardrum is often reddened during viral upper respiratory infections but viruses rarely cause painful otitis.

Clinical presentation and complications

The short incubation period of two to three days, nasopharyngeal irritation and sneezing, nasal discharge, mild sore throat, and sometimes cough are familiar features of a *cold*. Normally this subsides within a week but symptoms may persist longer in smokers or when complications arise. Postnasal drip often causes cough in children, especially at night. Fever, marked headache, and myalgia suggest influenza or a bacterial infection.

Signs of *sinusitis* are local pain, purulent nasal discharge, and often nasal obstruction which may progress with fever, localized headache, tenderness, and oedema over the affected sinus. Occasionally, infection spreads from the sinus to cause osteomyelitis, meningitis, brain abscess, or intracranial venous sinus thrombosis. Radiography or transillumination may show an opaque sinus or fluid level (Fig. 2.4).

The first sign of *otitis media* is often sudden onset of pain. The drum often ruptures within 48 hours with purulent discharge (Fig. 2.5). There is usually fever and auroscopy shows a red drum, bulging with loss of light reflex, and with loss of mobility when fluid or pus accumulates in the middle ear. Analgesics given for associated respiratory tract discomfort may mask initial pain. Mastoiditis (Fig. 2.6), osteomyelitis, and perforation into the meninges, formerly common complications, are now unusual with antibiotic therapy, but by preventing

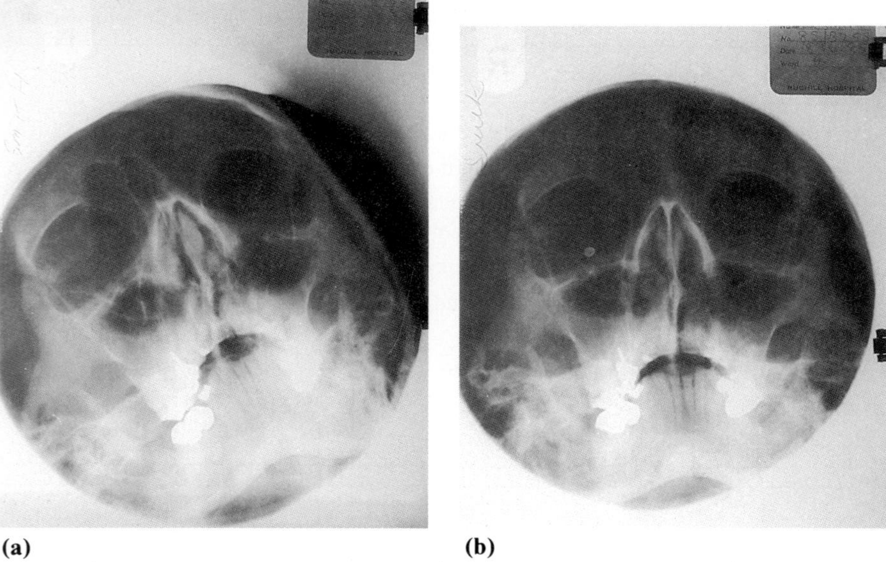

(a) (b)

Fig. 2.4 (a) and **(b)** Pus level in maxillary sinuses moves when head is rotated.

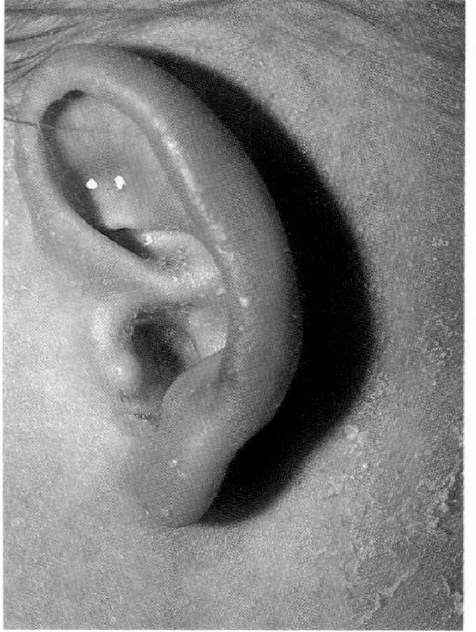

Fig. 2.5 Purulent discharge from the ear in otitis media after rupture of tympanic membrane in measles: note desquamation.

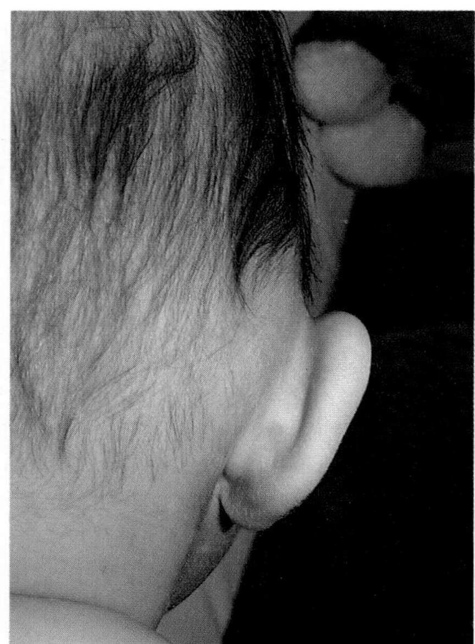

Fig. 2.6 Acute mastoiditis: erythematous, fluctuant swelling pushes the infant's ear inferiorly and anteriorly.

perforation this may encourage sterile fluid to persist in the middle ear ('glue ear'). Relapse may be due to inadequate or inappropriate antibiotics or reinfection through an unhealed perforation.

Management

Symptomatic treatment for *colds* may include the use of ephedrine nose drops for a few days to relieve nasal congestion, especially in babies who do not easily breathe through the mouth. Antibiotics are inappropriate unless there are bacterial complications. Antihistamines have little value unless there is associated allergy.

For *sinusitis* empirical antibiotic treatment (for example, with amoxycillin) is usually effective, perhaps together with flucloxacillin if staphylococcal involvement is suspected or metronidazole when the source is dental. Drainage is helped by decongestant nasal drops, occasionally by aspiration of pus.

Amoxycillin usually deals with bacterial *otitis media* in infants: co-trimoxazole and erythromycin are alternatives. Penicillin is usually sufficient over the age of five years when *H. influenzae* is less often involved. Improving drainage by ephedrine nasal drops is helpful; aspiration of pus through the drum is rarely required. Adequate analgesia should be given.

INFECTIONS OF THE EYE

The eye is exposed to airborne infection and to frequent introduction of organisms from contaminated fingers. Other sources of infection include flies, unsterile medical instruments, medicaments, and trauma. Spread of infection from local sepsis, or through the blood by cytomegalovirus or toxoplasma for example, can also cause disease.

Aetiology

Chlamydia trachomatis or *Neisseria gonorrhoeae* may infect the eye during birth (Chapter 11). Pyogenic cocci, pseudomonas species, *H. influenzae* and other bacteria and viruses causing respiratory infections (Tables 2.1 and 2.2) infect in later life. Several types of adenovirus (especially type 8) and enterovirus (especially type 70) cause epidemics affecting the conjunctiva and sometimes cornea. In many developing countries certain types of *Chl. trachomatis* damage the conjunctiva and cornea causing trachoma, an important cause of blindness, as also is measles in these areas.

Clinical features

Acute conjunctivitis. Local irritation, conjunctival inflammation (sometimes unilateral), and 'sticky eyes' in the morning are typical early features. Pus suggests bacterial aetiology. Possible trauma, foreign body, or allergy should be considered. Recurrence may result from a blocked naso-lachrymal duct.

Adenoviral infections, especially in epidemics, can also damage the cornea producing small opacities which resolve slowly ('keratoconjunctivitis'). Conjunctival inflammation may be severe and painful in acute haemorrhagic conjunctivitis due to enterovirus 70 (Fig. 2.7).

Acute keratitis. Corneal infection and ulceration may be secondary to conjunctivitis or trauma. Pain may be severe. In addition to adenovirus 8 (causing 'epidemic keratoconjunctivitis'), important viral causes are herpes simplex and varicella zoster (ophthalmic shingles). Herpes simplex can cause dendritic ulcers best seen by slit lamp examination after fluorescence staining. These are painful and like other herpes infections recurrent (Fig. 2.8). Measles is a major cause of keratoconjunctivitis in the malnourished, especially with vitamin A deficiency.

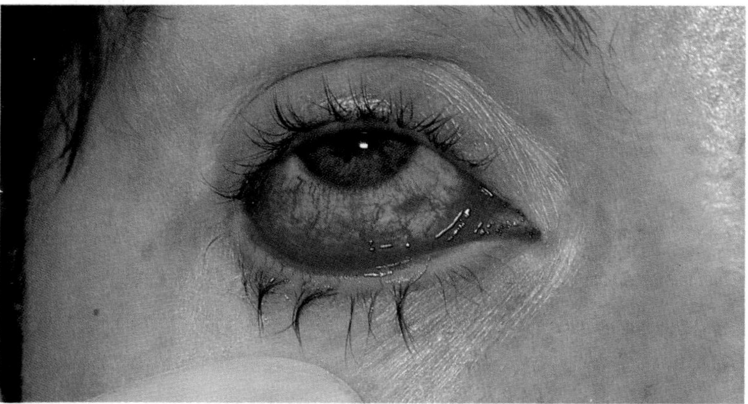

Fig. 2.7 Enterovirus type 70 can cause epidemic haemorrhagic conjunctivitis.

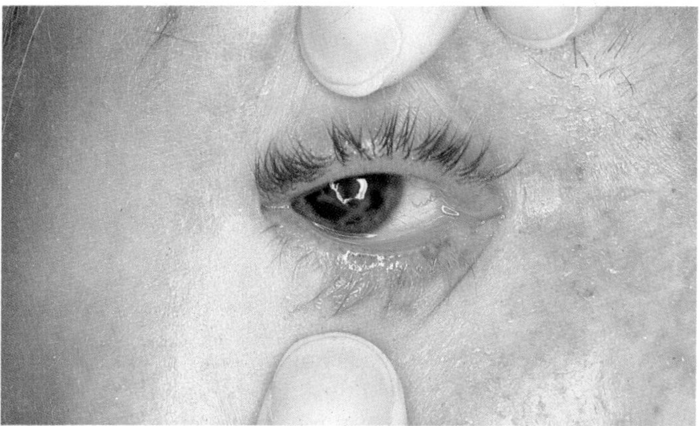

Fig. 2.8 A dendritic ulcer due to herpes simplex virus (stained with fluorescein).

Orbital cellulitis. This usually spreads from local sepsis of sinuses or skin, developing rapidly with pain and swelling of periorbital tissues (Fig. 3.8b). *Staph. aureus*, or in young children *H. influenzae*, are the usual causes.

Choroidoretinitis. Common causes are toxoplasmosis and, especially in AIDS, cytomegalovirus (Fig. 2.9) (Chapter 11). Fundal examination may show choroidal tubercles in disseminated tuberculosis or fluffy fungal deposits in intravenous drug misusers and immunocompromised patients. Toxocariasis (invasion of tissues by migrating larvae of *Toxocara canis*, the dog tapeworm) occasionally causes, mainly in children, retinal lesions resembling retinoblastoma.

Investigations

The responsible organism may be cultured from pus or conjunctival swabs, or in *Chlamydia* infections from scrapings of affected conjunctiva in which chlamydial inclusions may be visualized. Corneal scrapings are more helpful in herpes or adenovirus 8 keratitis. Identification of the organism is important in neonates, recurrent infections, and outbreaks. Identifying herpes infections can obviate using topical steroids which can seriously worsen herpes keratitis. Serological diagnosis can help in adenoviral outbreaks and toxoplasmosis.

Management

Antimicrobial drops or ointment are used for superficial bacterial infections, usually chloramphenicol and tetracycline for *Chlamydia* infections. Systemic

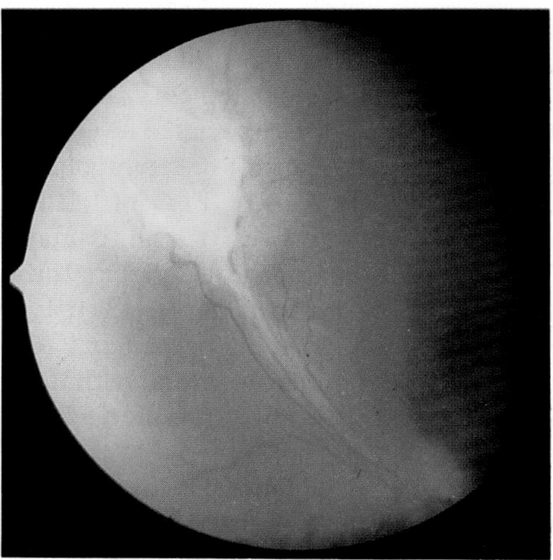

Fig. 2.9 Choroidoretinitis due to cytomegalovirus infection in a patient with HIV infection.

therapy may be required for infants, as indicated by isolation and sensitivity studies, and is required for orbital cellulitis and microbial chloroidoretinitis. Topical idoxuridine or acyclovir is useful for herpes simplex infections but less for zoster where oral acyclovir given early before the eye is involved is more effective.

Prognosis

This is usually good in conjunctivitis, but there can be serious damage or blindness if infection persists and if treatment is inadequate, especially in neonatal and *Chlamydia* infections. Herpes simplex keratitis often recurs. Deep infections and those with corneal involvement or inadequate response to therapy merit ophthalmological advice.

SORE THROAT SYNDROMES

Sore throat, with or without fever, is a common condition which can be caused by many different organisms, most of them viruses (Table 2.1). Often it is an early symptom of a cold, influenza, or systemic infection, but may be an isolated symptom in its own right. Local and symptomatic treatment deals with most sore throats, but it is important to recognize those bacterial infections which require specific antimicrobial therapy (Table 2.2).

Pharyngitis and tonsillitis

Clinical features and differential diagnosis

The intensity and distribution of inflammatory reddening of throat, tonsils, and soft palate, and the presence or absence of exudates, vesicles, ulcers, or petechiae vary and provide diagnostic clues. Associated coryza suggests a viral aetiology, often with little to be seen in the throat. Associated fever, myalgia, and headache suggest influenza (in which cough is usual) or streptococcal infection in which there is often purulent exudate on tonsils (Fig. 2.10) and painful swelling of jugulodigastric lymph nodes. Streptococcal tonsillitis may progress to quinsy (peritonsillar abscess) and the diffuse punctate erythematous rash of scarlet fever (Chapter 7) or postinfectious syndromes (Chapter 13) may develop. Vesicles or resulting ulcers on the soft palate suggest herpangina (Chapter 4). Insidious febrile illness not responding to antimicrobials and with thick exudate on tonsils and palatal petechiae suggests infectious mononucleosis (see below and Fig. 2.11).

Adherent membrane extending beyond the tonsils to palate and pharynx characterizes diphtheria (see below); inflammatory oedema around cervical lymph nodes can cause the 'bull neck' of severe diphtheria in which constitutional upset can be severe with pallor and hypotension. Diphtheria should be considered in patients recently arrived from abroad, especially if not immunized.

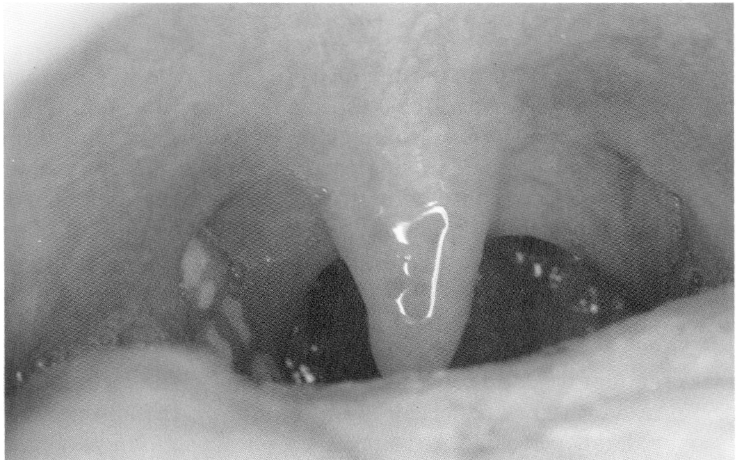

Fig. 2.10 *Streptococcus pyogenes* is the usual cause of follicular tonsillitis.

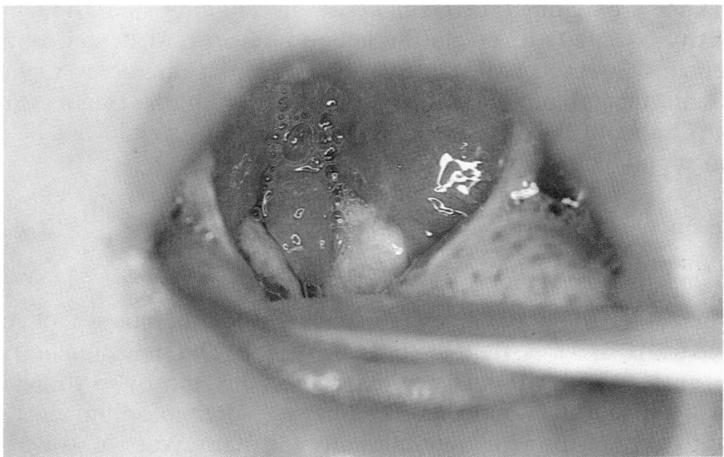

Fig. 2.11 The adherent exudate on the tonsils in infectious mononucleosis.

Lassa fever (Chapter 14) should also be considered in febrile patients with pharyngitis recently returned from tropical Africa. White plaques, often involving buccal mucosa and palate, are typical of *Candida albicans* infection (Chapter 4), which can be severe and recurrent in immunocompromise. Marked gingivitis, halitosis, and ulcerating membranes usually spreading from the gums typify the mixed anaerobic infection of Vincent's angina. In immunocompromised persons herpes simplex stomatitis may spread to involve the throat severely with ulceration and foetor.

Investigations

Throat swabs of exudate or from the inflamed area should be taken in severe cases and when the diagnosis is in doubt, especially to identify diphtheria which requires specific treatment urgently (see below). Fusiform bacilli, spirochaetes, and fungal bodies can be seen by microscopy. Viruses may be isolated from throat swabbings collected into viral transport medium (Chapter 19), and serological tests can help to diagnose influenza, adenoviral infection, and infectious mononucleosis. Blood can be examined for neutrophilia, typical of bacterial infection, or the atypical lymphocytes of infectious mononucleosis.

Management and complications

Saline gargles to remove debris, warm drinks, analgesics, and rest suffice for most viral infections, which are unaffected by antibiotics. In the absence of rhinitis, vesicles, or palatal petechiae to suggest a viral aetiology, penicillin or erythromycin should be given, preferably after taking a throat swab. Initial treatment can be parenteral in severe cases. The antibiotic can be discontinued if illness resolves quickly and culture is negative. Co-amoxiclav, chloramphenicol or cefotaxime are required for *H. influenzae* infection: this occasionally causes pharyngitis associated with epiglottitis and respiratory obstruction, septicaemia, and meningitis. Streptococcal infections may lead to rheumatic fever, glomerulonephritis, erythema multiforme, erythema nodosum or Henoch–Schönlein purpura (Chapter 13). Amoxycillin and related drugs should not be given in infectious mononucleosis (see below).

Infectious mononucleosis

Aetiology and epidemiology

This term, rather than 'glandular fever', is used for the illness caused by the Epstein–Barr (EB) virus (human herpesvirus 4). This infection is prevalent throughout the world with many primary infections, mostly in young children who have subclinical or mild febrile illnesses. After infection most persons continue to excrete the virus intermittently for many years in saliva and can spread infection later in life, for instance by kissing.

Clinical presentation and complications

The incubation period is around five to six weeks. Milder illnesses may present with unexplained fever and perhaps lymphadenopathy. The typical pharyngeal ('anginose') form usually seen in teenagers and young adults is characterized by fever which increases over several days with gradual onset of sore throat. After about a week the illness reaches its peak when the tonsils are covered with whitish-yellow adherent exudate (Fig. 2.11). There may be inflammation and oedema of the rest of the pharynx and soft palate. Involvement of adenoids can cause nasal obstruction and nasal intonation. Difficulty in swallowing often causes dribbling and in severe cases even breathing and drinking may be obstructed. These features are less marked after previous tonsillectomy. Palatal petechiae are common (Fig. 2.12).

Lymphadenopathy, especially of cervical and axillary nodes, and splenomegaly are common, also hepatitis which is usually without jaundice. Rarer complications include haemolytic anaemia, myocarditis, and central or peripheral neuropathy. A sparse macular rash may be seen (Chapter 7) distinct from the marked morbilliform rash that follows administration of ampicillin or related antibiotics to these patients.

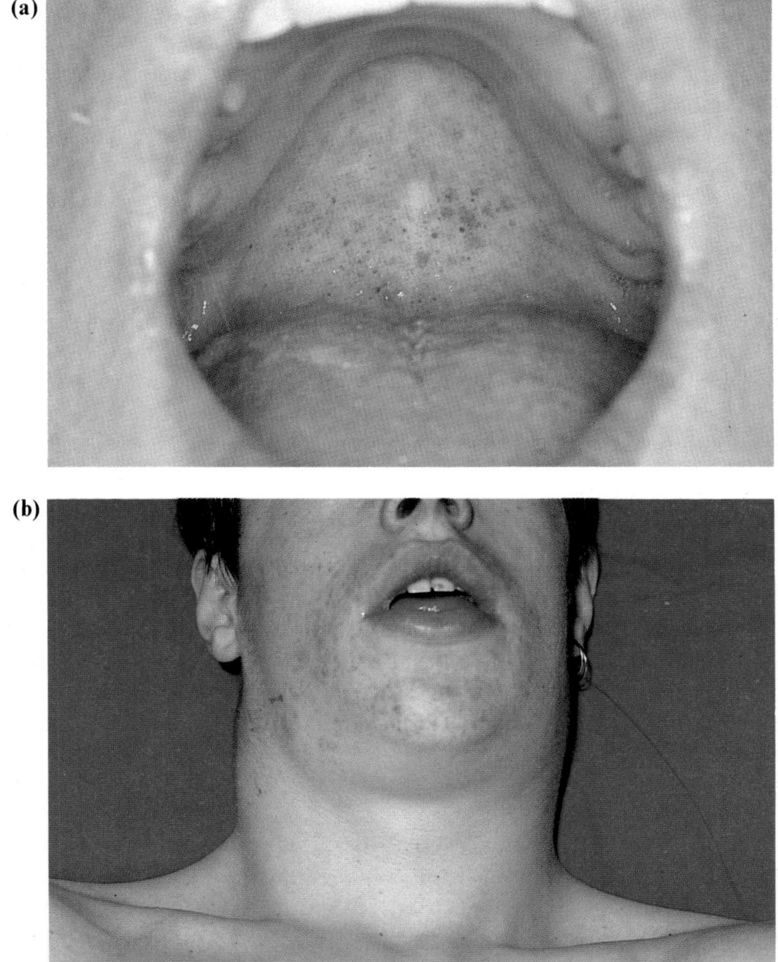

Fig. 2.12 Infectious mononucleosis. **(a)** palatal petechiae; **(b)** cervical lymphadenopathy.

Investigations

The total peripheral white-cell count may be normal but atypical lymphocytes are usually present. Heterophile antibody detected by the 'monospot' or more specific Paul Bunnell–Davidson test supports the diagnosis but tests may not become positive for some weeks and can remain so for up to a year. Current infection is indicated by IgM antibody to EB virus capsid antigen which is detectable for six to eight weeks after infection and in all patients whereas about 10 per cent do not have heterophil antibody.

Management and prognosis

Rest and analgesics are appropriate with care to maintain hydration. Respiratory obstruction may be helped by corticosteroids to reduce oedema but occasionally intubation or tracheostomy is necessary. Secondary bacterial infection is unusual. Ampicillin and similar drugs which cause a rash must be avoided. Some patients recover quickly but a few have prolonged malaise and lethargy (Chapter 13). Complications generally resolve completely.

Diphtheria

Aetiology and epidemiology

Infections with *Corynebacterium diphtheriae* are common in many under-developed and tropical countries, often as relatively harmless indolent ulcers on the lower limbs of barefoot children. These immunizing infections, from which absorption of toxin is slow, are less common in urbanized and developed areas where bacilli spread mainly by droplets from nasopharyngeal infections. Massive absorption of toxin from the larger colonized area of the throat causes both local tissue necrosis and life-threatening damage to the heart, nervous system, and other organs. The nasal mucosa, soft palate, pharynx, larynx, trachea, and bronchi can be involved, possibly with respiratory obstruction. Mild and silent infections predominate. Not all strains of the organism are toxigenic and visualization of Corynebacteria in stained smears from exudate does not prove the diagnosis. In Britain and most other countries where children are routinely immunized, diphtheria and circulation of *C. diphtheriae* in the community become rare, but occasionally the disease affects non-immunized children or adults, usually infected by a carrier arriving from abroad.

Clinical features

Toxicity is often severe in pharyngeal diphtheria. There is less absorption of toxin from disease confined to the tonsils or lower regions of the respiratory tract but laryngeal involvement causes obstructive 'croup' (see below), stridor, and eventual signs of asphyxia. In severe cases patients may die from circulatory collapse caused by cardiac arrest or sudden arrhythmia, usually during the second week; survivors may have residual cardiac failure, arrhythmia or heart block. Toxin may paralyse the palate, eyes, face, or laryngopharyngeal tract, rarely limbs, usually after several weeks. Diphtheria bacilli may also infect the nose or ear as well as skin ulcers, but with little toxicity.

Management

When diphtheria is suspected, antitoxin should be given without delay, from 10 000 to 80 000 units according to severity by intramuscular or cautious intravenous injection, after a preliminary test dose of 0.2 ml subcutaneously to detect sensitivity to horse serum. Adrenaline and hydrocortisone must be on hand to treat possible anaphylaxis. Delayed serum reactions with rash and joint swellings may develop after about 10 days. Penicillin or erythromycin help to eliminate infection and prevent production of further toxin. Tracheostomy to relieve respiratory obstruction and supportive care for circulatory complications and any late paralyses may be required.

Prevention

Close contacts should have throat swabs cultured and their immunization status determined. Non-immunized close contacts should receive prophylactic penicillin or erythromycin and be vaccinated. Erythromycin is more effective against nasopharyngeal carriage. Children should be routinely immunized against diphtheria (Chapter 16).

CROUP SYNDROMES

Croup is characterized by cough, inspiratory stridor, or low-pitched, crowing respiration caused by obstruction of breathing around the level of the larynx or epiglottis. It is commonest in young children whose narrow larynx is more easily obstructed by inflammatory swelling or exudates from diphtheria, *H. influenzae*, or other infections. It can also result from inhalation of a foreign body. In chronic or recurrent croup, tuberculosis, and congenital laryngomalacia should be considered.

Acute laryngotracheobronchitis

This classical croup mainly affects children up to four years old, particularly overweight males. It is usually caused by viruses, especially those of the parainfluenza group, but sometimes influenza, respiratory syncytial virus, and measles. It is commonest in winter.

Clinical presentation

There is often preceding coryza, pharyngitis, or cough. Fever is variable: marked fever in a 'toxic' child suggests bacterial infection. A harsh 'bovine' cough or cry may have been evident for some hours or days but stridor usually begins abruptly, often in the evening or at night. There is varying respiratory distress, causing anxiety in both child and parents. There is inspiratory indrawing of soft tissues of neck, abdomen, and even the lower rib cage in infants, with cyanosis in severe cases. Stridor often resolves within a few hours, but respiratory failure and sudden collapse may be precipitated, for example, by emotional upset. It is important to exclude epiglottitis as the primary or concurrent cause of stridor (see below).

Investigations

Virus isolation and serology may confirm the aetiology. Blood culture and throat swab may identify a bacterial cause. Throat swabbing is dangerous in epiglottitis because it may precipitate acute obstruction: X-ray examination of the larynx may aid assessment.

Blood gas estimation sometimes helps to assess the degree of anoxia but should not alone determine whether endotracheal intubation or tracheostomy should be undertaken. An exhausted child may deteriorate suddenly.

Management

Stridor often resolves spontaneously. Humidification is generally helpful by softening mucus and soothing inflamed mucosa, and may be carried out in the bathroom with a hot shower running or by steam inhalations or a steam kettle with care to avoid scalding. In hospital a humidification tent (Fig. 2.13) usually relieves symptoms rapidly. If hypoxaemia develops, oxygen should be given with humidification. Oxygen eases symptoms but does not treat the cause; the child may deteriorate rapidly if removed from oxygen for feeding. Secondary infection

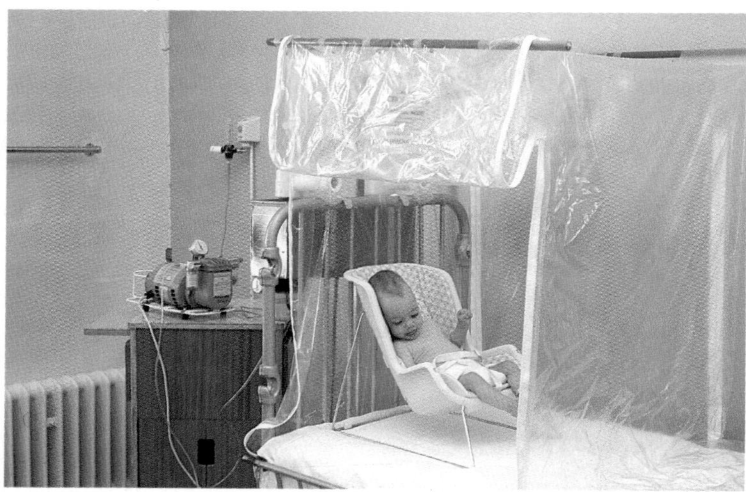

Fig. 2.13 The 'croupette', closed when in use, can provide a humidified atmosphere which may be oxygenated and/or cooled.

in viral croup is uncommon but antibiotics are indicated when bacterial infection is suspected; for example, from a high polymorphonuclear white-cell count. Some studies suggest that in severe cases corticosteroids reduce the need for tracheostomy.

Complications and prognosis

Persistent complications are unusual but croup may recur during later respiratory infections.

Acute laryngitis

This causes hoarseness or loss of voice, usually in older children or adults infected by various viral respiratory pathogens including those that cause croup. The diameter of the larynx and trachea in adults does not make obstruction a problem unless congenital abnormalities are present or the epiglottis is involved. Usually self-limiting, laryngitis may be prolonged in cigarette smokers or those who use their voice frequently such as singers. In AIDS, candidal laryngitis can be persistent.

Acute epiglottitis

Aetiology and epidemiology

This acute infection of the epiglottis and surrounding structures is almost always caused by *H. influenzae* type b. Like croup it is most frequent in winter and commoner in young male children. It sometimes affects older people, especially if immunocompromised.

Clinical presentation and differential diagnosis

The patient is usually toxic and febrile and prefers to sit upright with neck extended. There may be respiratory distress from partial obstruction which often becomes complete as a sudden emergency. The swollen red epiglottis may be seen when the mouth is open. Fever is usually absent after aspiration of a foreign body or in angioneurotic oedema which is usually preceded by a sting or drug intake.

Investigations

Radiography of the neck can help to distinguish epiglottitis from the subglottic obstruction of viral croup. There is generally neutrophilia and *H. influenzae* can usually be cultured from the blood.

Management and prognosis

Patients with suspected epiglottitis must be nursed intensively, usually with an endotracheal tube inserted prophylactically or tracheostomy performed. The throat of a non-intubated patient should not be examined or swabbed for fear of precipitating obstruction. In this life-threatening condition chloramphenicol or cefotaxime should be given without waiting for culture and sensitivity patterns. Recovery is usually complete.

INFLUENZA

Influenza can cause major morbidity and mortality world-wide in local or widespread epidemics. It causes a broad range of clinical manifestations at all levels of the respiratory tract, also affecting several other systems through complications.

Aetiology and epidemiology

The causal myxoviruses exist as three independent types. Type A strains infect many animal and bird species as well as man, with antigens varying by 'drift' and 'shift'. Antigenic drift results from selection pressure by immunity in the host population which favours minor variants; these tend to replace one another as prevalent subtypes between major epidemics (Fig. 2.1). Antigenic 'shift' appears to result from genetic recombination of human with avian or animal virus, providing major antigenic change which allows the new virus to spread with little or no hindrance by existing herd immunity. This can cause a major epidemic or pandemic involving most or all age groups. Influenza types B and C viruses infect only man and show little antigenic change except for 'drift' by influenza B. Type C infections are mainly confined to children; influenza B also has relatively greater impact on children but occasionally causes moderate or localized outbreaks involving older age-groups also. Recovery gives good immunity to symptomatic reinfection with the same strain of influenza virus.

Most infections occur in outbreaks during winter or spring. Mortality is highest in the elderly and debilitated, and those living in poor conditions with inadequate heating, also in the undernourished, overcrowded populations of developing countries. Much mortality is associated with deaths from secondary pneumonia, exacerbation of chronic lung, and cardiac diseases (see Fig. 17.4). This significantly raises recorded mortality 'from all causes' during epidemics. Fulminating, fatal illness in young adults has featured in several influenza A pandemics.

Clinical presentation and complications

After a brief incubation period of 24–72 hours the onset is typically abrupt with fever, shivering, headache, muscle pains, and malaise. As infection spreads there may be nasal, pharyngeal, and tracheal symptoms, dry cough, and burning substernal discomfort. Meningism is sometimes present. Mild and subclinical infections are common, often indistinguishable from a common cold, especially in young children, and some elderly patients may have little or no fever. Damage by the virus to respiratory mucous membranes and bronchi does not usually increase sputum or cause abnormal chest findings but can allow secondary bacterial invasion (Fig. 2.14). Secondary infections with *Strep. pneumoniae*, *H. influenzae*, and other organisms are especially common in smokers and those with chronic obstructive airways disease. Staphylococcal pneumonia is an important complication in these groups but also in young, otherwise healthy adults, typically with rapid onset and high mortality (Fig. 2.14). The fulminating 'influenzal pneumonia' of young adults seen in the devastating 1918–19 pandemic, of which occasional cases still occur, is severe and haemorrhagic with copious, blood-stained sputum obstructing airways and leading to rapid death. It is rarely due to the virus alone but is usually a mixed infection with staphylococci (Fig. 2.15), *Strep. pyogenes* or sometimes other bacteria. Croup is another manifestation of influenza in children during epidemics. Influenza type A, B, and C infections cannot be distinguished clinically in individual cases, though type C tends to be mildest, and type A the most severe; gastrointestinal upset is commoner in children with influenza A infection, and myalgia and leg weakness in children with influenza B.

Investigations

The specific diagnosis can be confirmed by demonstrating viral antigen in respiratory samples, by isolating virus, or demonstrating the antibody response (Chapter 19). This is impracticable in most cases and during epidemics, but it is

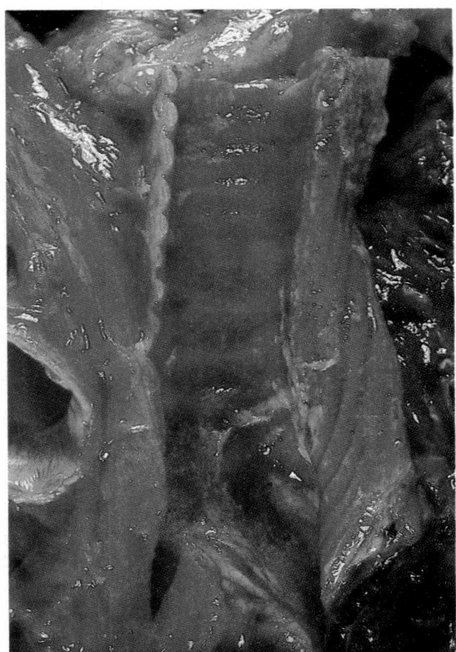

Fig. 2.14 An extensively inflamed trachea in a fatal case of influenza.

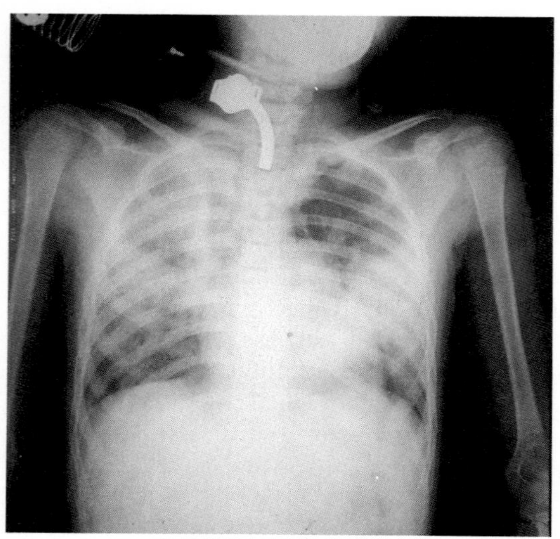

Fig. 2.15 Staphylococcal pneumonia complicating influenza. Tracheostomy was required.

important for virological diagnosis to be made in a sample of acute respiratory illnesses in order to give warning of epidemics and possible changes in antigens of influenza strains, an important element of surveillance and attempted control. One should not allow a casual diagnosis of 'flu' to delay recognition of some other important disease.

Management

Rest, fluids, and analgesics usually suffice. Appropriate antibiotics may be required for secondary infection. Amantadine, an antiviral drug (see below) has little effect once illness has started. Fulminating and staphylococcal pneumonia require vigorous supportive therapy with oxygen, physiotherapy, and maintenance of airways, sometimes by endotracheal intubation or tracheostomy, as well as antibiotics. Recovery is usual within a week, but many experience more prolonged malaise, sometimes with 'post-influenzal' depressive illness. Rarely, encephalomyelitis or Guillain–Barré syndrome may develop during or as a sequel of influenza. Reye's syndrome (Chapter 13) has been associated particularly with influenza B. Myocarditis can cause chest pain or heart failure, sometimes fatal.

Prevention

Isolation, where practicable, is important especially in hospitals and institutions for the elderly. Amantadine or rimantadine has limited prophylactic effect against influenza A (not B or C) if given before exposure and maintained for the period of risk—this is worth consideration for vulnerable groups (Chapter 18) for example, in a geriatric unit. Influenza vaccines when given before the epidemic season protect incompletely against the corresponding subtypes of virus

(Chapter 16). The formula of these vaccines is changed as required to match the antigenic pattern of currently active influenza A and B viruses as judged from continuing surveillance co-ordinated by the World Health Organization.

LOWER AIRWAY SYNDROMES

Acute bronchitis

This is usually an extension downwards of inflammatory infection of the upper respiratory tract, commonest in winter and spring and mainly due to viruses such as those causing the common cold, adenoviruses, influenza, and measles; *Mycoplasma pneumoniae* can be responsible, but the primary role of other bacteria is unclear.

Clinical presentation

Cough, dry or productive and mildly purulent, often follows upper respiratory tract infection. Rhonchi may be heard. Fever varies with the aetiology, being usual in measles and influenza but uncommon with rhinoviral infection.

Investigation and management

Management is symptomatic unless secondary bacterial infection requires antimicrobial therapy. Identification of the primary infecting agent is rarely helpful to management.

Acute bronchiolitis

Aetiology and epidemiology

This infection is rare after the second year of life and is mainly due to respiratory syncytial (RS) virus, although other viruses such as influenza and parainfluenza virus and *Mycoplasma pneumoniae* can also cause bronchiolitis. RS virus is highly infectious and reinfections are common, mainly as minor and upper tract syndromes. This virus causes large annual epidemics in winter with associated peaks of bronchiolitis causing pressure on paediatric hospital beds (Fig. 2.1). Almost every child experiences one or more infections with RS virus in early life.

Clinical presentation and complications

After a few days' incubation period fever, sometimes with coryza and cough, leads to respiratory distress with intercostal indrawing and subcostal recession on inspiration. These signs often worsen over two to five days, contrasting with the transient few hours' respiratory distress of viral croup. The marked rhonchi, inspiratory wheeze, and widespread fine crepitations can resemble asthma. There may be anoxia, cyanosis, and occasionally heart failure. The illness generally resolves within a week. Secondary bacterial infection is very unusual. Death is rare in otherwise fit infants, but those with developmental defects of heart are particularly at risk and some cases of 'cot death' (sudden infant death syndrome) have been associated.

Investigations

Chest X-ray usually shows hyperinflation, rarely consolidation (Fig. 2.16). RS virus can rapidly be identified by immunofluorescence tests (see Fig. 2.3) or isolated in culture; antibody response is slow in infants.

Management

Careful nursing is required, with oxygen if needed. Feeding can be given by a small nasogastric tube. Ribavirin given by aerosol may be helpful in severe cases. Antimicrobials are unhelpful. Although the clinical picture resembles asthma, response to corticosteroids is unimpressive. There is no obstruction to be relieved by endotracheal intubation, but intermittent positive pressure ventilation may save the life of an anoxic, exhausted child.

Prevention

Contact with highly prevalent RS virus is hard to avoid, but crossinfection should be minimized, especially in hospital where compromised children of the same vulnerable age may be exposed. No vaccine is yet available.

Chronic bronchitis

This condition is characterized by chronic or recurrent cough and increase in sputum for three months in each of at least three years. Cigarette smoking, atmospheric pollution, genetically-determined lung disease, and infection all play a part in the aetiology. It is sometimes associated with airflow limitation, wheezing, and emphysema as components of 'chronic obstructive airways disease'. Periodic relapses and exacerbations of illness are often precipitated by respiratory viral infections, especially by rhinoviruses and influenza, often involving bacteria such as *H. influenzae*, *Strep. pneumoniae*, and *Branhamella catarrhalis*. Exacerbations are treated by physiotherapy, relief of bronchospasm, and antimicrobials such as amoxycillin, oxytetracycline, or co-trimoxazole.

Pertussis

'Pertussis' is a more appropriate name for this condition than 'whooping cough' because not all those infected, particularly infants and adults, 'whoop'.

Aetiology

The usual cause is *Bordetella pertussis*, occasionally *B. parapertussis* which is more common in well-immunized populations. Some viral, chlamydial and *Mycoplasma pneumoniae* infections mimic the cough, and sometimes even the

(a)

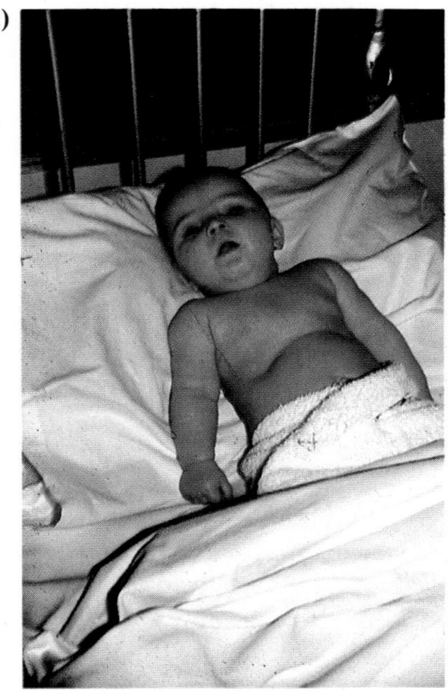

(b)

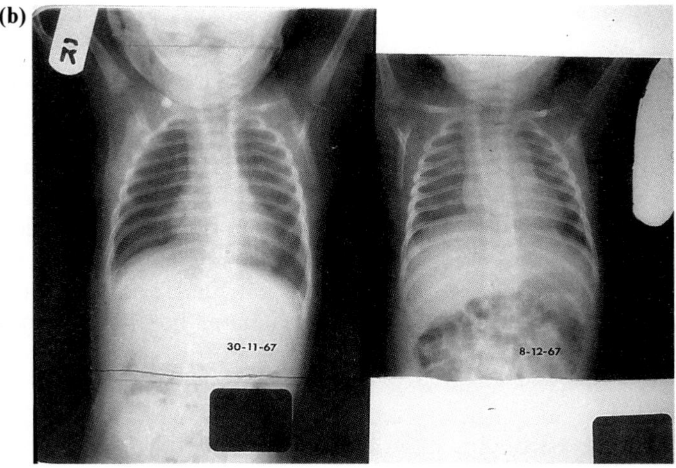

Fig. 2.16 Acute bronchiolitis: **(a)** subcostal recession on inspiration; **(b)** chest X-ray shows hyperinflation and increased translucency, then recovery to normal.

whoop, but not the prolonged illness with repeated spasms in a child well between spasms which typifies *Bordetella* infection.

Epidemiology

The explosive expiratory cough is highly effective in spreading infection to susceptibles. There is no transferred maternal immunity and all age-groups may be affected. Morbidity is greatest in those under five years, and mortality in those under one year of age, some of whom may not be diagnosed because they are too young to 'whoop' properly. Epidemics occur every three to four years, but in poorly-immunized populations many cases also occur during the intervening years (Fig. 2.1). Infants are mainly infected by their non-immunized older siblings who bring pertussis into the household after becoming infected at school or playgroups.

Clinical presentation

After incubation for 7 to 14 days a catarrhal cold with simple (but infectious) cough develops, progressing over several days to repeated expiratory paroxysms. These may be followed by rapid inspiration causing the 'whoop'. There is tenacious sputum, difficult to expel. Groups of paroxysms up to 30 or 40 times a day, more frequent at night, are often terminated by vomiting and may be triggered by feeding, crying, or manipulations such as examining the throat. Fever in uncomplicated illness is rare. The illness may be exacerbated by intercurrent viral infections (Fig. 2.17) during its long course which is commonly more than 30 and sometimes up to 100 or more days. Parents are often exhausted by giving constant attention.

Investigations

Swabs for culture are best taken from the nasopharynx either pernasally with a special thin wire swab or through the mouth (Chapter 19). Culture is usually positive early in the illness in infants who have not received antibiotics, and may remain positive for many weeks. Culture is less reliable in older children and adults, when serological diagnosis may be more helpful. There is commonly both absolute and relative lymphocytosis in peripheral blood with counts up to 30×10^9/litre or higher.

Complications

Apnoea during a paroxysm may cause cyanosis and/or convulsions. Most persistent neurological features can result from anoxia, but also from haemorrhage, metabolic or toxic factors. Secondary infection, atelectasis, and aspiration bronchopneumonia are common but sometimes recognized only by chest radiography. Fever suggests secondary infection of lungs. Umbilical or inguinal hernia and rectal prolapse from coughing may develop especially in the very young. Protrusion of the tongue during paroxysms can damage the frenum when teeth are present. Venous engorgement causes epistaxis, facial petechiae, and

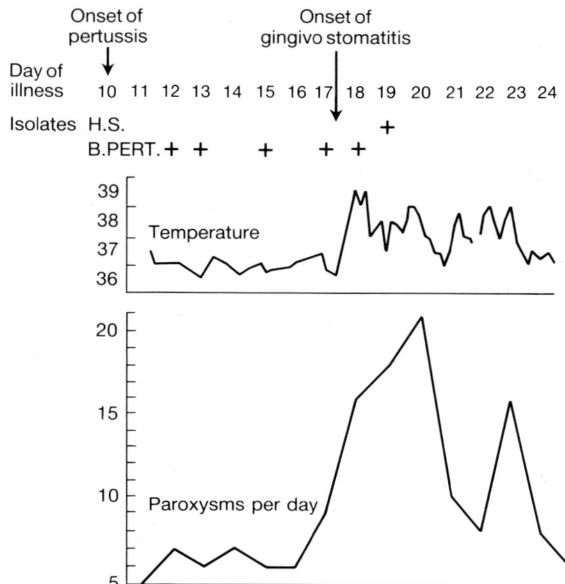

Fig. 2.17 Pertussis paroxysms increasing due to intercurrent primary herpes simplex stomatitis.

subconjunctival or occasionally intracerebral haemorrhages which may be fatal or permanently damaging. During the long illness, intercurrent respiratory and other infections like gastroenteritis, measles, or in the tropics tuberculosis, can add to the child's problems. Repeated vomiting can lead to malnutrition. Mortality in Britain is now low, but pertussis is one of the most lethal infections of malnourished, non-immunized infants in developing countries.

Management and prevention

Careful observation, lifting and comforting during paroxysms, cleaning away phlegm, minimizing inhalation of vomit, and re-feeding when necessary are important. Cough suppressants are generally unhelpful. Steroids may help but their side effects must be kept in mind. Rarely, intubation may be required for continual coughing or recurrent apnoeic episodes. Certain antibiotics such as erythromycin may shorten the duration of illness but only if given during the first five days of illness. They may not always eliminate organisms from the nose but are effective for treating infective complications. Isolating the patient from other susceptible children, especially the very young, should be attempted. Vaccination is discussed in Chapter 16.

PNEUMONIA

Infection may reach the lungs by inhalation of fine particles into terminal air spaces, or by downward movement of infected material in the muco-ciliary blanket. The normal cleansing action of cilia can be damaged by viral infection, chemical pollutants (for example, smoke), excess alcohol ingestion, or local disturbance from tumour or foreign body. Infection can also reach the lungs via the bloodstream (as in intravenous drug abusers) or by local extension from an infective focus. Pneumonia ('pneumonitis') was traditionally classified as 'lobar' (where inflammatory consolidation was mainly confined to one or more lobes), or as 'bronchopneumonia' (patchy involvement of the lungs, not confined to a single lobe). Later 'atypical pneumonia' was used to describe patchy pneumonias unresponsive to sulphonamides and later to penicillin in which clinical signs were scanty, radiological signs of pneumonia disproportionally great and no bacterial cause demonstrable—mainly due to mycoplasma, adenovirus, *chlamydia*, or rickettsial infections.

More useful for management is the distinction between (a) *primary pneumonia* affecting previously healthy persons, in which a limited number of pathogens are likely to be involved, and (b) *secondary pneumonia* in persons with previous respiratory or other disease including tumours and immune deficiencies, after surgical operations, in those very old, immobile, or unconscious, and patients receiving immunosuppressive therapy. In this second group a wide range of infecting agents may be involved including Gram-negative bacteria, viruses, protozoa, and mycotic infections which may require different specific therapies than those for primary pneumonias. Tuberculous pneumonia is described in Chapter 9.

Epidemiology

Pneumonia occurs worldwide as a major cause of death in the elderly ('the old man's friend'), in the unfit and malnourished and in children in crowded, deprived conditions. Antimicrobial treatment has greatly reduced mortality in developed countries in all but the oldest age groups. Environmental sources of infection include birds and other animals (sources of psittacosis, Q fever, and plague), contaminated mechanical ventilation systems (sources of legionellosis) and humidifiers and respirators (sources of *Pseudomonas* infection).

Clinical presentation (Table 2.5) *and complications*

Onset may be abrupt, more usually gradual with fever (not always present in elderly or very ill patients). Breathlessness and often rapid breathing—which may be shallow—are common. Cough, usually unproductive at first, may become purulent or blood-stained. Inspiratory crepitations are usually heard early over the affected area, often before radiographic changes appear. Bronchial breathing and dullness to percussion are common later. Breath sounds may be

Table 2.5. *Typical features of pneumonia*

	Lobar pneumonia (Fig. 2.18)	Bronchopneumonia (Fig. 2.15)	'Atypical' pneumonia (Fig. 2.21)
Chest movements	Reduced on affected side	No change or symmetrical reduction	No change
Percussion note	Dull over affected lobe	Variable	Variable often no change
Breath sounds	Bronchial breathing	Harsh or normal	Usually normal
Vocal resonance	Increased plus whispering pectoriloquy	Normal	Normal
Other sounds	Fine crepitations becoming coarse later	Coarse crepitations and rhonchi	Crepitations, rhonchi, or none
Pleural rub	Common	Occasional	Occasional
Fever	Intermittent, spiking	Continuous	Intermittent or continuous
Sputum	Small amounts early increasing later, may be bloodstained	Copious and purulent	Variable, usually little

reduced by pleural effusion or limited movements caused by pleural pain. This pain may be over the affected area or referred to the shoulder tip when the diaphragm is involved. Sudden resolution of pleuritic pain suggests developing pleural effusion. *Pleural effusion* in bacterial pneumonia is usually straw-coloured with predominantly polymorphonuclear cells but more lymphocytes in later stages. Primary tuberculous effusions are almost entirely lymphocytic but may contain some polymorphs when secondary to tuberculous bronchopneumonia (Chapter 9). Blood in the fluid suggests malignancy, or pulmonary embolus.

Empyema is a complication of bacterial pneumonia characterized by persistent fever, systemic upset, and failure to respond to therapy. It usually follows pneumonia due to *Staph. aureus* (Fig. 2.20), Gram-negative or anaerobic infection, occasionally pneumococcal or rare mycotic pneumonias. Aspiration and drainage of purulent effusion assist both diagnosis and treatment, supplemented by appropriate antimicrobial therapy and treatment of any other underlying condition.

Lung abscess can follow pneumonia, particularly staphylococcal. It can also follow aspiration into the lung of organisms from infected teeth or sinuses when coughing is impaired. It may be secondary to bronchial obstruction by inhaled vomit or food, other foreign bodies, or tumour. Staphylococci, Gram-negative and anaerobic organisms are mainly involved. Fever and cough are usual. Sputum may be abundant and foul smelling in anaerobic infection. Bronchoscopy is required to look for obstruction, for foreign bodies, and surgical drainage may be necessary but less often with adequate antibiotic therapy.

Investigations

Chest X-ray confirms the presence and distribution of lung consolidation and other abnormalities such as pleural effusion or cavitation. Before starting antibiotic therapy, blood culture and sputum, if available, should be taken to identify and check the sensitivity of the infection to antimicrobial drugs. Direct staining of sputum may reveal pneumococcal, staphylococcal, tuberculous, or Gram-negative organisms. *Legionella* antigen may be detected by immunofluorescence in sputum or urine. Blood should also be sampled for serological detection of mycoplasma, chlamydia, Q fever, or legionella antibodies. Polymorphonuclear leukocytosis is usual in non-tuberculous bacterial pneumonias, a mildly raised count with normal differential in *Mycoplasma* infection, normal or low counts in chlamydial, tuberculous, or Q fever infection. Cold agglutinins may be present in *Mycoplasma pneumoniae* infection. Organs other than lungs may be involved and a high creatine kinase level suggests Legionnaires' disease. Blood gas estimations are essential for management of severe cases. Follow-up chest X-rays and perhaps bronchoscopy are important if underlying pathology such as carcinoma is suspected.

Management and prognosis

In all but the mildest cases antibiotics (Table 2.6) should initially be given

Table 2.6. *Suitable initial choice of antimicrobials in pneumonia*

Infecting agents	Antimicrobials
Pneumococcus	Benzylpenicillin or erythromycin
Legionella	Erythromycin and/or rifampicin
Mycoplasma	Erythromycin or oxytetracycline
Staphylococcus	Flucloxacillin + an aminoglycoside
H. influenzae	Chloramphenicol, cefotaxime, or augmentin (in less severely ill)
Klebsiella pneumoniae	An aminoglycoside
Anaerobic organisms	Benzylpenicillin or amoxycillin + metronidazole (add aminoglycoside if source if subdiaphragmatic)
Unknown bacterial cause	Erythromycin + an aminoglycoside (+ metronidazole)
Cox. burnetii	Oxytetracycline or erythromycin
Chlamydia	Erythromycin or oxytetracycline
Varicella zoster	Acyclovir
Cytomegalovirus	Ganciclovir
Respiratory syncytial virus (severe)	Ribavirin
Pneumocystis carinii	Cotrimoxazole, or pentamidine

parenterally. Oxygen is often necessary and should be humidified. Physiotherapy should be given several times a day (and night) if there is significant sputum, although those very ill may need direct suction via an endotracheal tube or tracheostomy. Assisted ventilation can be life-saving in respiratory failure. Regular examination is essential to detect complications such as cardiac arrhythmias or failure, pleural effusion, abscess formation, or empyema. Prognosis depends on previous health, on underlying conditions such as emphysema or carcinoma, and on the effectiveness of antimicrobial treatment. Especially in previously fit patients, attentive and intensive management can assist complete recovery even of those extremely ill.

BACTERIAL PNEUMONIAS

Pneumococcal pneumonia

The *Pneumococcus* remains the commonest bacterial cause of classical lobar pneumonia (Fig. 2.18). The fully developed illness is common in less developed

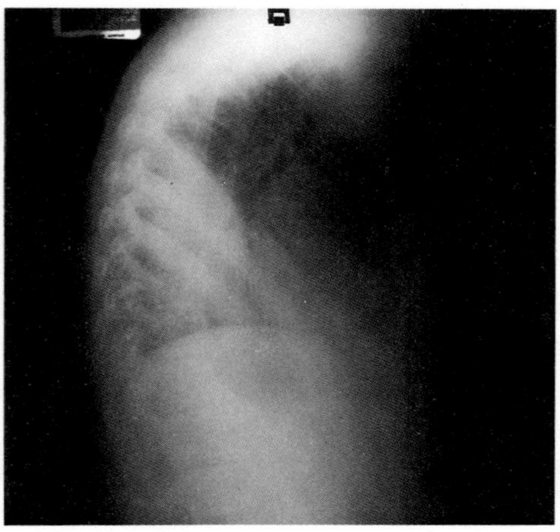

Fig. 2.18 Lobar consolidation in pneumococcal pneumonia.

and tropical regions but less so in countries such as Britain. This may be due to different intensity of exposure, nutritional status, or earlier use of antimicrobials; pneumococci are commonly found in the respiratory tract, with or without pneumonia, and belong to numerous serotypes with different degrees of pathogenicity. Pneumococcal pneumonia is a common complication in AIDS.

Clinical presentation

Onset is usually rapid with high swinging fever, rigors, cough, and small amounts of rust-coloured sputum. In very early stages, fine crepitations may be heard before full signs of pneumonia develop. The untreated illness progresses with increasing fever, consolidation, signs of respiratory distress, and delirium until the 'crisis', sudden fall of teperature and resolution or death around the tenth day. Lobular and bronchopneumonia can occur especially in the very young and elderly in whom symptoms may develop more insidiously.

Investigations and management

The organism can often be cultured from blood collected before treatment or from sputum. Counter-immunoelectrophoresis may reveal pneumococcal antigens in sputum, serum, or urine. Early use of benzylpenicillin or other appropriate antimicrobial markedly reduces morbidity and mortality. Mortality is highest in bacteraemic cases. Polyvalent polysaccharide vaccine is available to reduce the risk of pneumococcal infection in the elderly, those undergoing splenectomy or with other immunocompromising conditions.

Legionnaires' disease

Legionella pneumophila is the commonest member of the Legionellaceae causing pneumonia. These Gram-negative organisms are widely distributed in warm, wet habitats. Infection occurs by inhalation of airborne droplets or dust containing legionellas. They are often dispersed from taps, showers, or spray from cooling towers, causing outbreaks or sporadic cases. Air-conditioning systems have distributed the organisms in outbreaks in hospitals and hotels. Person-to-person spread is unknown. About one-third of reported cases in Britain are contracted abroad.

Clinical presentation

The illness may resemble pneumococcal pneumonia but with slower onset of malaise, myalgia, headache, increasing fever, and dry cough. Illness progresses, often with diarrhoea, abdominal and chest pain, disproportionate confusion or clouding of consciousness, and sometimes neurological features such as ataxia and peripheral neuropathy. There is often renal insufficiency with blood and protein in the urine. Occasionally pneumonia is absent. Serious disease is more likely in older adults, smokers, heavy drinkers, and those with underlying disease or immune compromise.

Investigation and management (Fig. 2.19a and b)

Early clinical suspicion of this disease is important to ensure prompt treatment with appropriate antibiotics (high dose erythromycin and/or rifampicin). The organisms may be demonstrated in respiratory secretions or lung biopsy material by immunofluorescence, or cultured on selective media, more easily from tracheal aspirate or bronchial washings than from sputum but diagnosis is

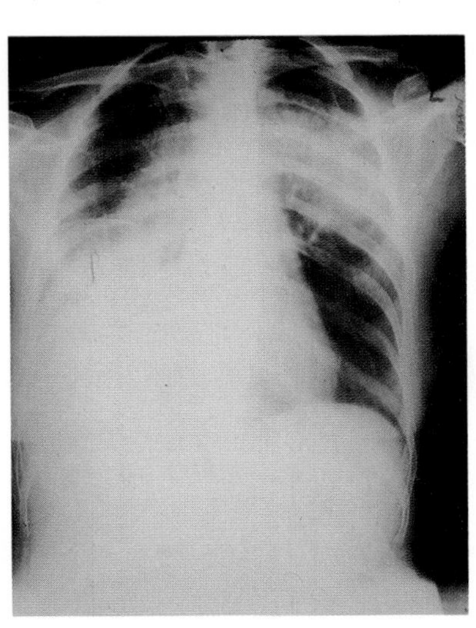

(a)

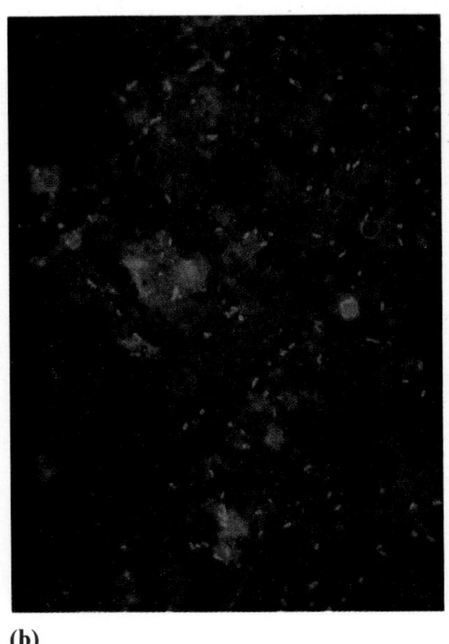

(b)

Fig. 2.19 (**a**) Multi-lobar consolidation in Legionnaires' disease; (**b**) Legionnaires' bacillus in lung biopsy material demonstrated by immunofluorescence.

usually achieved by detecting the rising titre of antibodies which may take up to six weeks. A high creatine kinase, low sodium and absolute lymphopenia suggest *Legionella* infection and are useful early findings.

Prognosis

In outbreaks mortality varies from 10 to 25 per cent. The long-term prognosis is usually good but there may be persistent respiratory embarassment and neurological sequelae, especially in those previously unfit.

Prevention

In localized outbreaks a common source should be sought in order to undertake preventative measures such as cleaning and chlorination of water-supply systems, cooling towers, and air ducts. Maintaining hot water supplies above 55 °C is advocated.

Staphylococcal pneumonia (Figs. 2.15 and 2.20)

This type of bronchopneumonia typically develops during viral infections such as influenza or measles, and is usually accompanied by septicaemia. Intravenous drug users are especially susceptible usually through infected emboli originating from right-sided endocarditis. It is also common in diabetes and those with cystic fibrosis.

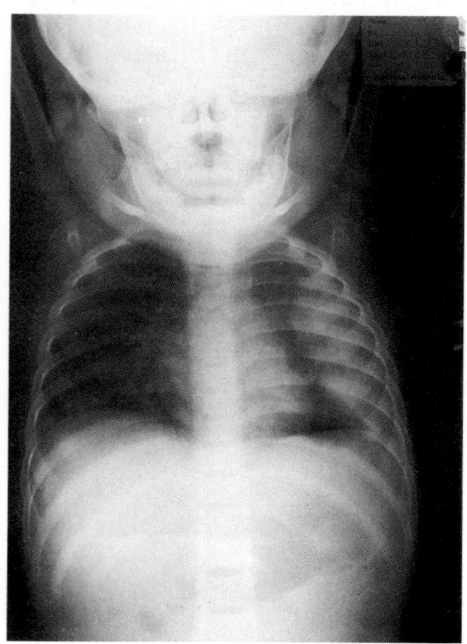

Fig. 2.20 Empyema following staphylococcal pneumonia.

Clinical presentation

The onset may be very rapid, even fulminant, causing much lung damage before antibiotics can be administered. There may be abundant blood-stained sputum, multiple lung abscesses, empyema and abscesses elsewhere in the body. Endocarditis may result from the septicaemia.

Investigations and management

Staphylococci are profuse in sputum and can usually be isolated from blood. Echocardiography should be performed when endocarditis is suspected. Treatment should preferably comprise two antistaphylococcal drugs, checked *in vitro* against the patient's own organism to exclude antagonism, and continued for about six weeks. Death may occur in a few hours in fulminant cases. Tracheostomy, clearance of airways, and intensive care may be required although the outlook in drug users with endocarditis if often good. Renal failure and the nephrotic syndrome can result from immune complex deposition in kidneys. Follow-up is important in those with cardiac involvement.

Haemophilus influenzae pneumonia

Haemophilus influenzae may cause exacerbations of chronic bronchitis or bronchiectasis with acute but rarely severe illness. Infection is much more serious in young children when generalized infection may also involve the epiglottis (see above), larynx, bloodstream, and meninges, in addition to causing bronchopneumonia. Treatment is by chloramphenicol or cefotaxime.

Klebsiella (Friedlander's) pneumonia

Klebsiella pneumoniae is in the normal respiratory flora of up to 25 per cent of people. It occasionally causes marked consolidation with abscess formation, prostration and thick bloody sputum, especially in debilitated patients and chronic alcoholics. Treatment is with aminoglycosides. Mortality is high.

Mycoplasma pneumonia

Mycoplasma pneumoniae causes epidemics lasting many months every few years (Fig. 2.1). Outbreaks in institutions and sporadic cases occur, mostly in cooler months of the year. All ages are affected, most often older children and young adults.

Clinical presentation

Many illnesses are mild. There may be associated upper respiratory tract symptoms, sore throat, or erythematous rash, and often pneumonia is revealed by chest radiography despite few abnormal signs on examination. The consolidation is usually lobular rather than lobar (Fig. 2.21).

Investigations and management

The organism cannot easily be cultured and diagnosis is usually made serologically. Cold agglutinins may be present and can precipitate haemolytic anaemia or thrombosis if the patient is exposed to cold. The sedimentation rate is often markedly elevated. Infections may resolve spontaneously but usually respond rapidly to erythromycin. Recovery is usually complete but convalescence may be slow. Erythema multiforme (Chapter 13) is occasionally associated.

Gram-negative bacterial pneumonia

Pneumonia caused by *Escherichia coli*, *Pseudomonas aeruginosa*, *Serratia marcescens*, and other aerobic Gram-negative bacteria has increased in recent years, particularly in elderly and ill patients both in hospital and in the community. These organisms may be commensals aspirated from the upper tract, inhaled from infected ventilators or humidifiers, or blood-borne from foci often in genito-urinary or gastrointestinal tracts. The use of broad-spectrum antibiotics in ill and elderly patients predisposes to colonization and infection. Isolation of the same organisms from blood as well as repeatedly from the respiratory tract supports the diagnosis. Antimicrobial therapy depends on the sensitivity of the

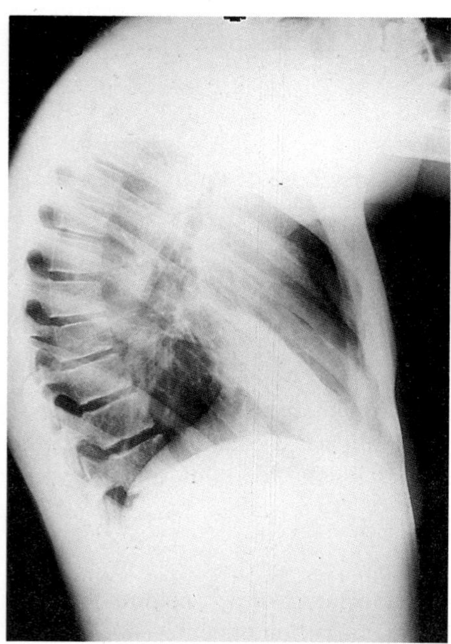

Fig. 2.21 Localized consolidation in *Mycoplasma* pneumonia may be associated with few or no abnormal physical signs.

organism—often an aminoglycoside. The prognosis depends largely on the underlying condition which is usually a serious illness receiving intensive care.

Anaerobic bacterial pneumonia

This usually follows aspiration of oral or upper respiratory tract material. Dental sepsis, foul sputum, and a history of unconsciousness with suppressed coughing suggest this infection which can extensively damage the lung. The organisms may be only isolated from blood or lung biopsy specimens.

OTHER BACTERIAL PNEUMONIAS

Uncommon causes of pneumonia include infections acquired from animal sources by inhalation, such as Q fever, chlamydial pneumonia, some of which (anthrax, Chapter 3; plague, Chapter 8) can cause fulminating disease with high mortality.

Rickettsial pneumonia: Q fever

Rickettsias are highly infectious by inhalation, usually causing systemic disease (Chapter 14) but pneumonia is usual in Q fever. *Coxiella burneti*, the cause of Q fever, is a common world-wide infection of sheep, cattle, and many other animals, often airborne over long distances. Sharp, febrile illness is usual in Q fever with possible acute or chronic infection of heart valves, liver and other organs. Diagnosis is usually serological, treatment by tetracycline.

Chlamydial pneumonia

These infections also are usually diagnosed serologically and treated by tetracycline or erythromycin.

Psittacosis/ornithosis

Systemic infection with fever and usually pneumonia can be caused by *Chlamydia psittaci* inhaled from birds which may be ill or apparently healthy. Psittacine birds (parrots, budgerigars) can cause the more severe disease *psittacosis*. The less severe *ornithosis* is acquired from other birds such as pigeons, ducks, turkeys.

Chlamydia pneumoniae (strain TWAR), a human adapted strain, has recently been recognized as a respiratory pathogen. It causes illness resembling ornithosis, sometimes involving myocardium, usually spreading from human to human.

Chlamydial pneumonia in infants

This is caused by human-adapted strains of *C. trachomatis* which commonly infect the genital tract or eye (Chapter 11). Contracted during birth, it causes diffuse pneumonia one to three months later.

Mycotic pneumonia

These infections are uncommon, but increasingly encountered in immunosuppressed patients and in AIDS. The organisms are common in the natural environment but rarely invade immunologically normal persons.

Aspergillosis

Aspergillus fumigatus can cause allergic fever and bronchospasm but also, in the immunocompromised, acute and often fatal illness with pneumonia and dissemination to other organs. A fungus ball (aspergilloma) may develop in lung damaged by bronchiectasis or in a tuberculous cavity. Culture of tissue is more reliable for diagnosis since the organism may be a commensal in sputum. Antifungal drugs tend to be toxic, unpredictable in effectiveness, and generally reserved for more severe infections (Chapter 18).

Cryptococcosis

Cryptococcus neoformans, usually originating from pigeon droppings, likewise causes systemic disease including pneumonia. Granulomas containing the organism cause radiological opacities in the lung. Serological tests can aid diagnosis and treatment is with amphotericin B or fluconazole.

Pneumocystis pneumonia

Pneumocystis carinii, formerly considered a protozoon but now recognized as a fungus, is mainly encountered as one of the commoner opportunistic infections in AIDS or immunosuppression; for example, after organ transplants (Chapter 12).

Viral pneumonia

Many viruses which infect the respiratory tract can occasionally cause pneumonia, most often in young children. Immunocompromised patients and sometimes the elderly may develop pneumonia from such viruses as parainfluenza and respiratory syncytial virus.

Influenzal pneumonia

This is rare, but a few of the fulminant pneumonias encountered mainly in young adults during epidemics are due to virus alone, though more often associated with bacteria (see above).

Adenoviral pneumonia

Adenoviruses occasionally cause severe pneumonia in young children, often with distressing cough resembling whooping cough. Certain types have caused outbreaks of atypical pneumonia in communities of young adults such as military recruits.

Measles

This causes viral bronchitis as a feature of the usual illness, but can also cause severe giant-cell pneumonia in the immunodeficient child or immunocompromised adult.

Varicella

This likewise can involve the lung during chickenpox in immunocompromise or pregnancy, especially in smokers, causing pneumonia with high mortality. Early acyclovir treatment is essential.

Cytomegalovirus

This virus of the Herpes group can cause pneumonia as a primary manifestation in infancy, and is one of the commoner infections reactivated by immunocompromise, e.g. after renal transplantation, as life-threatening pneumonia.

3

Localized skin and tissue infections

Infections remain localized when generalized or systemic spread has not occurred, as in cutaneous anthrax, when effective immunity has prevented spread or, occasionally, when the infecting organism is not invasive. In addition to localized infections of skin and subcutaneous tissue, this chapter deals with those of muscles, lymph nodes, bones, and joints.

SKIN AND SUBCUTANEOUS TISSUE INFECTIONS

These infections usually show the classical quartet of heat, swelling, tenderness and redness, sometimes with additional distinctive features. Diagnostic assessment requires both history and examination of the character of the lesion, its distribution and course of evolution both of its appearance and distribution. Local rashes can be roughly divided into groups (Table 3.1). Some, like scabies, may be initially localized before becoming more widespread.

Erysipelas

Aetiology and epidemiology

Streptococcus pyogenes causes this erythematous intradermal infection. It is commonest in older females. The organisms spread particularly along lymphatics, causing their fibrosis. Minor degrees of lymphoedema may result and predispose to recurrences at the same site.

Clinical features and differential diagnosis

An area of local erythema and tenderness develops at the site of infection, usually without any obvious preceding skin lesion or wound. Redness and tenderness rapidly spread within the dermis and, for example, within 24 hours can spread from the ankle to involve all skin below the knee. Local oedema is best demonstrable at the edge of the lesion, which is sharply demarcated and readily palpable (Fig. 3.1) as oedema is trapped within the dermis. Systemic toxicity is usually marked with fever, chills and sweats, and confusion or delirium in the elderly. The commonest sites are lower leg and face. On the face it usually starts on one cheek and spreads rapidly to the other to give a typical butterfly distribution (Fig. 3.2). Draining lymph nodes may be enlarged and tender.

Table 3.1. *Localized rashes due to infection*

Condition	Key local diagnostic features
Erythema	
Erysipelas	Lower limb or face, sharply demarcated palpable border.
Cellulitis	
streptococcal	Poorly demarcated, border not palpable.
anaerobic	Poorly demarcated, crepitus, gangrene.
Lyme disease	Erythema chronicum migrans—slowly enlarging erythema from site of bite with resolution in centre.
Vesicles	
Herpes simplex	Perioral or genital, may be autoinoculated elsewhere. Sometimes identifiable provocator if reactivation.
Shingles	Unilateral dermatomal distribution.
Pustules	
Boils/carbuncles/ abscesses	Fluctuant, pointing, purulent discharge.
Orf } Paravaccinia }	Animal contact; on hands or arm, larger than herpetic lesions.
Anthrax	Contact; black centre, daughter vesicles, massive surrounding oedema.
Crusts	
Impetigo	Golden crusts, starting around nose. Children.
Dried vesicular or pustular rashes	See above.
Scaling	
Cutaneous mycoses	Often circular with active edges, in warm and moist areas.
Pityriasis versicolor	Slow enlargement of lightly pigmented, slightly flaky lesions, especially on trunk. Hypopigmented when sun-tanned.
Pityriasis rosea (Chapter 7)	Initial single herald patch with scaly border before more generalized rash, otherwise symptomless.
Others	
Infected wounds	Primary trauma; spreading cellulitis, discharge.
Scabies	Distribution, burrows, pruritus.
Pediculosis	Excoriation, pruritus, nits (with hair lice), poor hygiene (with body lice).
Septicaemias	Haemorrhagic lesions, especially peripheral.
Warts	Commonly nodular or flat. Usually localized. May be genital or plantar.
Molluscum contagiosum	Hard, pale papules with umbilication. Not itchy.
Tuberculosis (Chapter 9)	
Lupus vulgaris	Chronic grouped nodular lesions, with surrounding red/brown discoloration. Apple jelly appearance on pressure. Results in scarring.
Cold abscess	Discharging sinus usually with underlying induration but lacking heat, tenderness, and erythema.

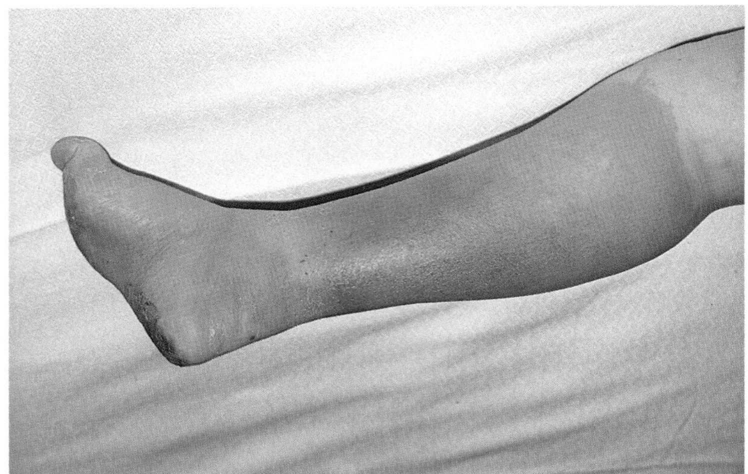

Fig. 3.1 Erysipelas.

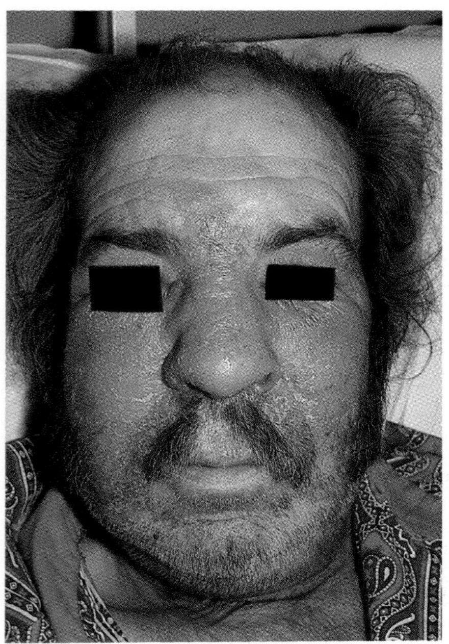

Fig. 3.2 Erysipelas. Two days after starting treatment the fiery erythema has begun to fade and some desquamation is evident.

The main differential diagnosis is from infection within subcutaneous tissues (cellulitis—see below) which is also caused mainly by *Strep. pyogenes* but is distinguished by the lack of a well-demarcated, palpable border.

Investigations and management

It is often difficult to isolate *Strep. pyogenes* from skin swabs but any moist areas should be sampled. Pharyngeal carriage is often present. Blood cultures should be taken but are rarely positive. Neutrophil leucocytosis and a rise in streptococcal antibodies are usual.

Parenteral benzylpenicillin causes steady improvement in fever and toxicity but erythema fades only slowly from bright red to a browner appearance and from the periphery inwards. When severer symptoms have settled, oral penicillin may be substituted and continued for 10–14 days.

Prognosis

Septicaemic complications are rare. Recurrences are common and increasingly likely with each recurrence. If, in addition, the patient becomes sensitized to commonly used drugs such as penicillin, erythromycin, and co-trimoxazole, others such as tetracycline, ciprofloxacin or clindamycin may be necessary. With each recurrence more lymphatics become fibrosed, eventually with chronic lymphoedema and an 'elephantiasis'-like appearance (Fig. 3.3). Patients prone to recurrence should maintain good skin hygiene, keep cuts and abrasions clean and covered and report early evidence of infection.

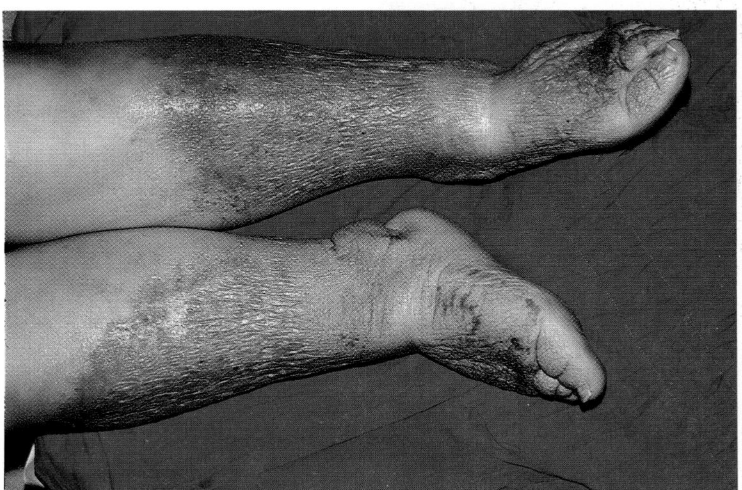

Fig. 3.3 Chronic lymphoedema after several attacks of erysipelas.

Cellulitis

Aetiology and epidemiology

Cellulitis is inflammation of subcutaneous tissues caused usually by *Strep. pyogenes*. An important minority of cases is caused by other organisms, particularly anaerobes like *Clostridium perfringens*. Infection is usually secondary, entering through minor skin lesions such as cuts or chickenpox lesions. Anaerobic infection is most likely in poorly perfused tissue, especially after deeper wounds, fractures, or surgical procedures such as amputation. Such infections are often caused by multiple organisms.

Clinical features

Streptococcus pyogenes cellulitis, like erysipelas, rapidly spreads as a localized warm erythematous area; tense and tender to touch. As infection is beneath the dermis local oedema can diffuse more readily and the edge of the lesion is neither well-defined nor palpable (Fig. 3.4). High fever is likely and the patient feels unwell. A scarlet fever rash can develop if the responsible strain produces erythrogenic toxin to which the patient has no antibodies. Septicaemia sometimes results. If, as on the hands of butchers and abattoir workers, *Strep. pyogenes* infection of minor injuries recurs frequently, then chronic, local, slightly exudative lesions develop and systemic symptoms and spreading cellulitis become less common, suggesting development of some immunity.

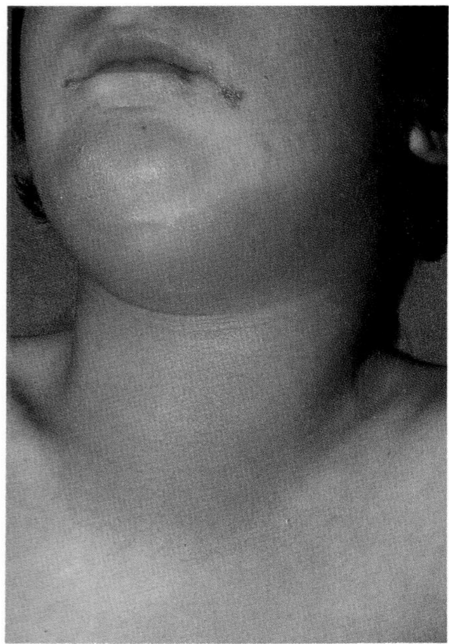

Fig. 3.4 Cellulitis. Local spread from streptococcal cervical adenitis.

Anaerobic cellulitis has the above features but is usually accompanied by local interruption of blood supply within subcutaneous tissues, leading to dusky overlying erythema and eventual patches of gangrene (Fig. 3.5). Crepitus in involved tissues may be present if gas-producing anaerobes are responsible. The patient is usually very ill and rapidly becomes shocked and comatose. Clinical categories of anaerobic cellulitis (Table 3.2) vary with the organisms responsible, the overlying skin appearance, and the depth and tissue plane along which infection spreads. However, many of these features overlap. In necrotizing fasciitis, rapid necrosis spreads along fascial planes involving fascia much more extensively than overlying, visible skin lesions suggest.

Infection secondary to trauma, fractures or surgery may cross fascial planes to involve underlying structures such as muscle and bone. Classical *gas gangrene* caused by *Cl. perfringens* involves muscle and subcutaneous tissues. Septicaemia is likely.

Investigation and management

Neutrophil leucocytosis is present. Blood should be cultured and any moist lesion swabbed for culture. If surgery is undertaken the wound should be swabbed and necrotic tissue sent to the laboratory. If there is no major underlying precipitant such as recent surgery or trauma, and no evidence of gangrene, crepitus or severe toxicity, it is reasonable to diagnose *Strep. Pyogenes* infection and treat with benzylpenicillin parenterally.

If anaerobic or mixed infections are suspected, necrotic tissue should be

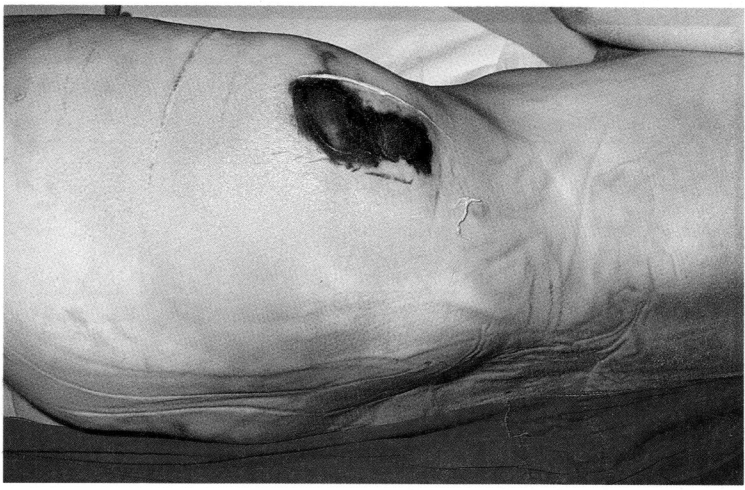

Fig. 3.5 Anaerobic cellulitis. Surgery revealed necrotizing fasciitis. Organisms spread via a rectal sinus resulting from Crohn's disease. Eight different anaerobes or facultative anaerobes were isolated.

Table 3.2. *Examples of gangrenous skin and soft tissue infections*

Variety	Main features
Progressive bacterial synergistic gangrene	Slow development and spread in skin and superficial subcutaneous fat. Central necrosis, surrounded by dusky and then erythematous zones. Caused by non-haemolytic *Staph. aureus.*
Meleney's ulcer	Slow development and spread in skin and subcutaneous fat. Enlarging ulcer with undermined edges resulting in fistulae and further ulcers. No gangrene. Caused by haemolytic streptococci.
Fournier's gangrene	Rapid development and toxicity. Superficial, usually scrotal or perineal. Often caused by coliforms and *Bact. fragilis.*
Necrotizing fasciitis	Rapid; toxicity. Affects deeper subcutaneous fat and fascia. Gangrene often remote from source. Usually caused by mixed coliforms and anaerobes, occasionally *Strep. pyogenes.*
Synergistic necrotizing cellulitis	Slow onset, rapid progression. Can involve underlying muscle. Gangrene patchy with foul smell and crepitus. Usually caused by coliforms and *Bact. fragilis.*
Clostridial cellulitis	Slow onset and progression. Muscle not involved. Slight, foul-smelling, serous exudate; crepitus. Usually caused by *Cl. perfringens.*
Clostridial myonecrosis (gas gangrene—see also Table 3.4)	Rapid onset and progression. Predominantly muscle involvement. Profuse serous exudate with sweet smell. Toxicity. Usually caused by *Cl. perfringens* but also *Cl. oedematiens* and *Cl. septicum.*

surgically exposed and excised and a combination of antibiotics such as ampicillin, gentamicin, and metronidazole given for wide cover against anaerobes and facultative anaerobes. Hyperbaric oxygen may be helpful and shock requires resuscitation.

Prognosis

Streptococcal cellulitis usually resolves uneventfully with benzylpenicillin, but as cellulitis resolves suppuration of draining lymph nodes may occasionally become evident and require surgical drainage. Mortality is high in well-established, anaerobic soft tissue infections. Large areas of excized tissue require skin grafting after healthy granulation tissue has formed.

Impetigo

Aetiology and epidemiology

Staphylococcus aureus is the usual cause, sometimes associated with *Streptococcus pyogenes* which is occasionally isolated on its own. *Staphylococcus aureus* phage type II is particularly associated with bullous impetigo. Most frequent in young children, it is spread by contact or via fomites, commonly within a household.

Clinical features

Lesions usually start around the nares and spread locally and wherever inoculated on other sites, particularly hands. Irregular, golden-yellow crusts are typical (Fig. 3.6), without preceding vesicular or pustular stages. There may be surrounding erythema, and exfoliation of skin may give the appearance of a burst bulla. There is no systemic illness unless deeper tissue is invaded. Draining lymph nodes may be enlarged and tender, but septicaemia is rare. The main diagnostic differentiation is between primary impetigo and secondarily impetigenized lesions of, for example, chickenpox, scabies, pediculosis, or eczema, reflecting the distribution of the underlying disease. Differentiation may be difficult where the lesion is around the nose as with impetigenization of reactivated herpes simplex lesions.

Management and investigation

Swabs of lesions readily yield *Staph, aureus.* Unless there is systemic spread, shown by fever, underlying cellulitis, or enlarged, tender lymph nodes, treatment with a topical antistaphylococcal antibiotic is sufficient. As topical preparations

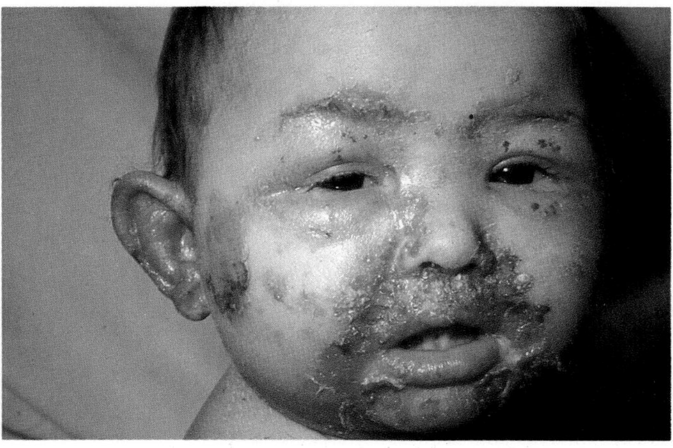

Fig. 3.6 Impetigo. Golden crusts with some exfoliation around mouth and on cheek.

can be allergenic, agents less likely to be used systemically such as fusidic acid are preferable. Healing follows without scarring.

Boils and carbuncles

Aetiology and epidemiology

These are usually caused by *Staph. aureus*. Boils usually originate in hair follicles, particularly in warm, moist areas of skin such as neck or axilla, or in areas of previous trauma such as injection sites (Fig. 3.7) or insect bites. They are commoner in diabetics and the immunocompromised, particularly those with abnormal neutrophil function.

Clinical findings

A boil begins as a firm, tender papule which then suppurates. Systemic features such as fever and malaise may be present, particularly if there is local spread as shown by underlying cellulitis or enlarged, tender draining lymph nodes. Carbuncles are larger, deeper, and also involve underlying subcutaneous tissue. Commonest on the nape of the neck or back (Fig. 3.8a) they discharge eventually through multiple sinuses.

Investigation and management

Pus should be cultured and diabetes excluded. Suppuration and 'pointing' is encouraged by local heat. Larger collections of pus should be incised and drained. Antistaphylococcal antibiotics, such as flucloxacillin, are advisable for all but the simplest boils to prevent spread, especially on the face, and given parenterally to those severely ill.

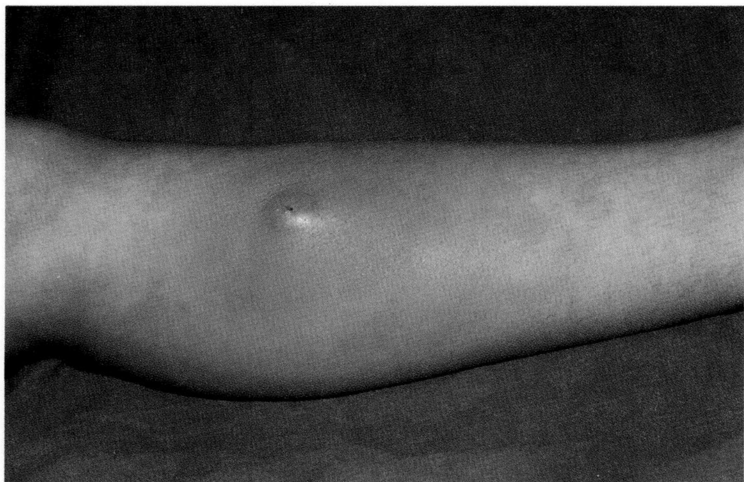

Fig. 3.7 Boil. Staphylococcal infection of lower leg at injection site in drug abuser.

(a)

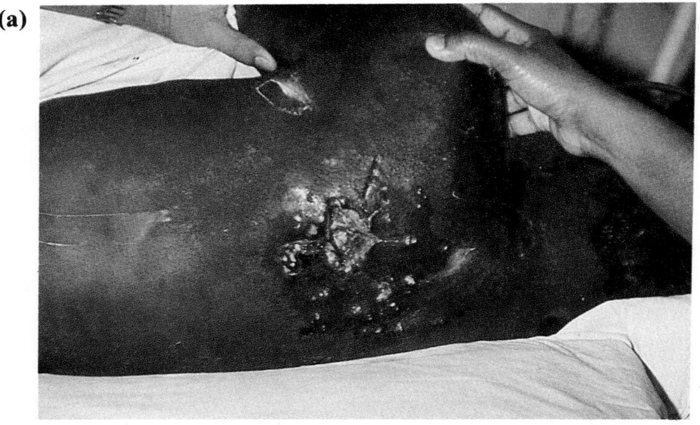

(b)

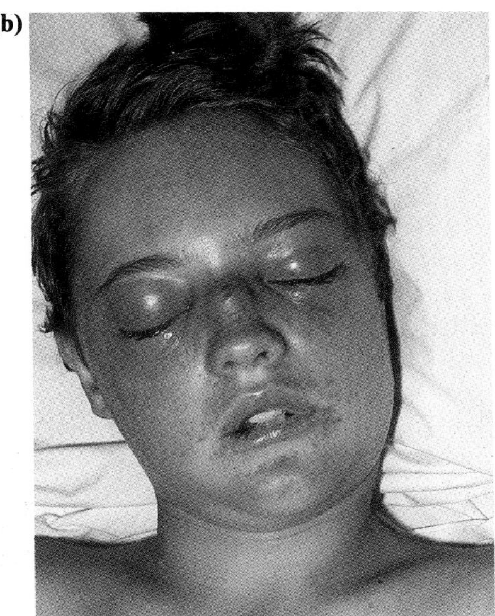

Fig. 3.8 **(a)** Carbuncle; **(b)** Staphylococcal boil on the bridge of nose has led to orbital cellulitis.

Prognosis

Septicaemia and local spread may complicate even simple skin lesions, particularly in the immunosuppressed. Spread along facial veins may lead to periorbital cellulitis (Fig. 3.8b) or septic thrombosis of the cavernous sinus. Sufferers from recurrent staphylococcal abscesses often have low serum iron without anaemia. Whether this is causative or reflects increased iron utilization is unclear but iron supplements lessen the likelihood of further recurrence.

Breast abscess

Aetiology and epidemiology

Staphylococcus aureus is responsible for most cases of mastitis and abscess in the lactating breast. These usually develop in the second or third week after delivery and affect approximately 5 per cent of women.

Clinical presentation and management

Fever is accompanied by local heat, redness, tenderness, and swelling in one breast quadrant. At this early stage, oral flucloxacillin should be given and breast-feeding may be continued. However, should a fluctuant area be detected, drainage is also necessary and breast-feeding from the affected breast should be temporarily discontinued to avoid precipitating staphylococcal sepsis in the infant. Milk should be removed from that breast by pump and discarded. Breast-feeding can be resumed after resolution.

Herpes simplex

Aetiology and epidemiology

Primary herpes infection usually involves the mouth (Chapter 4), occasionally with dissemination, or the genitalia (Chapter 10), or sometimes is inoculated locally into skin by trauma ('scrumpox', 'herpes gladiatorum') or contamination causing paronychia (Fig. 4.11).

After primary infection, non-replicating herpesviruses persist in sensory ganglia. After reactivation by suitable stimuli (commonly sun, cold, trauma, menstruation, fever, or viral respiratory infections), the virus travels along sensory nerves to cause skin lesions. Bacterial meningitis, pneumonias, and malaria are potent but less common reactivators; depression of cell-mediated immunity, especially by steroids, leukaemia, or lymphoma, often provokes reactivation. Intervals between reactivations vary, commonly from several months to a year or so for type 1 herpes virus reactivations that usually involve the face or mouth. The usual site for reactivation of type 2 is the genital area and recurrences are more frequent initially (Chapter 10).

Clinical manifestations

Reactivation usually affects the border of normal and vermilion skin on the lateral aspect of the lips (Fig. 3.9). It is often preceded by a tingling sensation but no systemic upset (except for any reactivating illness). Lesions are sometimes seen solely on cheeks and forehead; other non-genital sites are occasionally involved. Individual lesions progress through papular, vesicular, pustular, and crusting stages (Fig. 3.10a–d). Lesions are closely grouped and develop in synchrony. Intra-oral lesions alone are uncommon but rarely follow a stimulus such as prolonged sucking of ice. In immunocompromised patients more extensive and

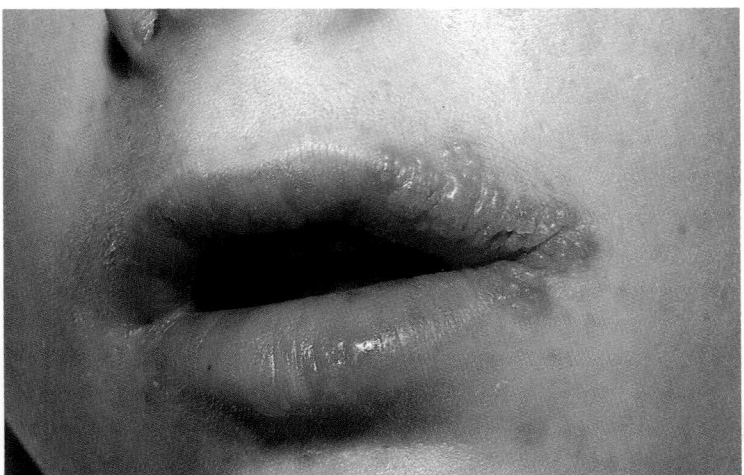

Fig. 3.9 Reactivation of herpes simplex. Cluster of confluent vesicles.

deeper lesions appear with slower resolution or even progression, and spread into the oral cavity is common. Unilateral grouped lesions must be distinguished from shingles; this may be impossible clinically with a sparse shingles rash although zoster lesions are usually scattered over a dermatomal distribution whereas those of herpes tend to be closely grouped.

Investigation

Diagnosis is usually clinical, but electron microscopy of material from the lesions can reveal herpes-type virus particles and their identity as herpes simplex can be shown by culture or serological tests.

Management and complications

In immunocompetent patients herpes skin lesions resolve in 4–5 days. At the stage of prodromal paraesthesia, acyclovir cream or oral tablets may prevent development of overt lesions. In the immunocompromised all lesions should be treated with acyclovir orally or, if disseminated, intravenously.

Secondary infection of skin lesions with *Staph. aureus* or *Strep. pyogenes* may supervene. Herpes simplex infection may precipitate erythema multiforme or Stevens–Johnson syndrome (Chapter 13). These syndromes may reappear with every recurrence of herpes and in these patients it is particularly important to treat early prodromal features with acyclovir.

Shingles (herpes zoster)

Reactivation of latent varicella-zoster virus in sensory ganglia usually results from waning immunity with age. It occasionally affects the young however, even

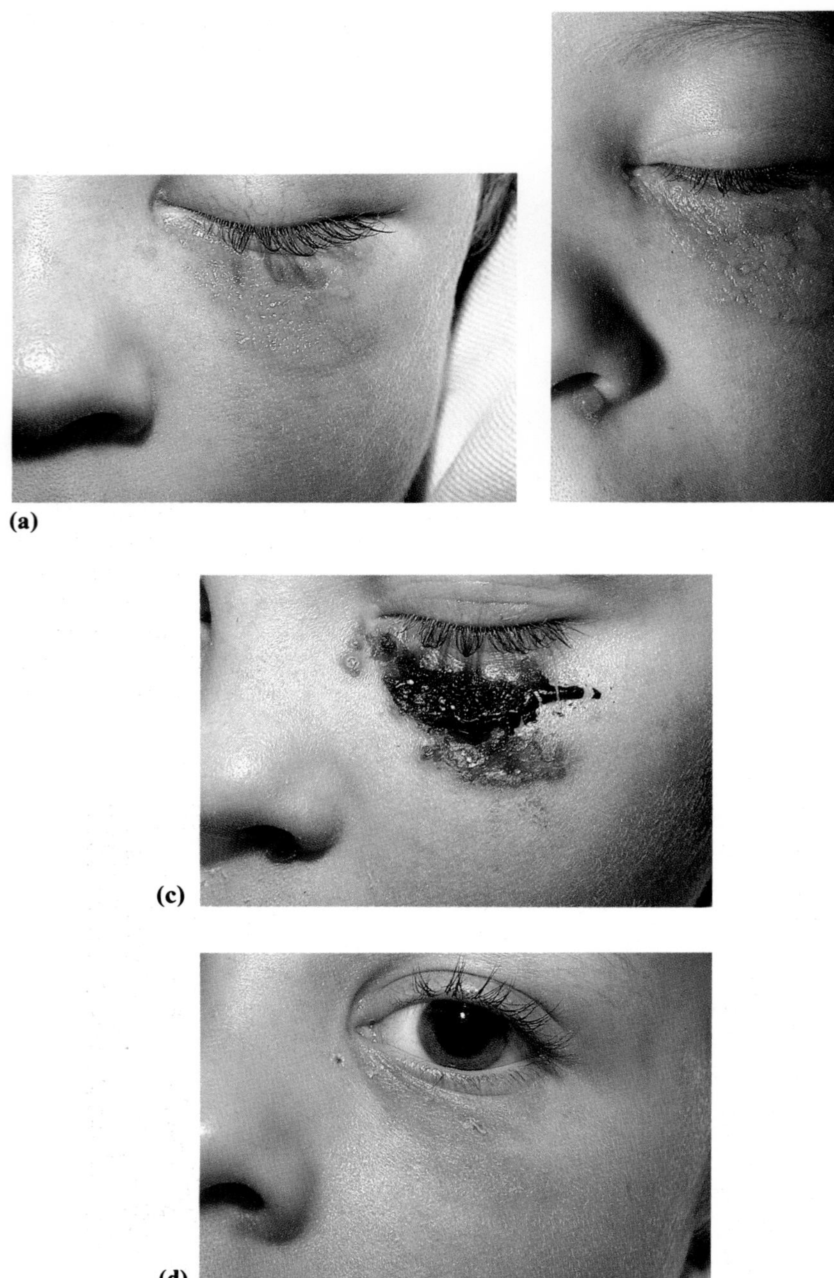

Fig. 3.10 Evolution of reactivated herpes simplex. **(a)** vesicles; **(b)** pustules; **(c)** crusts; **(d)** heals without scarring.

in the neonatal period following maternal varicella during pregnancy. Reactivation is commoner in immunocompromised persons. Second and third attacks of shingles can occur.

Clinical features

The rash may be preceded by a painful, burning sensation or severe pain in the distribution of the affected dermatome. It develops through the same stages as varicella (Chapter 7) which individual lesions resemble. Classically they are closely grouped with unilateral distribution limited to one or two adjacent dermatomes. The commonest single nerve root involved is the ophthalmic division of the trigeminal nerve (Fig. 3.11), although the commonest body area involved is the thorax (Fig. 3.12). Ophthalmic shingles may affect the eye via the long ciliary nerves, often shown by development of skin lesions on the side of the tip of the nose, an area also supplied by a branch from the ciliary ganglion. Oral lesions are present in both maxillary and mandibular shingles. Severe shingles can be accompanied by paresis of muscles supplied by nerves of the same spinal root; for instance, weakness of triceps with shingles over the distribution of C6 and C7. The Ramsay Hunt syndrome is thought to represent reactivation of virus within the geniculate ganglion. It causes cutaneous lesions within the pinna of the ear, oral lesions on the side of the tongue and hard palate, and facial palsy from involvement of the facial nerve as it runs past the geniculate ganglion within the bony facial canal. All forms of shingles are usually accompanied by background pain with intermittent sharp stabs superimposed, but some patients are pain-free. Some, especially the elderly, feel generally unwell and may appear 'toxic'. Slight viraemia occasionally causes a few disseminated lesions scattered elsewhere over

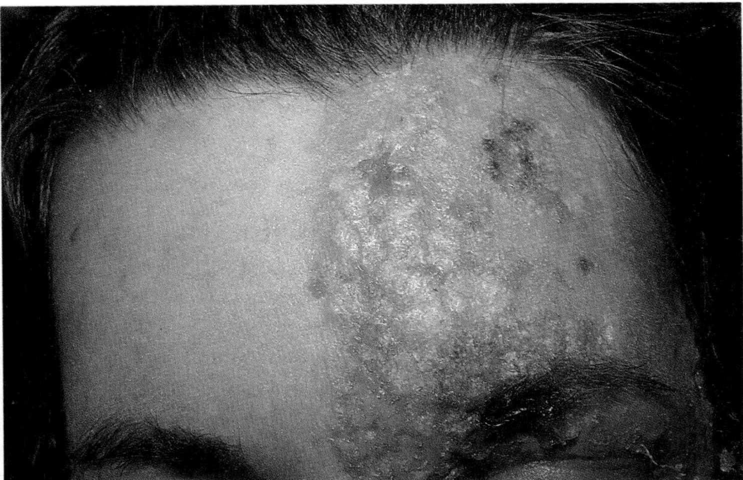

Fig. 3.11 Shingles. Mainly vesicular lesions on erythematous base in distribution of ophthalmic division of left trigeminal nerve.

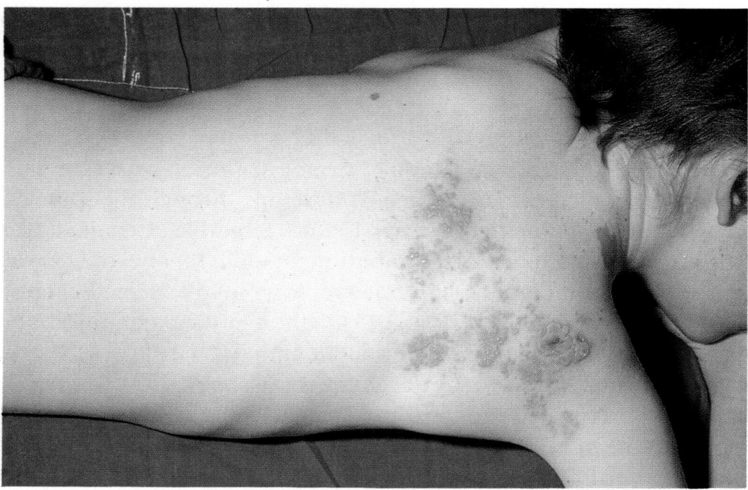

Fig. 3.12 Shingles. Pustular lesions in upper thoracic dermatome.

the skin. Immunocompromised patients may develop larger, more slowly evolving, or progressive lesions, and viraemia often leads to a widespread varicella rash (Fig. 12.8); post-infectious encephalitis is rare.

Investigation and differential diagnosis

Unilateral lesions with dermatomal distribution are unlikely to be confused with other diseases except when a small cluster of reactivated herpes simplex lesions develop at an unusual site such as thorax; viral culture and serology may be necessary to distinguish these from a sparse shingles rash. Severe pain may precede the rash. Migraine, nerve root compression, or emergencies like myocardial infarction or appendicitis may enter initial diagnosis depending on the site of the affected dermatome. Before the rash appears the pain is often thought to be a muscular strain or Bornholm disease.

Management

Pain relief may require strong analgesics. The potentiating action of chlorpromazine may help. Acyclovir is indicated intravenously for those immunocompromised and for severe or spreading lesions. Acyclovir orally and/or topically may speed resolution of mild shingles if used within the first five days of the rash. Pain may persist even when skin lesions have healed; stabbing pains may be very difficult to ease. If conjunctivitis and chemosis from involvement of the ophthalmic branch of the trigeminal nerve are marked, antibiotic eye ointment helps to prevent secondary bacterial infection. Atropine eye drops and oral acyclovir should be used if keratitis is present and ophthalmological advice sought if this is severe. Post-herpetic neuralgia is commoner in elderly females and can be difficult to control, though it usually resolves within six months to two years. Carbamazepine in high dosage may help.

Orf and paravaccinia

These are primarily poxvirus infections of animals now known to be caused by the same virus. Orf (contagious pustular dermatitis) naturally infects sheep and goats causing vesicular lesions that readily ulcerate. Infection of man is by contact, usually direct or from contaminated objects such as fences or clippers, most likely in farm and abattoir workers. A single painless lesion is usual, commonly on the hand, initially a 'meaty' papule slowly enlarging up to 1–1.5 cm. A flat vesicle (Fig. 3.13) may develop on this, often with haemorrhage into the base. There is no systemic upset and slow resolution follows over 3–4 weeks, without scarring. Erythema multiforme can complicate recovery.

Paravaccinia (pseudocowpox) causes vesicular lesions on the udders and teats of cattle. Lesions in man (milkers' nodes) are usually multiple painless papules on the hands that may enlarge to 1 cm and become bluish in colour. These resolve over a few weeks. They do not usually vesiculate or provoke systemic symptoms except for occasional allergic rashes.

Electron microscopy of fluid from early lesions can define the virus morphology and distinguish it from orthopoxviruses such as vaccinia or variola.

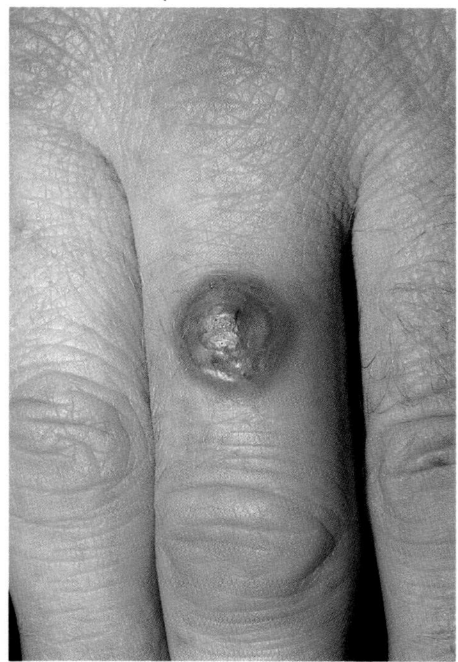

Fig. 3.13 Orf. Lesion developed after freeing sheep from barbed wire.

Anthrax

Aetiology and epidemiology

Anthrax is primarily a septicaemic infection of herbivorous animals that is much less prevalent in Britain than in warmer climates. *Bacillus anthracis* readily forms spores which are responsible for most transmission. Spores resist drying and freezing, may persist for years, and heavily contaminate the carcass of an animal dead from anthrax. Human infection mainly results from cutaneous inoculation or inhalation of spores during contact with animal products such as imported hides, bones and bone-meal, wool, and hair. More rarely, undercooked infected meat may be eaten or there may be direct occupational exposure, e.g. of abattoir workers or veterinary surgeons. Person-to-person transmission is rare.

Clinical features and diagnosis

Cutaneous anthrax. Occupation dictates the common sites where spores are inoculated, usually on hands and arms, shoulders or neck of those carrying infected products. A reddened papule develops rapidly into a larger vesicular lesion 1–2 cm in diameter surrounded by erythema and a wider zone of marked non-pitting oedema. The lesion is not painful but may itch. Central necrosis of the 'malignant pustule' occurs with drying to form a black crust or eschar (Fig. 3.14). 'Daughter' vesicles may develop at the periphery of the original lesion. Life-threatening septicaemic spread may follow, particularly involving the meninges, but in its absence the patient may suffer only moderate malaise, fever, and headache. Clinical diagnosis may be confirmed by microscopy of pustular material to show the Gram-positive bacilli and by culture.

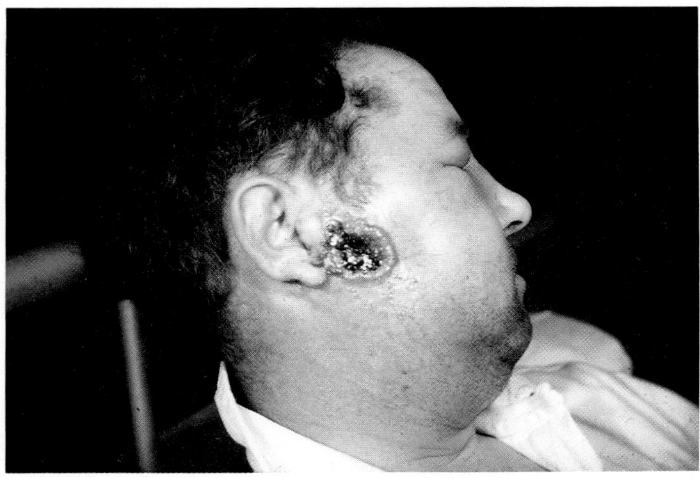

Fig. 3.14 Anthrax ('malignant pustule').

Non-cutaneous anthrax. Septicaemia may lead to severe generalized infection or meningitis. Pulmonary (from inhalation) and gastrointestinal (from ingestion) forms of anthrax are usually fatal.

Management and prognosis

Benzylpenicillin prevents complications in milder cutaneous forms but does not speed resolution of the lesion. High doses should be given intravenously for septicaemia, meningitis, pulmonary, or gastrointestinal anthrax. Erythromycin may be used in penicillin-hypersensitive patients. Hydrocortisone is used when oedema causes respiratory embarrassment.

Long-lasting immunity results. High-risk persons such as bone-meal workers can be vaccinated with killed *B. anthracis.* Control of infection in animals can be attempted by vaccinating those at risk and by examination and cremation of those suspected of dying of the disease. Transmission via animal products in the industrial setting can be reduced by good working practices (for example, dust control), or by irradiating or sterilizing products before handling.

Lyme disease

This zoonosis is caused by *Borrelia burgdorferi*, a spirochaete transmitted by ixodid ticks (*Ixodes ricinus* in Britain).

Clinical features

The most characteristic appearance is development of an erythematous papule at the site of the tick bite, that enlarges and spreads peripherally over several weeks to involve a large area of the body, with central clearing ('erythema chronicum migrans'). There may be flu-like symptoms and malaise. With or without this preceding rash, after some weeks neurological signs may appear, such as subacute lymphocytic meningitis, isolated cranial nerve palsies (usually facial), and peripheral neuropathy. Weeks or months after infection, with or without previous symptoms, flitting arthropathy may affect particularly the larger joints, and cardiac abnormalities (usually heart block) and chronic fatigue may develop.

Diagnosis and management

Diagnosis is difficult because serological tests are not usually positive early in infection, current tests lack sensitivity and specificity, and effective therapy may abort development of antibodies. Clinical diagnosis is therefore important for early treatment, and the possibility of Lyme disease should be suspected especially in areas of known endemicity of the tick vectors.

Oral penicillin, tetracyclines or erythromycin are effective in early disease but intravenous penicillin or cefotaxime are needed if serious complications develop, possibly followed by prolonged oral therapy; for example, with doxycycline.

Scaling cutaneous mycoses

Aetiology and epidemiology

Three main genera are responsible for human infection, *Epidermophyton* spp., *Trichophyton* spp., and *Microsporum* spp. Many animal fungi from domestic, farm, and wild animals can also produce lesions in humans and a history of contact should be sought. Infection can affect most sites but with preferences; for instance, epidermophyton for feet, trichophyton and microsporum for skin and scalp. Cutaneous mycoses can develop at any age, particularly in childhood; the incidence is higher in hot, moist climates.

Fungal skin infections cause itchy, scaling lesions with a small surrounding zone of erythema. Scaling must be distinguished from desquamation during the healing of any severe erythematous rash, especially after scarlet fever, toxic shock syndrome, and Kawasaki disease; fine desquamation sometimes follows measles. Fungal lesions slowly spread centrifugally, often healing at the same time in the middle. They are normally categorized according to their site.

Tinea pedis (Athlete's foot). Usually seen as scaling with occasional moist fissures between outer toe webs, this spreads locally and can involve nails. Sometimes it causes small, itchy vesicular and pustular lesions particularly over the lateral border of the foot but also over the sole. Cheiropompholyx sometimes accompanies this condition.

Tinea unguium (of the nails). Usually infected by local spread, involved nails become thicker, more opaque, and harder (Fig. 3.15).

Tinea cruris. Often seen in people with tinea pedis, this causes typical spreading lesions on contiguous areas of upper thigh, perineum, or lower abdomen (Fig. 3.16). The scrotum is generally spared.

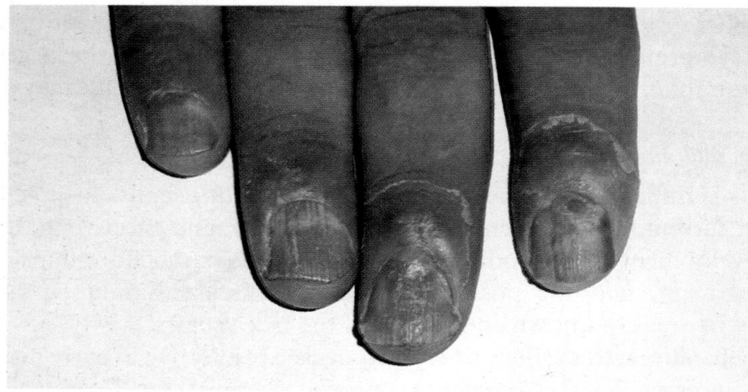

Fig. 3.15 *Tinea unguium.*

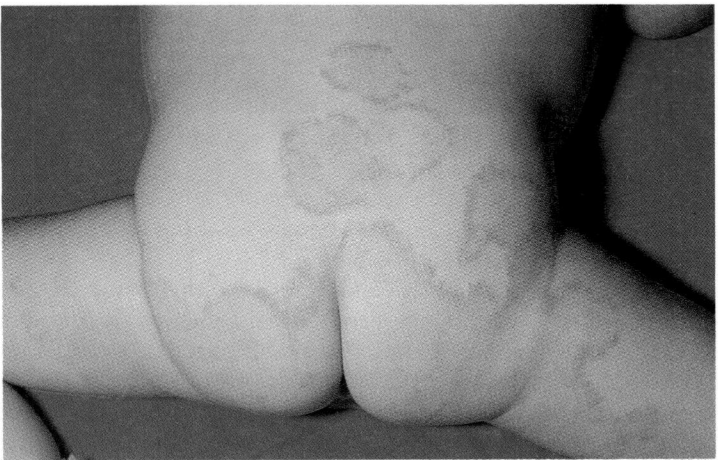

Fig. 3.16 *Tinea cruris.*

Tinea capitis. Circular lesions develop with broken off hairs in affected areas (Fig. 3.17). A boggy, well-demarcated raised mass with weeping surface called a kerion may develop. Animal fungi are rarely involved.

Tinea corporis (ringworm). This usually appears as spreading circular lesions on exposed parts of the body; sometimes these are plaque-like without healing in the centre.

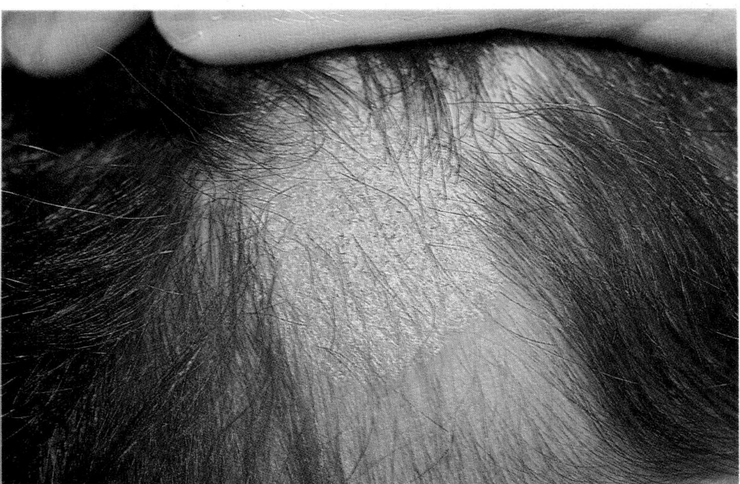

Fig. 3.17 *Tinea capitis.*

Diagnosis and management

Although readily recognizable when typical, unusual forms may be confused with many dermatological conditions including psoriasis, nummular eczema, seborrhoeic dermatitis, and lichen planus. Pityriasis rosea and some cutaneous lesions of secondary syphilis can also be similar.

Diagnosis is confirmed by microscopy and culture of scrapings or clippings from the lesion. Skin scrapings can be cleared by potassium hydroxide so that hyphae and spores become readily visible.

Most cutaneous mycoses resolve satisfactorily with topical treatment, but tinea capitis and tinea unguium may need systemic therapy. Keratolytic agents like Whitfield's ointment are cheap and reasonably effective, as are more specific agents like tolnaftate. However, the most effective agents are the imidazole derivatives such as clotrimazole, miconazole, and itraconazole. Itraconazole can be given orally. Oral griseofulvin can be used for tinea capitis or tinea unguium and continued for four weeks to avoid relapse. More resistant nail infections may require oral treatment for six weeks to six months; whether this is warranted for an infected toe-nail must be weighed against the side-effects of the drugs.

Prognosis

Relapses and re-infections are common. The commoner infections, tinea pedis and cruris, are suppressed by reducing the hot, moist conditions they favour. Thorough drying after bathing and the use of cotton socks and open sandals to encourage evaporation of sweat are helpful.

Pityriasis versicolor

Aetiology and epidemiology

The cause of this common mycosis of young adults is *Malassezia furfur*.

Clinical

Lesions start usually on the upper trunk as small circular areas with minimal flaking and just the slightest brown pigmentation. They occasionally cause mild pruritis but are normally symptom-free. The areas slowly enlarge and coalesce and often are not noticed until next summer as relatively hypopigmented areas against tanned uninfected skin.

Diagnosis and management

Malassezia furfur is not readily cultured but is easily seen in skin scrapings in potassium hydroxide. It responds readily to topical applications such as Whitfield's ointment or imidazoles such as clotrimazole. The ability to tan may not return for some months. Infection can recur.

Cutaneous candidiasis

Candida is often present in intertrigo and nappy rashes in which ammoniacal dermatitis is also a contributory factor. *Candida* involvement typically gives a large, moist, patchy, reddened area with a distinct, scalloped margin and a few, smaller satellite lesions nearby (Fig. 3.18). In obese females candidiasis is common in the axilla, under the breast, under abdominal folds of fat, and in the groin. Treatment is by exposure to air and anticandidal cream, such as nystatin or clotrimazole.

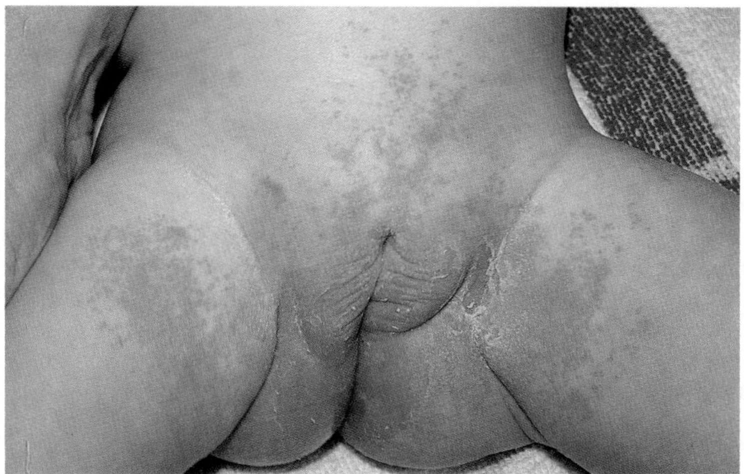

Fig. 3.18 Napkin dermatitis. *Candida* was isolated from perineum and oral thrush was also present.

Louse infestations (pediculosis)

Aetiology and epidemiology

Three species of lice infect humans: *Pediculus humanus corporis* the body louse, *Pediculus humanus capitis* the head louse (Fig. 3.19), and *Phthirus pubis* the crab louse (Chapter 13). Infection with body lice is related to poor hygiene, especially infrequent washing and laundering of clothes, and is usually found in neglected or elderly persons. The lice visit the body to feed, returning to the surrounding clothes where their eggs are deposited. Body lice may also transmit agents such as rickettsias or borrelias. Head lice remain in scalp hair, to which their eggs (nits) are firmly attached. They are particularly common in children, irrespective of cleanliness, and have become more prevalent perhaps because of longer hair-styles.

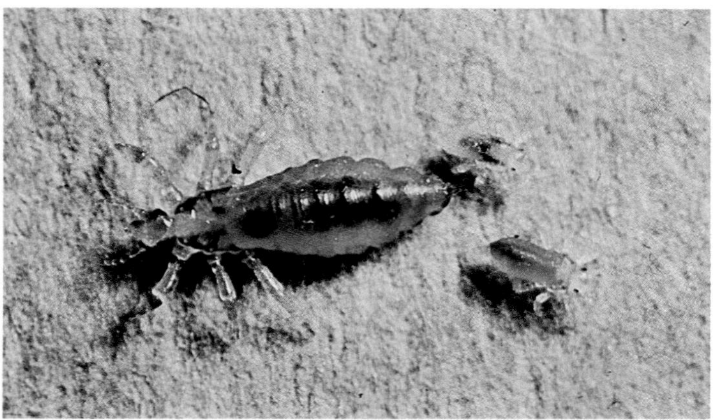

Fig. 3.19 Head louse: *Pediculus humanus capitis.*

Clinical features

The visits of the body lice leave an itchy erythematous maculopapular rash. Scratching can cause excoriation or secondary infection. Head lice are not seen as readily as their 'nits' which are laid on scalp hairs close to the skin, especially around the nape of the neck. They appear as 1–2 mm pink or white dots and, as the hair grows, become more distant from the scalp. Irritation by lice may induce scratching with secondary infection and lymphadenopathy.

Management

Head lice and nits are killed by lotions or shampoos of malathion or carbaryl, repeating application after one week. Remove nits and lice with a fine-toothed comb. The whole family should be treated. Clothes and linen should be give a hot wash or dry-cleaned to kill body lice and their eggs. A hot bath is sufficient for the patient.

Scabies

Aetiology and epidemiology

Scabies is caused by a mite, *Sarcoptes scabiei*, which is spread by direct bodily contact or, less importantly, via fomites such as blankets. The female mite burrows in the epidermis, laying eggs as she proceeds.

Clinical presentation

After an incubation period of about six weeks, pruritus results from hypersensitivity to the mite and its products. Itch is worse at nights and after hot baths. Generalized rash may develop, particularly in babies, with small, scattered papules, and occasional small vesicular elements (Fig. 3.20a–d). These are often

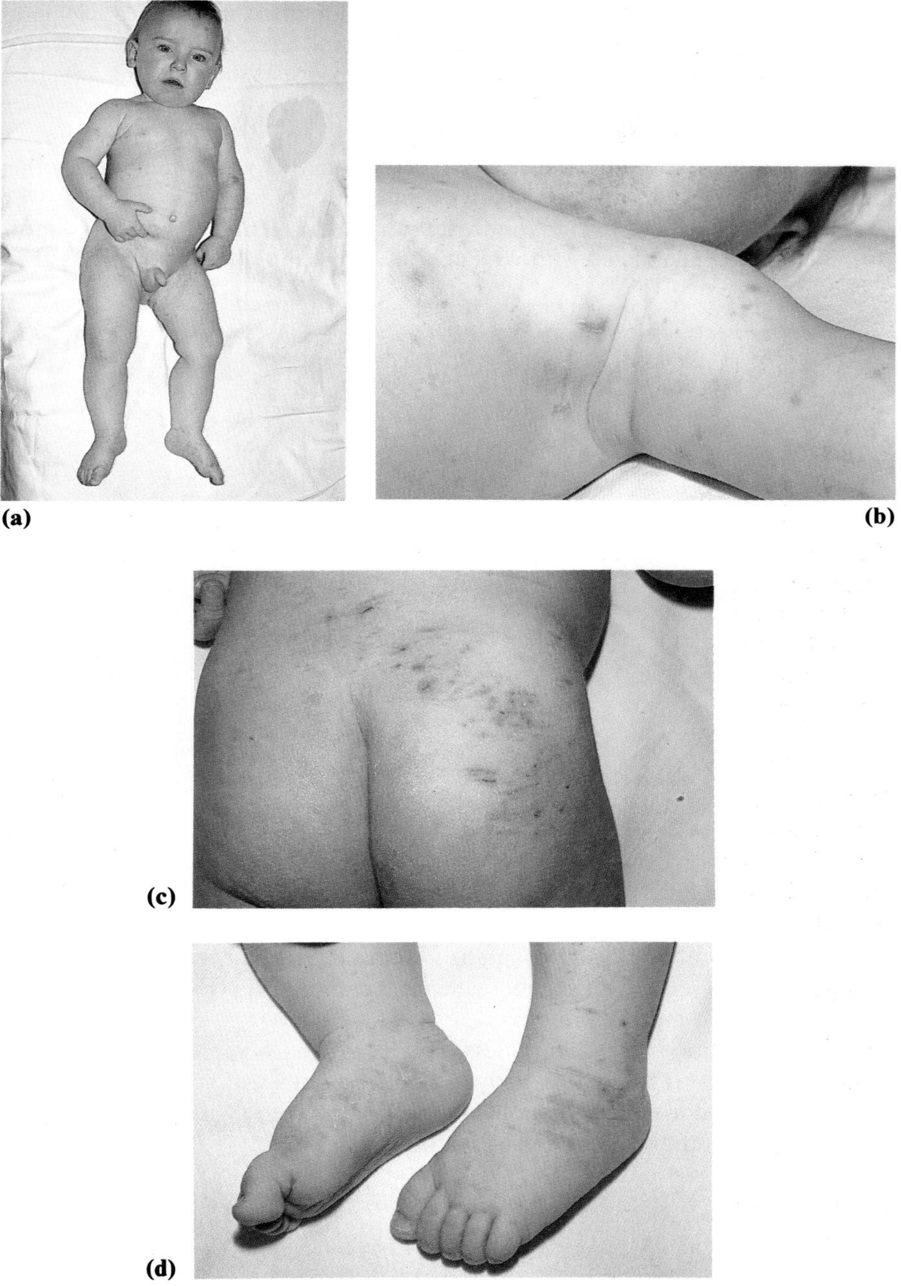

Fig. 3.20 Scabies. **(a)** An itching child with sparse scattered rash with excoriated lesions; **(b)** in axilla; **(c)** on buttocks; and **(d)** on feet.

excoriated and in older persons lichenified by scratching. Amidst these elements are found relatively few burrows, small linear streaks, often serpiginous, with a little flaking of skin at the entrance and a tiny grey dot at the end. In adults they are most often found on the webs of fingers, axillae, groins, penis, and in babies on the feet and neck. Scabies should be considered in any rash of slightly unusual appearance or distribution, especially if scratch marks are evident or if other family members itch. In the highly contagious 'Norwegian scabies', seen particularly in elderly or immunocompromised patients, the rash is more prominent and generalized with vesiculation and scaling but less pruritus (Fig. 3.21).

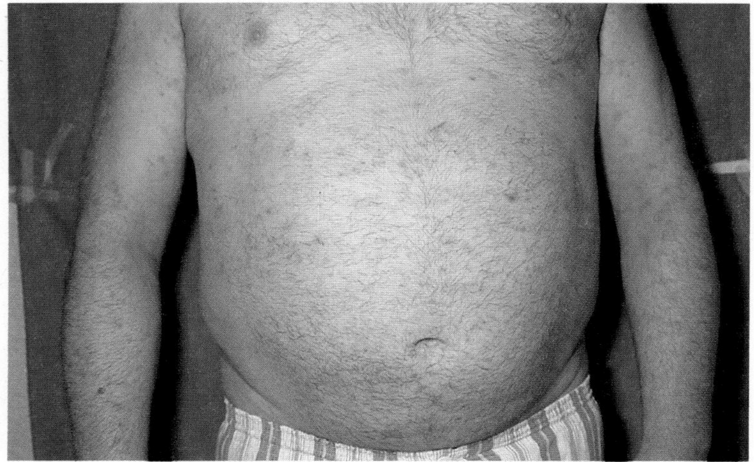

Fig. 3.21 Generalized scabies in adult.

Investigation and management

The adult mite or her eggs may be found by scraping a burrow with a needle point. If the grey dot of the adult can be visualized, 'tickling' with a needle point often causes it to seize the needle tip whence it can be visualized by microscopy Mild eosinophilia may be present.

Benzyl benzoate or the slightly less irritant gamma-benzene hexachloride is applied as lotion to the whole body, except face and scalp, after a hot bath. Treatment should include the whole family and be repeated after three days. All bed-linen and clothes should be given a hot wash.

Warts

Aetiology and epidemiology

Warts are caused by papillomaviruses of which there are over 50 human types. Different types are associated with different sites. Spread is by direct contact or via floors and fomites. Warts can appear at any age although flat warts (Fig. 3.22) and plantar warts (verrucas) (Fig. 3.23) are commonest in children. Genital warts (condylomata acuminata) are usually spread by sexual contact (Chapter 10). Virus transmitted during birth from the genital tract may cause neonatal laryngeal papillomas. Immunocompromised persons are more susceptible to warts.

Clinical features and investigations

Warts can develop on any area of squamous epithelium, keratinized or mucosal, into which the virus is inoculated. On normal skin, plane (slightly elevated) warts

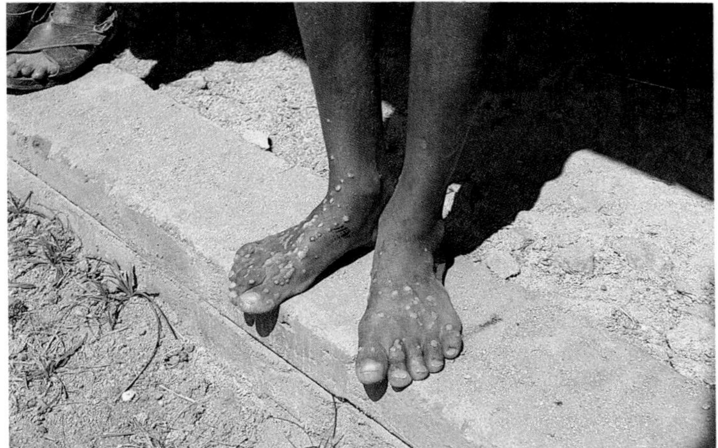

Fig. 3.22 Warts, widespread on feet.

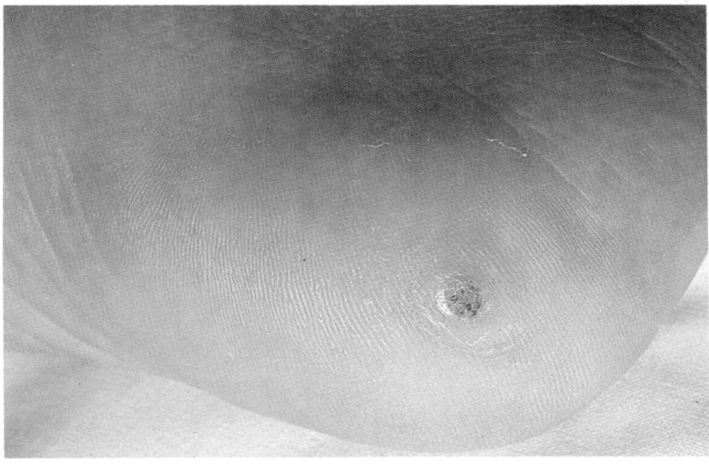

Fig. 3.23 Verucca on heel.

or smaller nodular warts develop; in the thickened epidermis of the sole, plantar warts are usually level with the surface, with deeper underlying growth. In warm, moist areas such as the mouth or genital area more profuse papillary warts may be seen. There are no associated systemic or local symptoms, except local pain from walking on plantar warts. Histology of warts is characteristic but rarely required. Routine serology or culture is not available but the virions can be seen by electron microscopy.

Management and prognosis

Warts resolve spontaneously, although this may take a few years. More rapid resolution can be achieved by local freezing by 'dry ice' or liquid nitrogen, or by applications of podophyllin or salicylic acid.

Cervical dysplasia is associated with papillomavirus types 6, 11, and 31 although it is not yet known whether these are causative. Similarly, types 16 and 18 are associated with cervical carcinoma-*in-situ* and invasive carcinoma of cervix and other genital areas.

Molluscum contagiosum

This rash caused by a poxvirus normally infects children and is endemic world-wide. Transmission is by direct inoculation from sexual or non-sexual contact or infected fomites. Outbreaks have involved children attending swimming pools and users of Turkish baths. The rash is not typical of other poxviruses and develops as hard, pearly papules with central umbilication commonly on the trunk and exposed skin (Fig. 3.24). Untreated, they slowly increase in size for a few weeks or months before resolving. There are no

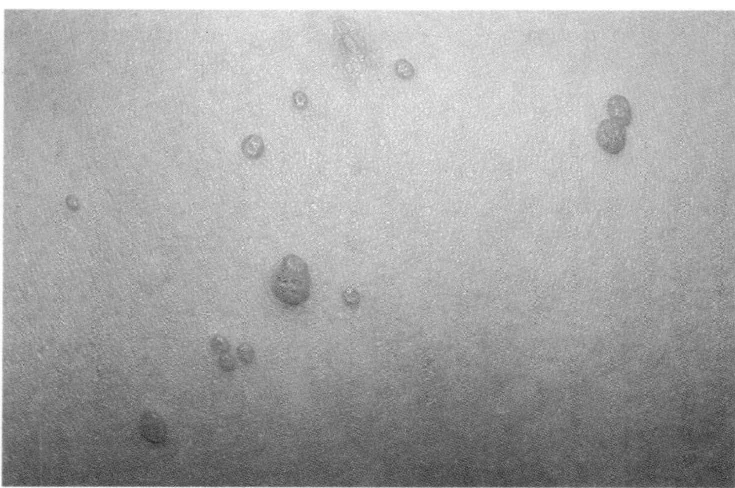

Fig. 3.24 Molluscum contagiosum.

associated systemic symptoms and individual lesions are not itchy. Lesions may be widespread in the immunocompromised.

The appearance is usually diagnostic but can be confirmed by squeezing firmly from the sides: cheesy white material is expressed and light microscopy confirms the presence of rounded 'molluscum bodies', or the virus may be visualized by electron microscopy.

Treatment, if necessary, is based on local damage to lesions, commonly by a silver nitrate stick, needle puncture, or liquid nitrogen, in order to present higher levels of antigen to the immune system and encourage immune response and speedy resolution. Iodine solution applied for 20 minutes twice daily may help.

INFECTIONS OF DEEPER TISSUES

Table 3.2 summarizes various necrotizing infections which generally affect damaged or devitalized tissues.

Wound infections

Aetiology and epidemiology

Traumatic wounds or bites are likely to be seeded with organisms when inflicted. Surgical wounds are cleaner, but after bowel has been opened about one-quarter of patients later develop significant wound infection. Other breaches of skin such as burns or varicose ulcers often allow infection to enter. Depending on the nature and depth of the wound, muscle, joints, or bone may also be involved. The organisms responsible vary with circumstances: multiple faecal organisms are likely to be responsible for wound infection after perforated appendix; skin or environmental organisms like staphylococci or *Clostridium tetani* are likely after trauma; *Pasteurella multocida* is commonly inoculated by cat or dog bites.

Clinical features and management

Signs of infection are usually evident as serous or purulent discharge, local erythema, and tenderness. Lymphangitis may be evident as reddened linear streaks draining wounds infected by *Strep. pyogenes* (Fig. 3.25). If superficial layers have already healed, a fluctuant abscess may be palpable in pyogenic infections that remain localized. Spread within deeper layers may give distant signs such as dusky erythema and gangrene as the first indications of, for instance, necrotizing fasciitis. If infection is limited to the wound tract with any discharges draining freely, generalized symptoms and signs are often absent. With tissue invasion systemic features are often marked. Management of all wounds must follow sound surgical practice: this may include debridement, delayed closure, haemostasis, insertion of drains, suturing that allows adequate tissue perfusion (i.e. not too tight), and prophylactic antibiotics.

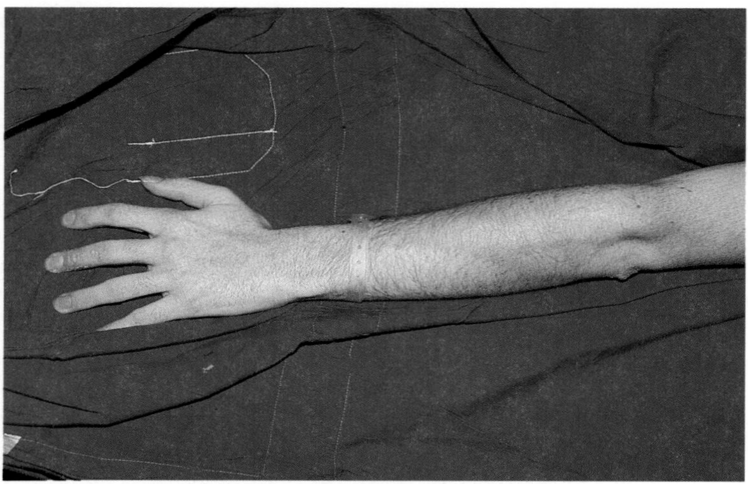

Fig. 3.25 Lymphangitis. Secondary to untreated streptococcal infection of middle finger.

Tetanus

Aetiology and epidemiology

Tetanus is caused by the neurotoxic exotoxin produced by the anaerobe *Cl. tetani* which is present in the bowel flora of many vertebrates, including man. It readily forms spores which are widely distributed in the environment. Tetanus usually follows inoculation of these spores into wounds at the time of injury and in Britain is most common in outdoor athletes and gardeners. Often the injury is minor or even unnoticed. In many areas of the developing world poor hygiene in cutting and keeping clean the umbilical cord results in neonatal tetanus with very high mortality.

Clinical presentation

After an incubation period usually from 3–21 days, slowly increasing muscle rigidity develops. To the patient this may be more noticeable as trismus (inability to open the mouth wide) but on examination rigidity of spinal muscles and board-like firmness of abdominal muscles are marked. Painful spasms, usually generalized, then become superimposed on the muscle rigidity, resulting typically in facial grimacing (Fig. 3.26) and arching of the back. Spasms may be provoked by sudden sensory stimuli like loud noises or abrupt handling. Later autonomic nervous system involvement may cause arrhythmias and highly labile blood pressure. Severity is greater when the incubation period is short and the evolution of illness is rapid.

The clinical features are usually distinctive but trismus can also be caused by the pain of mumps and by muscular dystonias as a side-effect of phenothiazines.

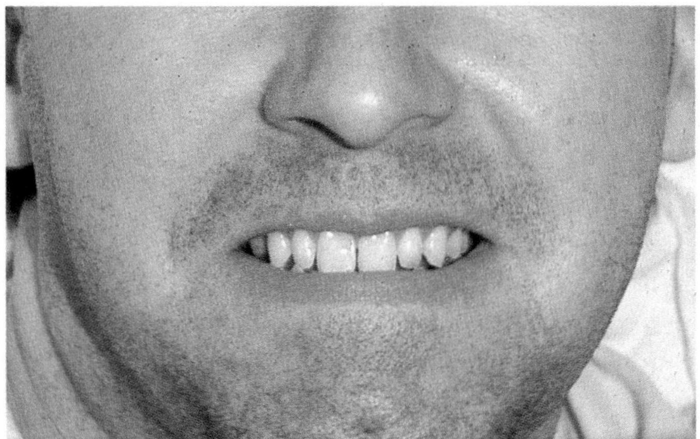

Fig. 3.26 Risus sardonicus—involuntary grimacing in tetanus.

Management

Human tetanus immunoglobulin should be given to neutralize circulating toxin. It may also be of benefit given intrathecally. Benzylpenicillin should be given, but since *Cl. tetani* is a strict anaerobe, the site of infection will have a poor blood supply and any evident wound should be surgically opened and debrided. In mild cases spasms can be controlled by diazepam, but more severe cases require management in intensive care units with assisted ventilation and muscle relaxants. Recovery should then follow within weeks as the effects of the already fixed toxin wear off, although mortality can result from autonomic instability. Tetanus does not reliably confer subsequent immunity, so after recovery immunization should be performed (Chapter 16).

Prevention

A primary course of tetanus toxoid and subsequent boosting every 5–10 years or so will prevent a person from developing tetanus even if wounds become infected with *Cl. tetani*. Neonatal tetanus is mainly prevented by ensuring immunity of pregnant mothers. Scrupulous cord hygiene is also important, where attainable, for preventing neonatal tetanus. The likelihood of tetanus developing from accidental trauma must always be considered and, depending on the nature of the wound and the immunization status of the patient, appropriate action taken (Table 3.3) in addition to any surgical attention.

Muscle involvement in infections (Table 3.4)

Muscles may be affected directly by the infecting agent, as in clostridial myonecrosis or acute trichinosis, or indirectly by toxin as with Legionnaires' disease or toxic shock syndrome. The mechanism by which many viruses cause

Table 3.3. *Tetanus immunoprophylaxis after wounds*

Immunization status	Clean and minor wounds	All others*
Primary vaccination course in past with		
Booster <5 years ago	—	—
Booster 5–10 years ago	—	Booster
Booster >10 years ago	Booster	Booster
Primary course uncertain or incomplete	Primary course	Primary course + human tetanus immunoglobulin

*High risk if >6 hours delay, heavily contaminated, or penetrating wound, or necrotic tissue.

myalgias is uncertain, probably by infection of muscle cells but in some cases by immunological processes. Clinically, the involved muscles are tender to touch and may be swollen. As the muscle involvement is usually only one part of the whole illness, most of the conditions in Table 3.4 are mentioned elsewhere in the book.

Lymphadenitis

Infections can cause either localized (Table 3.5) or generalized (Table 3.6) lymph node enlargement. These tables are limited to conditions which may occur without such obvious diagnostic features as a local skin or throat lesion or generalized rash. Infections are included, like infectious mononucleosis, that often produce a clinically distinct picture of fever, confluent tonsillar exudates, palatal petechiae, generalized lymph-node enlargement, splenomegaly, or 'ampicillin rash', but which can also often present with fever and generalized lymph-node enlargement alone.

INFECTIONS OF BONE AND JOINTS

The older child and adult can usually indicate precisely the site of pain of an infected joint or bone, but pain may be referred from an infected hip to the knee. In the younger child or infant evidence may be less direct, for instance a limp, reluctance to move a limb, or irritability when it is passively moved.

Table 3.4. *Muscle involvement in infections*

Condition (and cause)	Features
Clostridial myonecrosis (gas gangrene—see also Table 3.2) (*Cl. perfringens*)	Predisposing cause. Crepitus. Severe toxicity.
Acute trichinosois (*Trichinella spiralis*)	Eating uncooked meats. Periorbital oedema, painful swollen muscles, high fever, eosinophilia.
Cysticercosis (*Taenia solium*—Chapter4)	Eating uncooked pork. Radiological calcifications.
Legionnaires' disease (*Legionella pneumophila*—Chapter 2)	Severe pneumonia, diarrhoea, confusion. High creatine kinase.
Toxic shock syndrome (*Staph. aureus*—Chapter 8)	Erythematous rash, fever, myalgia, vomiting and diarrhoea. High creatine kinase. Usually during menstruation.
Influenza (virus—Chapter 2)	Headache, myalgia, cough. Fever. Occasional temporary severe calf pain during convalescence from type B.
Bornholm disease (coxsackievirus)	Sudden onset, usually intense lower intercostal muscle pain, sometimes pleuritic, sometimes abdominal. May relapse over several weeks.
Myalgic encephalomyelitis (often coxsackievirus B—Chapter 13)	Chronic relapsing course after initial infection characterized by extreme fatigue.
Poliomyelitis (poliovirus—Chapter 6)	Asymmetrical paralysis secondary to lower motor neurone destruction.

Osteomyelitis

Aetiology and epidemiology

The agents responsible (Table 3.7) usually reach the bone asymptomatically via the bloodstream, but can also invade from a contiguous focus of infection or by direct inoculation from trauma or surgery. Osteomyelitis is commoner in males (2.5:1) and develops most often in childhood and adolescence involving long bones of lower limbs in about 75 per cent of cases. Patients with sickle-cell disease are prone to *Salmonella* osteomyelitis.

Clinical features

Local bone pain and tenderness are usually accompanied by fever, malaise, and

Table 3.5. *Localized infective lymph node enlargements which may present without specific features*

Condition	Features
Streptococcal cervical adenitis (*Strep. pyogenes*)	Large, tender nodes, fever, often 1–2 weeks after streptococcal sore throat or skin infection. May suppurate or lead to spreading cellulitis (Fig. 3.13).
Toxoplasmosis (*Toxoplasma gondii*—Chapter 11)	May be localized. Often cat contact or ingestion of undercooked meat. Atypical mononuclear cells. Characteristic biopsy. Positive serology.
Scrofula (Atypical mycobacteria—Chapter 9)	Tonsillar and submandibular nodes, not usually tender, and usually no systemic upset. Suppurates and discharges via sinus.
Tuberculous lymphadenitis (Chapter 9)	Usually posterior cervical and supra-clavicular nodes. Often TB contact.
Cat scratch diseae	History of exposure to cat, with indolent papule at inoculation site. Positive skin test. About 10 per cent suppurate. Pleomorphic Gram-negative bacillus sensitive to cefotaxime or aminoglycoside.
Persistent generalized lymphadenopathy (HIV virus infection—Chapter 12)	Lymph nodes must be larger than 1 cm at two separate sites for 3 months to be classified as PGL. Positive serology.
Lymphogranuloma venereum (Chapter 10)	Inguinal node enlargement with suppuration. Initial lesion usually overlooked. Usually imported. Positive serology.

anorexia. In the infant, infection readily spreads from the metaphysis to adjacent joint and soft tissues, often with obvious erythematous swelling. In the young child, infection does not spread so readily although overlying oedema may cause non-erythematous swelling. At older ages surface features are not visible.

Investigation

In the neonate and young child X-rays may show increased soft tissue shadowing. At older ages, radiological changes do not develop for at least 10 days but a technetium bone scan can demonstrate increased uptake earlier. Bacteriological diagnosis is usually made by blood culture, by subperiosteal needle aspiration in the younger child or from material removed at operation. Neutrophil leucocytosis and high ESR are present.

Table 3.6. *Generalized infective lymph node enlargements which may present without specific features*

Condition	Features
Infectious mononucleosis (Epstein–Barr virus infection—Chapter 2)	Fever for 5–14 days. May have characteristic enanthem, hepatitis, splenomegaly, 'ampicillin rash'. Atypical mononuclear cells. Positive serology.
Cytomegalovirus infection (Chapter 11)	Fever, splenomegaly, hepatitis. Atypical mononuclear cells. Positive serology.
Toxoplasmosis (Chapter 11)	Cat contact or raw meat consumption. Splenomegaly, fever. Atypical mononuclear cells. Positive serology.
Rubella (Chapters 7 and 11)	Over 50 per cent have no rash, but lymph nodes prominent especially occipital.
Syphilis (Chapter 10)	At secondary stage—patient often asymptomatic (may have rash, condylomata and oral lesions).
HIV virus infection (Chapter 12)	Often prominent feature in PGL and AIDS. May be fever, sweats, weight loss, diarrhoea, Kaposi's sarcoma, and opportunistic infections.

Table 3.7. *Aetiology of osteomyelitis*

Infecting agent	Approximate frequency (per cent)	
	Children	Adults
Staph. aureus	80	60
Strep. pyogenes/Strep. pneumoniae	6	10
Gram-negative bacilli	10*	25
Tuberculosis and other	4	5

*Especially *H. influenzae* and *Salmonella*.

Management

As *Staph. aureus* is the most likely pathogen, two antistaphylococcal agents such as flucloxacillin, fusidic acid, or rifampicin should be given. After severe symptoms have resolved, oral therapy is continued for 4–6 weeks with at least two months' treatment for vertebral osteomyelitis. Collections of pus or sequestra should be surgically removed.

Prognosis

In a few patients, infection relapses on stopping treatment and chronic osteomyelitis may develop. Sinuses to skin may discharge purulent material. Any sequestra should be removed; long-term treatment (up to six months) with antibiotics is often curative.

Arthritis and arthralgia

Aetiology and epidemiology

Infective causes of arthritis (Table 3.8) are usually bacteria or viruses, although sterile 'reactive' arthritis may develop during an infective illness, possibly from immune complex deposition. This may also explain arthralgia (joint pains without joint signs) in many infective conditions. Similarly, some patients are predisposed to develop reactive arthritis after infections, sometimes associated with eye lesions and other features as in Reiter's syndrome (Chapter 13). Most joint infections are blood-borne, although there may be direct inoculation by surgery or trauma, or spread from contiguous structures such as bones. The organisms causing childhood osteomyelitis (Table 3.7) are also responsible for septic arthritis, though *H. influenzae* normally infects only children under three years of age. In adults the possibility of gonococcal aetiology must be considered.

Up to 4 per cent of prosthetic hip and knee joints become infected after insertion. Early infections are probably established at the time of operation; later infections are likely to be blood-borne. When a prosthesis is replaced infection develops in up to a third: *Staph. epidermidis* is isolated in about 40 per cent and *Staph. aureus* in 20 per cent.

Clinical features

A swollen, painful joint with restricted movement is typical, usually accompanied by fever and systemic upset. Bacterial arthritis normally involves only one joint, usually the knee, although shoulder, wrist, hip, interphalangeal, and elbow joints are commonly involved.

Migratory polyarthritis with skin lesions can be caused by the gonococcus, mainly with sparse pustules on limbs (Chapter 10), or the meningococcus, with haemorrhagic lesions especially on fingers and feet (Chapter 6), and may complicate Lyme disease.

Joints affected by rheumatoid arthritis are prone to infection, usually by *Staph. aureus*, which should be suspected if an acutely inflamed joint does not respond to anti-inflammatory agents.

Arthralgia or arthritis accompanying viral infections involves many joints. In hepatitis B, joints are involved only briefly in the prodromal period at the same time when urticaria may be noted. With other viruses listed in Table 3.8 the joint involvement usually develops during the major illness and with rubella and parvovirus may persist for some months. Rubella virus has been isolated from the

Table 3.8. *Infective causes of arthritis*

Infecting agent	Monarticular arthritis	Polyarticular arthritis
Bacterial (pyogenic)		
Staph. aureus	+	−
Strep. pyogenes	+	−
H. influenzae	+	−
N. meningitidis	+	+ (sometimes)
N. gonorrhoeae	+	+ (sometimes)
Mycobact. tuberculosis	+	−
Mycoplasma pneumoniae	+	−
Borrelia burgdorferi	+	+
Viral		
Rubella (and some other togaviruses)		+
Parvovirus		+
Hepatitis B (prodrome)		+
Mumps		+
HIV (primary infection)		+
Reactive		
N. meningitidis		+
N. gonorrhoeae		+
Bacterial endocarditis		+
Post-infective		
(a) Reiter's syndrome		
Shigella		+
Campylobacter		+
Salmonella		+
Yersinia		+
Non-specific urethritis		+
(b) Rheumatic fever		
Strep. pyogenes		+

synovia of patients with chronic arthritis such as juvenile rheumatoid arthritis. The significance of this is not yet clear.

Investigation and differential diagnosis

Joint aspiration can differentiate bacterial arthritis from other infective or non-infective causes like gout. In bacterial arthritis synovial fluid typically shows organisms by Gram stain (culture is more sensitive), the leucocyte count is over 50×10^9/litres (over 90 per cent polymorphs) and the glucose level is low. Blood cultures should also be performed and other relevant sites sampled—for instance urethra, cervix, and rectum if gonococcal infection is suspected. A raised ESR is usual but not always present. X-rays often show only soft tissue swelling and radioisotope scans do not consistently differentiate the causes of joint inflammation. Most infected prostheses show radiological signs of loosening of the prosthesis.

Treatment

Appropriate antibiotics should be given for bacterial infection. If the initial Gram stain is negative, *Staph. aureus* should be suspected and treated as for osteomyelitis. Gonococcal disease responds rapidly to benzylpenicillin within 48 hours (unless due to a penicillinase producer). *H. influenzae* infection in a young child should be treated by chloramphenicol. Intra-articular antibiotics are not usually necessary since antibiotic levels in synovial fluid are as high as those in blood. Joints should be immobilized initially and aspirated to dryness, repeating this if fluid re-accumulates.

Prognosis

This depends on the cause. Septic (bacterial) arthritis, especially in adults, usually leaves permanent damage, limited movement and pain. Arthrodesis or joint replacement may be necessary after infection is eradicated.

4

Infections of the gastrointestinal tract

Diarrhoeal diseases are extremely common, especially in developing countries and wherever hygiene and sanitation are poor and excrement contaminates the environment, water, and food. In world terms these diseases are estimated to kill about six million persons each year and to incapacitate many more. Throughout the developing world they rival acute respiratory infections as the main killers of children, especially those under five years old and those who are malnourished. In those under two years, one tropical study showed that mortality from diarrhoea was 1 in 3 in the severely malnourished and 1 in 10 in those moderately malnourished. In Mediterranean countries diarrhoea causes from 10–25 per cent of paediatric admissions, with serious mortality especially under two years of age.

Most diarrhoeal infections are acquired by ingestion, being transmitted through food, water, contaminated fingers, feeding utensils, or even surgical instruments and medicaments. The respiratory route also is important for rotaviruses which are among the most important causes of diarrhoea in children world-wide in both developing and developed countries.

Factors favouring 'food poisoning' include warm climates which encourage rapid multiplication of bacteria in foods and beverages; urbanization and travel, which facilitate local, national, and international spread; and problems with food production and preparation which ensure that even in developed countries these diseases remain common.

In Britain in recent years, acute diarrhoea and vomiting with severe dehydration have become less common in babies. This is probably due to factors such as improved hygiene of bottle-feeding, increase in breast-feeding (which is markedly protective), earlier oral rehydration, decreased use of antibiotics (which are mainly ineffective), earlier referral to hospital, and possibly a change in virulence of some common pathogens. Among adults, however, the picture is different: *Campylobacter* and *Salmonella* infections reach epidemic proportions in summer when more cold meals are consumed (Fig. 4.1). Mass catering as in canteens provides better opportunities for *Clostridium perfringens* as well as salmonellas, pathogenic *Escherichia coli* and enterotoxic staphylococci to cause outbreaks.

Many pathogens which enter through the gastrointestinal tract do not cause predominantly gastrointestinal symptoms. Examples include *Salmonella typhi*, hepatitis A, other enteroviruses, and the 'intoxications' shown in Table 4.1. Non-diarrhoeal intestinal infections with helminthic infections such as hookworms,

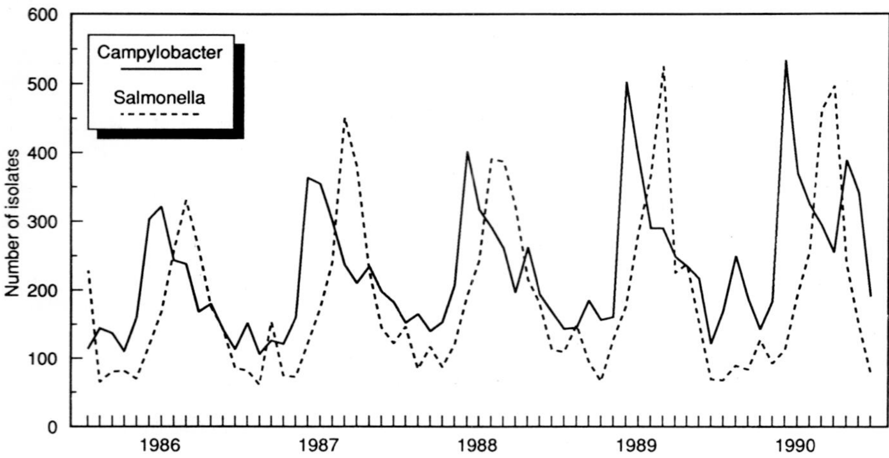

Fig. 4.1 The annual summer increase in campylobacter and salmonella infections (Scotland: laboratory reports by 4-week periods, 1986–90).

round worms and strongyloides are also common where methods of faecal disposal are primitive.

ROUTES OF INFECTION AND EPIDEMIOLOGY (Fig. 4.2)

Person-to-person spread

Transfer of faecal contamination is common among children, in families and close communities, especially where facilities for personal hygiene such as hand-washing are poor or sanitary standards low. Hepatitis A virus, shigellas, *Giardia lamblia*, and threadworms commonly spread in this way.

Water-borne infection

Water is a classical vehicle for transmission of cholera (Chapter 14), hepatitis A, amoebiasis, giardiasis, and many other infections of which the causal agents are excreted in faeces or urine. Provision of uncontaminated water for drinking and for washing is one of the most important communicable disease control measures. Separation of water supplies from sewage and related pollutants made a major contribution to reducing the impact of diarrhoeal diseases in developed countries. In poorer countries sewage collection and provision of disposal plants and water purification plants are often difficult or impossible to provide. In many tropical countries faeces are left exposed to dry and contaminate land and water. Untreated excreta are used in some countries to fertilize crops. Protection of water sources is important, since removal of infectious particles by filtration is

Table 4.1. *Main food and water-borne infections and intoxications*

(*a*) *Infections causing diarrhoea*	
Campylobacter jejuni Salmonellae	Infected poultry, milk, eggs, and less commonly meat
Escherichia coli *Shigella*	Food, water or utensils contaminated by human faeces
Entamoeba histolytica	Food or water contaminated by human faeces
Cryptosporidia	Water or food contaminated usually by animal faeces or slurry
Giardia lamblia	Water contaminated by human faeces
Vibrio cholerae	Water, occasionally food contaminated by human faeces
Vibrio parahaemolyticus	Infected seafoods
Bacillus cereus	Rice, contaminated, reheated
Yersinia enterocolitica	Food contaminated by human or animal faeces
Clostridium perfringens	Meat contaminated by animal intestinal flora
Small round structured viruses	Infected seafood or food contaminated by human faeces
(*b*) *Intoxications*	
Bacterial enterotoxins	Preformed in food (staphylococcal, clostridial, *B. cereus*)
Botulism	Heat-labile neurotoxin from *Cl. botulinum* in non-acid anaerobic conditions, e.g. canned vegetables, fish products
Paralytic shellfish poisoning	Saxitoxin–from 'red tide' plankton dinoflagellates
Scombroid poisoning	From histaminic decomposition products of mackerel, tuna fish
Ciguatera poisoning	Neurotoxic lipid in Pacific and Caribbean fish
Tropical neuropathy	Cyanide poisoning from cassava
Solanine poisoning	From green potatoes (and some other alkaloids)
Mycotoxins	'Mushroom poisoning'; aflatoxins

not always reliable. Some organisms such as hepatitis A and other enteroviruses, cysts of amoebae, cryptosporidiae, and *Giardia lamblia* resist normal chlorination. Sterilization by boiling is effective but expensive. Recycling water for re-use by dense, urban populations in industrial countries increases the organism load, particularly of viruses. Drought conditions can increase the concentration of pathogens and the probability of related infections.

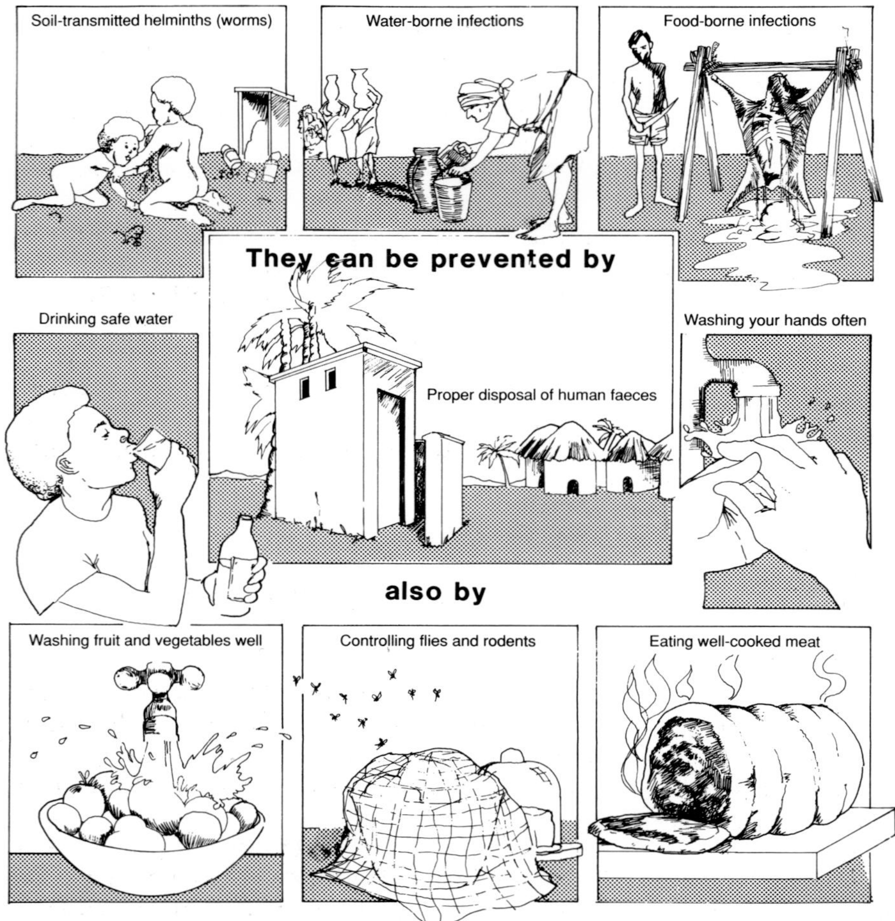

Fig. 4.2 Food-, soil-, and water-borne infections (adapted from *World Health*, March 1984, World Health Organization).

Food-borne infection

Provision of uncontaminated food is also essential for preventing infections, as recognized by legislation and arrangements to provide relatively uninfected food materials in most countries. Freshly cooked food consumed hot is usually safe. Paradoxically, the sophisticated organization for mass catering involving 'food-handlers' (often literally) and the use of bulk deliveries of foodstuffs produced by intensive methods in developed countries have made possible outbreaks on a bigger scale than ever before. Some infections may be inherent in the foodstuffs themselves; for example, salmonella infection of poultry, cattle, and derived products. Intensive production methods and recycling of offals as animal feeds can amplify and perpetuate these infections of farmed animals (see Fig. 4.7). Seafoods, particlarly molluscs, can be infected from polluted sea water, since organic particles which may contain faecal bacteria or viruses provide food for many of these animals and are concentrated within them. Microbial infections such as *Cl. perfringens*, staphylococci and *E. coli* can reach food as contaminants, often in small amounts from unsafe preparation methods. Time lapse and warm temperatures may then allow them to multiply up to dangerous levels or produce pathogenic concentrations of toxin. Education of those preparing and cooking food is most important, together with supervision of those employed in the kitchens of restaurants, hospitals, and canteens: good hygiene and hand-washing are essential.

Suspicion of a possible food-borne infection or intoxication requires immediate action to attempt to determine the source and institute preventive measures. Many of these infections are therefore legally notifiable. Regional and national surveillance arrangements monitor trends and possible sources of new infections and outbreaks.

HOST FACTORS AND PATHOGENESIS

Host resistance

The baby is exposed to gastrointestinal infections from birth. Passive immunity derived from the mother, and possibly factors in breast milk, confer useful protection; for example, most rotavirus infections are symptomless in the newborn. Breast-feeding also protects the infant from pathogens contaminating prepared and artificial feeds.

Gastric acidity gives some protection against acid-labile organisms, and infection by these is more likely when acidity has been lost through pernicious anaemia, gastric surgery, or medication. Acquired, specific immunity may help to speed elimination of organisms from the gut. It may give useful protection against symptomatic disease due to, for instance, coliform organisms. This immunity tends to be incomplete and temporary and gives little or no protection against the numerous other serotypes of these organisms, such as those which cause travellers' diarrhoea (see below).

Pathogenesis

Disease may result from invasion of gastrointestinal tissues or from the action of toxins. Many important diarrhoeal diseases are mediated by enterotoxins, such as the heat-labile toxin of *E. coli*, immunologically related to cholera enterotoxin.

A specific region of this enterotoxin attaches to host cell-membrane receptors and sets up a chain of reactions which culminate in formation of excessive amounts of c-AMP. A cascade of events then causes intestinal epithelial cells relatively to over-secrete electrolytes and water as the fluid, diarrhoeal stool.

In order to establish damaging infection, viruses and most other organisms must make effective contact with enterocytes, often by combination with specific receptors. This is not easy because of competition by the indigenous gastrointestinal flora and because of mucous secretions, peristaltic flushing, and antagonism by bile salts, lysozyme, lactoferrin, and perhaps immunological factors. Thus to cause disease the infecting dose must usually be large. The sites of infection vary. Some agents such as *Giardia lamblia* are principally small bowel pathogens. *Vibrio cholerae* and *E. coli* may involve the whole intestine but with major toxic effects in small bowel. Salmonella and campylobacter infections are often confined to the colon.

Oral rehydration operates by replacing electrolytes and water depleted by diarrhoea. Absorption of the electrolytes through the intestinal mucous membrane is facilitated by sugar. Provision of simple oral rehydration preparations is making a major contribution to reducing mortality from childhood diarrhoea in the developing world (Table 4.2).

CLINICAL PRESENTATION

'Gastroenteritis' is used to decribe illness with both vomiting and diarrhoea, but is usually a misnomer since the stomach itself is rarely infected (except for colonization of antral mucosa by *Helicobacter pylori* which has a role in the pathogenesis of peptic ulcers).

Diarrhoea is very common but not always present. An unformed or watery stool is generally considered as diarrhoea, likewise unusually frequent passing of soft stools. In babies a greenish watery stool suggests infection, although some dried cow's milk preparations containing iron can mimic this colour. The breast-fed baby's normal stool is yellow, looks 'curdled' and is often passed frequently and explosively (Fig. 4.3).

Frank blood and mucus in the stools ('dysentery') (Fig. 4.4) and colicky abdominal pain relieved by defecation suggest large bowel involvement. Persistent pain suggests possible ulceration, perforation, or toxic dilatation. Salmonella and campylobacter infections especially can cause localized pain which may be confused with appendicitis or diverticular disease. Vomiting is common and may be predominant. Bile in vomit suggests intestinal stasis. Abdominal distention may be due to infective ileus, perhaps aggravated by over-using antimotility drugs. Dehydration (Table 4.3), is inevitable if fluid loss, including bowel contents (often substantial with ileus), exceeds intake.

Before concluding that diarrhoea and vomiting are primarily due to gastrointestinal infection, more generalized disease or disease of other systems

Table 4.2. *Solutions for oral rehydration*

1. *Prepared powders for dissolving in boiled water*
   (a) Compound sodium chloride and dextrose oral powder, B.P.
   Dissolve to contain per litre (mmol):

Na^+	35
K^+	20
Cl^-	37
HCO_3^-	18
Dextrose	200

   (similar preparations are available commercially).

   (b) Oral Rehydration Salts—WHO Diarrhoeal Control Programme
   ORS–citrate. Reconstitute in water to give the following (g/litre):

sodium chloride	3.5
trisodium citrate, dihydrate	2.9
potassium chloride	1.5
glucose, anhydrous	20.0

   (ORS–citrate replaces previous ORS-bicarbonate as being more stable in storage as dry powder in sealed packets in tropical climates. Supplied widely in developing countries for giving by non-medical persons including parents.)

2. '*Home-made*' *solution*

Boiled water	one pint (half-litre)
Salt	two pinches (half a 5 ml spoonful)—no more
Sugar	one handful (four 5 ml spoonfuls)
	Flavour with fruit juice as required

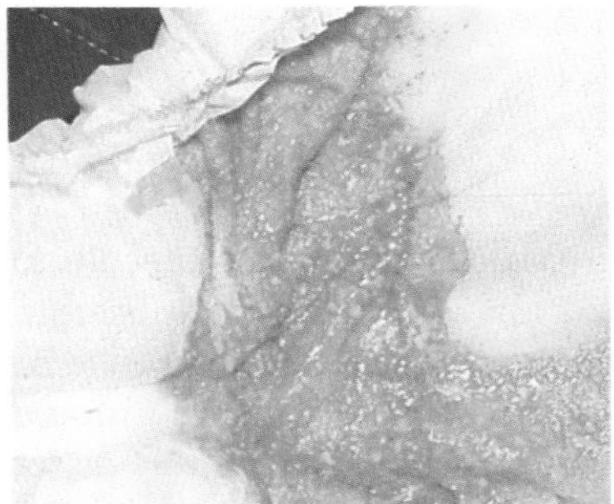

Fig. 4.3 The normal yellow stool of the breast-fed baby.

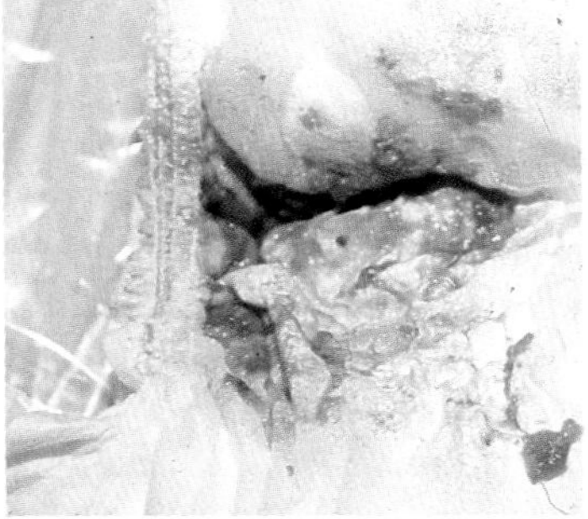

Fig. 4.4 Watery diarrhoea with blood and mucus as in dysentery.

Table 4.3. *Features of dehydration*

1. *Hypo- and isonatraemic*		
	Mild	Increased thirst
		Reduced urine output
		Presume present when there is vomiting or diarrhoea
	Severe	Loss of skin turgor and elasticity
		Sunken appearance of eyes; dry mucous membranes
		Apathy
		Sunken fontanelle (in infants)
2. *Hypernatraemic*		
	Marked thirst	
	Restlessness and irritability, convulsions	
	Skin turgor may appear normal	
	Conjunctival injection	

must be considered; for example, otitis media in children can present with vomiting. Diarrhoea in exanthematous infections may reflect direct involvement of the alimentary tract or merely febrile upset.

Broad-spectrum antibiotics especially can cause loose or watery stools and may precipitate pseudomembranous colitis (see below). Although not a major feature of infective diarrhoeal illnesses, fever is common at the onset of rotaviral infection in infants and in shigella dysentery. Fever may also indicate infection elsewhere in the body or complicating septicaemia. Convulsions may be provoked in children especially by shigella infections—not always 'febrile', sometimes apparently due to toxin. Differential diagnosis of chronic or recurrent diarrhoea includes cystic fibrosis, coeliac disease, disaccharidase deficiency, and inflammatory bowel disease (for example, Crohn's disease and ulcerative colitis).

INVESTIGATIONS

Stool examination

By a combination of electron microscopy, virological and bacterial culture methods, and of immunofluorescent, serological, and toxin-identification techniques applied to the stool, the cause of gastrointestinal infection can often be determined. Immediate examination of stools is necessary to identify motile trophozoites of amoebae and *Giardia lamblia*. Cultures should preferably be set up on the same day as the specimen is taken. Ova and cysts can be looked for at leisure. Specific diagnosis allows appropriate action to prevent spread and control outbreaks, assists with surveillance and the monitoring of routine preventive methods, but does not often alter the management of individual patients.

Blood culture

This should be performed on all toxic or persistently febrile patients. Salmonella in particular but also campylobacter and other organisms can cause bacteraemia.

Serology

Serology if available may help identify the responsible organisms (for example, *Campylobacter*, *Yersinia*, *Giardia*).

Other tests

The presence of reducing substances in stool suggests complicating disaccharidase deficiency (see below). Investigations for milk 'allergies' are less useful.

MANAGEMENT OF DIARRHOEA AND VOMITING

The decision whether a patient with diarrhoea and vomiting should be admitted to hospital depends in part upon the facilities available at home. Persistent fever or vomiting, severe dehydration and collapse, rigors, abdominal pain or distension are indications for admission, for parenteral rehydration, resuscitation, special investigations, or surgical opinion.

Dehydration

Prevention and correction of dehydration is the key to management of diarrhoeal illness. Unless vomiting is persistent, rehydration can usually be achieved by oral fluids given frequently in small amounts (Fig. 4.5). Absorption of electrolytes is helped by the presence of monosaccharide such as glucose in the rehydration fluid. Solid food is best avoided in the acute stages but nutrition should be maintained in young children, especially if already malnourished. Breast-feeding should be continued. Fresh cows' milk and reconstitued milk powders may be poorly tolerated. Milk substitutes such as those made from soya beans are useful when illness is thought to be prolonged by disaccharidase deficiency or other secondary causes of malabsorption. Parenteral rehydration should be given when repeated vomiting prevents oral intake, when intestinal transit is too rapid to allow oral replacement, in severe dehydration with impending 'shock', and when intra-abdominal complications are suspected.

It is useful to monitor serum biochemistry during assessment and correction of dehydration. Osmolarity and serum urea are usually increased. Dehydration from diarrhoea and vomiting is usually '*hyponatraemic*', with lowered sodium, bicarbonate, and potassium levels. Hypokalaemic metabolic acidosis is usual when fluid and electrolytes have been lost mainly through diarrhoea; acidosis may cause rapid respiration. *Hypernatraemia*, with raised sodium but normal or

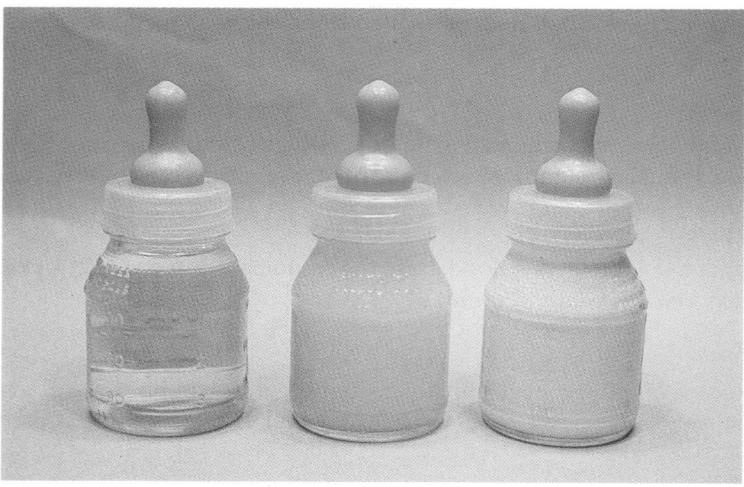

Fig. 4.5 Oral rehydration in infants: initially with oral rehydrating solutions, then diluted milk, finally normal feeds. Breast-feeding should continue during rehydration with ORS alone.

reduced bicarbonate and potassium levels, is a serious complication, commonly due to excessive salt in feeds or rehydration fluids. Usually found in children, it is suggested by irritability, convulsions, a doughy feel to the skin without marked loss of elasticity, often extreme thirst, and can lead to cerebral damage, shock, and renal failure. Fluid replacement should be cautious in hypernatraemia to avoid cerebral oedema.

Rehydration fluids

Table 4.2 shows several preparations for oral rehydration. Great care is required in preparing 'home-made' solutions, especially for children, to avoid excessive salt causing hypernatraemia. Fluids should not taste noticeably salty. When oral fluids are not retained, in the absence of ileus they can be given through a small Ryle's tube passed into the jejunum if facilities are not available for parenteral (usually intravenous) administration. Effective rehydration requires an estimate of fluid and electrolyte losses and of the appropriate normal daily fluid intake for the patient's age. Acidosis can be relieved by bicarbonate but this is usually unnecessary since, unless severe, acidosis usually corrects itself as dehydration resolves.

Drugs

Antidiarrhoeal agents are not usually helpful in the first few hours of illness. They can be used for prolonged symptoms especially in older children and adults. Codeine phosphate, loperamide or Lomotil can control diarrhoea and relieve colic but sometimes aggravate nausea and vomiting or even cause ileus. Milder diarrhoea may be eased by kaolin preparations. Antidiarrhoeal agents can cause constipation after recovery.

If necessary, vomiting can be controlled by phenothiazines; if persistent it requires investigation and usually hospital admission. Phenothiazines (including prochlorperazine and metaclopramide) may produce extra-pyramidal side-effects and oculogyric crises, especially in children in whom they are best avoided. The limited role of antibiotics in gastrointestinal infections is discussed below under specific diseases.

COMPLICATIONS AND SEQUELAE

Malabsorption

In children, and occasionally adults, diarrhoeal may 'grumble on' or relapse after reintroduction of solid food and milk. Gastrointestinal hurry or the 'irritable bowel' syndrome discussed below may be responsible, but intolerance, usually temporary, to milk or other foods may have developed. Reducing substances in stool suggest deficiency of disaccharidase which is required for normal absorption of lactose in cows' milk. Allergy to various food substances including milk may be present but intestinal biopsy to detect villous atrophy is rarely appropriate. Tests for allergic antibodies to food are usually unhelpful since they may be present in healthy children. When malabsorption is suspected, cows' milk can be empirically replaced by a lactose free substitute usually based on soya bean or chicken. Lack of response suggests underlying disease such as immunodeficiency. Normal feeding can usually be re-introduced after an interval of several weeks with careful monitoring for relapse.

Haemolytic uraemic syndrome

This serious complication with listlessness, persistent vomiting, pallor, or spontaneous bruising is commonest in children and is discussed in Chapter 13. One common cause is infection with *E. coli* serotype 0157.

Septicaemia

This is more likely in those who are elderly or immunosuppressed, after ingestion of a high dose of organisms and when there is achlorhydria. Certain agents such as salmonellas cause septicaemia more often than other bacterial invaders of the deeper layers of mucosa. If present or suspected, septicaemia should be treated by appropriate antibiotics and a watch kept for complications such as shock, disseminated intravascular coagulation, and remote septic foci (Chapter 8).

Paralytic ileus and perforation (Fig. 4.6)

Prolonged ileus can develop in association with extensive intestinal inflammation or ulceration and may be aggravated by anti-spasmodic drugs. Milder degrees

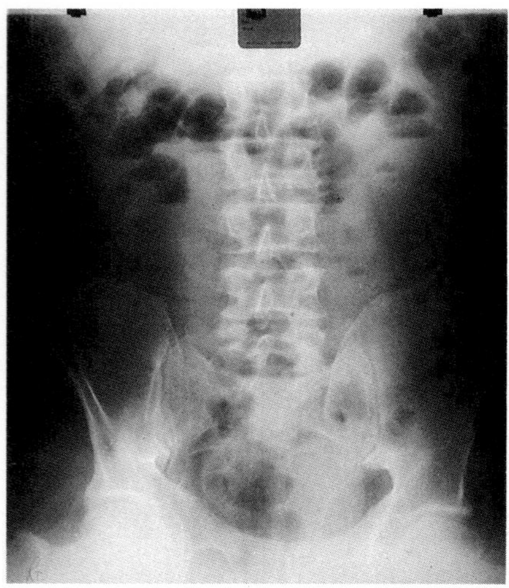

Fig. 4.6 Small intestinal levels in infective enteritis.

are common and may pass unnoticed or with only slight abdominal distension. Diarrhoea does not exclude ileus since it may be simply an 'overflow'. Intravenous fluids, gastric suction, and occasionally total parenteral nutrition may be necessary. It is important to monitor for perforation, peritonitits, or toxic dilatation of large and sometimes small bowel which may require surgery.

Reactive arthritis

Arthritis affecting principally the larger joints can follow various gastro-intestinal infections (for example, *Salmonella*, *Shigella*, *Campylobacter*, and *Yersinia* infections), especially in patients with HLA-B27 genetic determinants. It is usually self-limiting but may persist for six months or longer. If it is incapacitating, corticosteroids can be helpful.

Irritable bowel syndrome

This syndrome, in which the usual investigations fail to demonstrate structural disease but normal peristaltic function is disturbed, presents with various combinations of nausea, diarrhoea, flatus, and abdominal pain. It can be exacerbated by or follow bowel infections, such as amoebic dysentery or giardiasis. It may be related to food allergies and is often associated with anxiety states or fatigue syndromes.

Other inflammatory bowel disease

There are possible relationships between infection and such diseases as ulcerative colitis, gastric ulceration, and Crohn's disease. *Salmonella* and *Campylobacter* infections can mimic the gross mucosal changes and radio-graphic appearance of ulcerative colitis and may precipitate its onset or relapse. Treatment of

Helicobacter pylori by bismuth and antibiotics such as erythromycin and metronidazole can cure or produce prolonged remission of peptic ulcers.

SPECIFIC BACTERIAL INFECTIONS

Chapter 14 describes cholera, enteric fever, and amoebic dysentery and mentions several other diseases usually acquired in tropical or subtropical areas that can cause diarrhoea.

Salmonellosis

Aetiology and epidemiology (Fig. 4.7)

Approximately 2000 serotypes of *Salmonella* have been associated with disease in humans and animals. *Salmonella typhimurium*, *S. enteritidis*, and *S. virchow* have particularly often caused disease in man. Those which usually cause enteric fever (*S. typhi* and *paratyphi A*, *B*, and *C*—not to be confused with *S. typhimurium*) are discussed in Chapter 14.

Most of these human infections are contracted from foods such as poultry and milk, but occasional cases have been traced to domestic animals and reptiles.

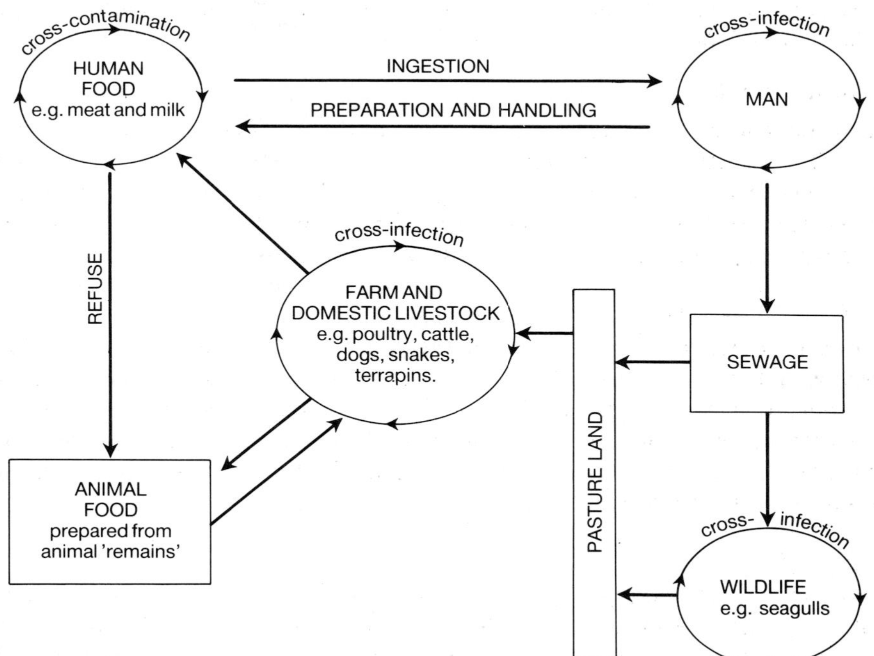

Fig. 4.7 Main transmission routes of salmonellas involving man.

Direct person-to-person spread is unusual, unless from diarrhoea in infants or incontinent adults, because a large number of organisms is required to cause disease in an otherwise healthy individual. Salmonellosis is most prevalent during summer months (see Fig. 4.1) when meat products, especially poultry, improperly cooked or stored, are eaten cold. At least three-quarters of the chickens intensively reared in Britain are infected at time of slaughter. Organisms can multiply and spread through a joint of cooked meat in just a few hours. Cross-contamination via utensils or preparation surfaces is also a hazard. Illness can be severe when frail or ill persons are affected (for example, in hospital outbreaks).

Clinical features and investigations

The incubation period is usually between 12 and 36 hours. Vomiting, diarrhoea and colicky pain are common. Symptoms can settle in a few days or may last for two or more weeks. More severe infections usually present with fever and prostration; blood culture is important to detect septicaemia. Rarely, they mimic enteric fever, even with 'rose spots' (Chapter 14). Disease can be especially severe where persisting faecal impaction prevents evacuation of bowel contents.

Complications and management

In addition to non-specific complications (above), septicaemia can cause disseminated abscesses, arthritis, endocarditis, meningitis, pyelonephritis, orchitis, or pneumonia.

Antibiotics are reserved for those in whom septicaemic spread is suspected or confirmed; they have no proven value in uncomplicated cases and tend to prolong carriage.

The 'carrier' state

Carriage of organisms for more than six weeks after acute infections is unusual but about 1 per cent of infected adults and 5 per cent of young children excrete them for a year of more. Infection rarely spreads from person-to-person if there is no watery diarrhoea and if hands are well-washed after defecation. Asymptomatic patients should not be prevented from resuming normal activities unless handling food for public consumption: the decision about returning to work then depends on individual circumstances. Courses of antibiotics for non-typhoid *Salmonella* carriers are generally unhelpful and may promote resistant strains, although ciprofloxin has recently shown promise and may be tried where carriage is potentially harmful as in food handlers or the incontinent.

Campylobacter enteritis

Aetiology and epidemiology

Campylobacter jejuni, the usual species causing disease in man, was only recognized as a cause of acute gastrointestinal symptoms in 1977. It is at least as common as salmonellosis and is likewise an important cause of animal disease. Human infection originates particularly from poultry, sometimes from farm and domestic animals or pets, and outbreaks have been caused by unpasteurized milk.

Clinical features

The incubation period is from 1 to 7 days, usually 3. Fever, malaise, and myalgia often precede diarrhoea by 24 hours or so, a useful feature to distinguish this infection from salmonellosis. Colicky pain is usual and blood may be present in the stool.

Diagnosis

The organisms can usually be cultured from the stool and occasionally from the blood. If stool cultures are negative and confirmation of the diagnosis is important, serological tests may be helpful.

Complications and management

These are in general as discussed above for other causes of diarrhoeal conditions. Erythromycin or ciprofloxacin can shorten illness but are normally reserved for severe or prolonged cases.

Shigella dysentery

Aetiology and epidemiology

'Dysentery' signifies the presence of blood and mucus in diarrhoeal stool, classically distinguished as 'bacillary' or 'amoebic'. Bacillary dysentery is caused by four species of shigella: *Sh. dysenteriae*, *Sh. flexneri*, *Sh. boydii*, *Sh. sonnei*. As social circumstances improve, *Sh. sonnei* becomes predominant. Bacillary dysentery is endemic worldwide, commonly spread from person to person but sometimes causing water-borne epidemics. Unlike salmonellas, shigellas primarily infect man. Amoebic dysentery is discussed in Chapter 14.

Clinical features and complications

Shigellas primarily infect the large intestine. Diarrhoea predominates over vomiting after an incubation period of usually 24–48 hours and sometimes prodromal fever and anorexia. Colicky pain and blood or mucus in the stools are common but less so with *Sh. sonnei* infections. Fever is usual, especially in *Sh. flexneri* and *Sh. dysenteriae* infections. Febrile convulsions can affect young children. Some illnesses persist for several weeks but prolonged carriage is unusual. Disease can be spread by asymptomatic infections. Septicaemia is rare and death is uncommon when rehydration is adequate.

Management

Antibiotics are more effective in shigella than salmonella infections but are unnecessary in mild illnesses. If required, the drug should be one to which the organisms are sensitive and which reaches adequate levels in the bowel lumen. Co-trimoxazole or amoxycillin can shorten both illness and carriage.

Escherichia coli enteritis

Aetiology and epidemiology

Pathogenic strains of *E. coli* are currently classified in five groups: (a) *Enteropathic* (EPEC), particularly incriminated in outbreaks of infantile diarrhoea, which act by adhering to and damaging enterocytes probably by local toxin production; (b) *Enteroadherent* (EAEC), causing diarrhoea but not producing recognized toxins; (c) *Enterotoxigenic* (ETEC), important causes of infant diarrhoea in developing countries and 'travellers' diarrhoea' in adults (see below), producing toxins of which one resembles that of cholera; (d) *Enteroinvasive* (EIEC), closely resembling shigellas and causing food-borne outbreaks and sporadic infections; (e) *Enterohaemorrhagic* (EHEC), causing haemorrhagic colitis by the action of shiga-type serotoxins, sometimes followed by the haemolytic uraemic syndrome in children (Chapter 14). Specific 'O' antigens are distinguishable by agglutination and there are many serotypes of which some are more likely than others to cause disease. Spread is usually faecal–oral causing sporadic cases or local outbreaks; hospital outbreaks occasionally cause temporary closure of wards for infants.

Clinical features

The incubation period is only a few days. The degree of illness varies with the age of the patient and the pathogenic properties of the infecting organisms. More severe illness is usually seen in babies, when vomiting and refusal of feeds usually precede diarrhoea which can be prolonged and liable to relapse often because of disaccharidase deficiency. Large bowel involvement may cause blood in the stools. There is often discomfort before defecation and sometimes 'toxic' or febrile convulsions.

Management and prevention

Adequate hydration and nutrition are important. Antibiotics are usually reserved for suspected tissue invasion or septicaemia. Prevention depends upon strict isolation and hygienic precautions during outbreaks and encouraging breast-feeding of infants.

Travellers' diarrhoea

This commonly afflicts recently-arrived travellers to another country, especially travellers from north and west Europe and north America to warmer countries with lower standards of hygiene and sanitation. Many infectious agents can contribute to its causation, including viruses and protozoa, but mainly enterotoxigenic *E. coli* of serotypes not previously encountered by the travellers.

When due to *E. coli*, the incubation period is short (one or two days), diarrhoeal illness often temporarily incapacitating but with quick recovery in a few days. Clinical experience suggests that acquired immunity to the strain involved lasts about six months. Treatment is symptomatic in milder illnesses. Antibiotics such as co-trimoxazole or ciprofloxacin may shorten the illness in more severe cases or provide prophylaxis where the brief illness would be particularly inconvenient.

Vibrio parahaemolyticus

This organism causes outbreaks of acute, watery diarrhoea, resulting from infected seafood (crustacea, fish, and molluscus) eaten raw or inadequately cooked. The incubation period is two to three days. The diarrhoeal illness varies in severity but is rarely life-threatening. Diagnosis is by isolating the organisms from stool or the responsible food; prevention is by adequate cooking of seafoods.

Bacillus cereus

This aerobic, sporing organism produces enterotoxins which can cause mainly vomiting with an incubation period of up to 5 hours or mainly diarrhoea with an incubation period of up to 16 hours. It is widespread in soil and in food materials; for example, rice husks. Keeping cooked food at room temperature and then re-heating is commonly responsible for outbreaks; for example, from fried rice. The onset is acute, colic may be severe, but illness rarely persists for more than 24 hours. Diagnosis is by isolating the bacillus from faeces and food residues.

Yersinia enterocolitica

This organism causes acute diarrhoea and sometimes terminal ileitis which can be confused with other causes of pain in the right iliac fossa. Reactive arthritis and erythema nodosum are more frequent than in other causes of diarrhoea. Transmission is mainly from foods of animal origin, commonly pigs, or water contaminated by pigs. Diagnosis is by isolating the organism from faeces or by serological tests; it is probably under-diagnosed. Tetracyclines may speed recovery.

Clostridium perfringens

This anaerobic organism is widespread in soil and in the gastrointestinal tracts of man and animals. Spores survive normal cooking temperatures and multiply if food is stored without refrigeration. The heat-labile toxin is formed when infected food is inadequately cooked or inadequately re-heated. The incubation period is about 8–18 hours and mainly diarrhoea is usual, sometimes with vomiting or fever of acute onset resolving within 12–24 hours. Abdominal cramps are prominent and severe cases may develop necrotizing enteritis. Bacteria of serotype A are generally responsible. Laboratory diagnosis is by semi-quantitative stool culture. Antibiotics are unhelpful.

Staphylococci

Staphylococci contaminating food either before or after cooking may multiply and produce heat-stable enterotoxins. Subsequent heating may sterilize the food but not destroy the toxin. Septic lesions on the skin of the food-handler may be responsible, and milk may be infected from a cow with mastitis. After a short incubation period (up to six hours) there is abrupt onset of vomiting, muscle cramps, and sometimes diarrhoea; the patient may rapidly become shocked and require parenteral fluids but recovery within 24 hours is usual and fatality is rare. Staphylococci which produce enterotoxin may be cultured from vomitus, faeces or the responsible food. Prevention depends upon food-handlers being aware of the risks of skin sepsis and minimizing the time between preparation and cooking of foods. Antibiotics are inappropriate.

Clostridium difficile

Pseudomembranous colitis is caused by toxin produced by this anaerobe in the intestine. It is occasionally present asymptomatically but, although sometimes no initiating factor is observed, previous use of antibiotics is a recognized precipitant. Many antibiotics have been implicated, notably clindamycin, ampicillin, and the cephalosporins. Occurring at all ages, symptoms vary from mild diarrhoea to fulminating toxic illness that can be fatal especially in debilitated patients. Marked pain and bleeding are unusual. The diagnosis is made by a combination of antibiotic history, the presence of *Cl. difficile* and/or toxin in the stools, mucosal appearance (mucoid or with yellow plaques–pseudo-membranes) and mucosal histology or radiographic appearances (Fig. 4.8).

Supportive treatment is sometimes necessary with intravenous fluids. Vancomycin or metronidazole given orally can be helpful. Symptoms usually subside within a week or so, but relapses occur and the therapy may need to be prolonged.

PROTOZOAL INFECTIONS

Giardiasis

Aetiology and epidemiology

The protozoon *Giardia lamblia* (Fig. 4.9a) colonizes the upper small intestine where flagellated trophozoites line the mucosa and damage the epithelium.

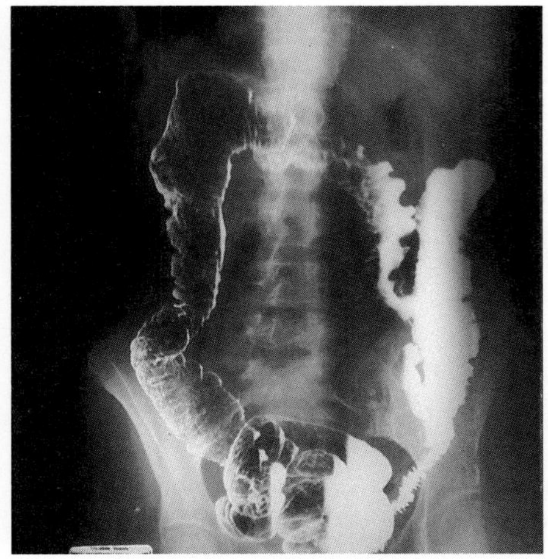

Fig. 4.8 Extensive ulceration of the colon in pseudomembranous colitis.

Infection is spread by cysts which resist normal levels of chlorination of drinking water. The disease exists world-wide, an often unrecognized cause of intestinal symptoms. Asymptomatic carriage is found in from 1 to 5 per cent of persons in developed countries and up to 30 per cent or more in developing countries. Dogs and other animals may also be infected.

Clinical features and investigations

All age groups are affected but repeated exposure give some immunity. Children and travellers from low to high prevalence areas most likely to be affected. Profuse diarrhoea is unusual; more usually one or two, foul-smelling motions a day are passed, often only in the morning. There may be abdominal discomfort, anorexia and nausea, alcohol intolerance and unusual travel sickness induced by food. Symptoms may be protracted or intermittent and mild infections can be confused with the irritable bowel syndrome, gall-bladder disease, or duodenal ulceration. Heavy infestation can cause malabsorption. To visualize trophozoites, stools or duodenal contents should be examined immediately. Usually only cysts are recognized, but often no parasites are seen and the diagnosis is made presumptively from rapid response to treatment. Immunofluorescence tests can increase the sensitivity of stool examination. In complicated or prolonged illness a 'string test' or villous biopsy can be helpful. The string test involves swallowing a capsule from which a string unwinds as it passes into the intestine. The end is tethered to the face with adhesive tape. It is left overnight, withdrawn and examined under the microscope when giardia trophozoites (Fig. 4.9a) may be seen.

Management and prognosis

Metronidazole or tinidazole usually relieve symptoms which may recur and necessitate several courses. Mepacrine hydrochloride and quinacrine are more

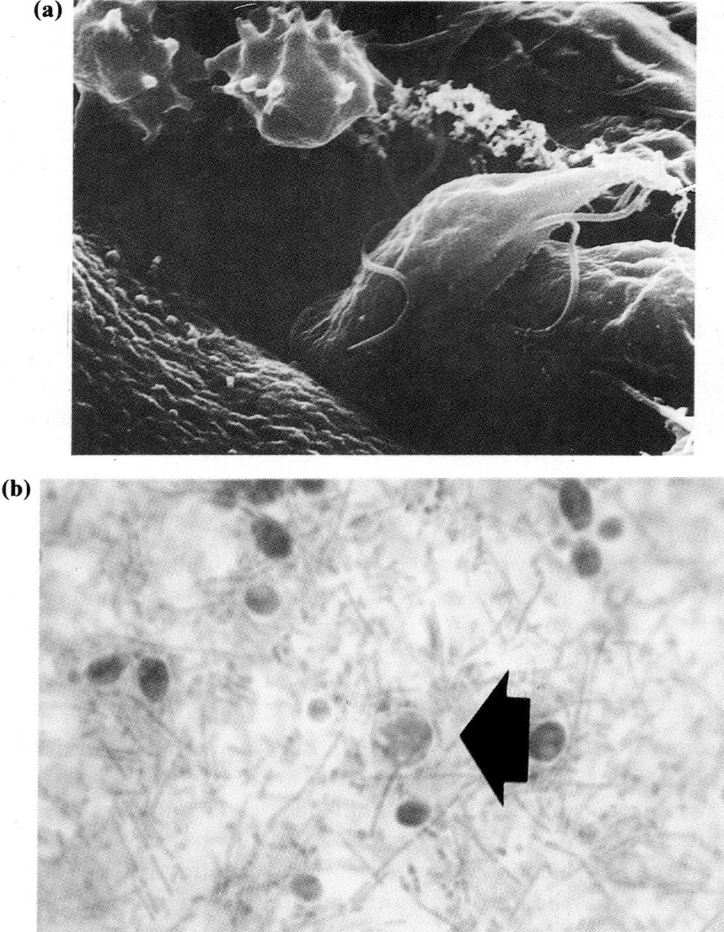

Fig. 4.9 **(a)** The flagellate trophozoite of *Giardia lamblia*; **(b)** the cryptosporidium—a recently recognized cause of diarrhoea.

toxic alternatives. Recurrence may be due to reinfection from asymptomatic family members who themselves can be treated. Although sometimes rather protracted and debilitating, espeically if diagnosis is delayed, complete recovery is usual.

Cryptosporidiosis

The protozoon *Cryptosporidium* can cause diarrhoea and vomiting, usually self-limiting but sometimes prolonged with severe and watery diarrhoea, especially in those immmunosuppressed (for example, in AIDS), malnourished or homosexual with the 'gay bowel syndrome' from faecal contamination during anal intercourse. It occurs world-wide, sporadically and in outbreaks. Spread may be person-to-person, from animals or animal products or sometimes via contaminated water. Special staining (for example, modified Ziehl–Neelsen) of faecal material for oocysts is required (Fig. 4.9b). Treatment with spiramycin may be helpful in severe cases but generally specific treatment is ineffective.

VIRAL INFECTIONS

Numerous viruses have been discovered in recent years, largely by electron microscopy, in normal and diarrhoeal faceces, especially from infants and young children (Table 4.4). Many of them are important causes of diarrhoeal disease. Several other important viral infections of the gastrointestinal tract rarely, if ever, cause diarrhoea.

Table 4.4. *Viruses in diarrhoeal stools*

Rotaviruses	Cause diarrhoea in infants, sometimes in older children and occasionally adults.
Astroviruses	Cause diarrhoea mainly infants.
Fastidious adenoviruses	Cause diarrhoea mainly in infants.
Norwalk group ('small round structured viruses')*	Cause sporadic or epidemic illness in adults and older children, e.g. winter vomiting disease.
Calciviruses Coronaviruses Enteroviruses ('small round viruses')*	Some possibly related to diarrhoeal disease.

**Nomenclature for reporting by electron microscopy.*

Rotaviral enteritis

Aetiology

The most widespread and important viruses causing diarrhoea are rotaviruses (Fig. 4.10). At least five groups of rotavirus have been distinguished in human infections. Group A has 11 serotypes and is responsible for most outbreaks.

Epidemiology

All age groups may be involved but older infants are most often and most severely affected. In the newborn, infection is usually asymptomatic (suggesting temporary passive immunity). After the first few months of life rotaviruses cause

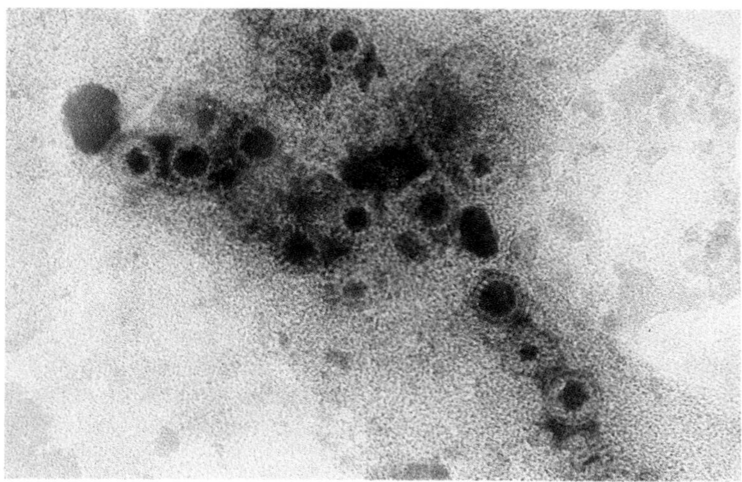

Fig. 4.10 Rotavirus showing the 'cartwheel' appearance.

many of the more severe attacks of diarrhoea and vomiting in babies. Antibodies are acquired in the first two years or so of life and family outbreaks rare, probably because adults have considerable immunity to illness though not necessarily to silent infection. Disease occurs throughout the world, in temperate climates more often in winter when bacterial diarrhoeas are less prevalent. Spread is faecal–oral but apparently also by the respiratory route.

Clinical features

No specific clincal features distinguish rotaviral from other forms of infective diarrhoea, but in children fever, vomiting, and dehydration are commoner than in other diarrhoeas and there may be mild respiratory symptoms. The onset in severe cases is abrupt after an incubation period of 24–48 hours. Often in children dehydration requires correction after which recovery is usual within a week. Immunodeficient children and elderly or immunosuppressed adults may suffer severe, prolonged, and life-threatening diarrhoea. The disease is especially severe and dangerous in malnourished infants in tropical countries. Rotavirus infection can be diagnosed by detecting virus particles in faecal extracts by electron microscopy, viral antigen by immunological tests or characteristic viral ribonucleic acid by gel electrophoresis. There is no specific treatment.

Prevention

Cross-infection should be minimized in hospitals and among babies in close contact. It is most likely while diarrhoea is present.

Winter vomiting disease

This self-limiting illness is common at all ages. Vomiting predominates over diarrhoea, myalgia, and malaise. The syndrome is attributed to the 'Norwalk agent' and related viruses. Intra-family spread is common and recovery is usually rapid within 12–24 hours.

Primary herpes simplex gingivo-stomatitis

Aetiology and epidemiology

This form of herpes simplex infection mainly affects young children. It has world-wide distribution. Most primary infections are asymptomatic. The incubation period is around 4–5 days but may be longer. There is sometimes a history of exposure to someone with 'cold sores'. Neonatal infection can be acquired from the birth canal. Adults may be affected, often by sexual contact, with pharyngitis commoner than stomatitis.

Clinical features and complications

The onset is fairly rapid with marked fever, irritability, and sore mouth. After 12–48 hours shallow ulcers appear (the vesicles rupture quickly due to mouth movements) and spread rapidly over the buccal mucosa, gums, tongue, and usually lips (Fig. 4.11a and b). Occasionally the pharynx is involved. Dribbling infected saliva down the face causes local spread, usually vesicular, and sometimes lesions elsewhere such as paronychia of a sucked finger. Pain causes difficulty in opening the mouth and swallowing. Anterior cervical nodes are enlarged and tender. Feeding and drinking are painful and there may be dehydration.

Investigations and differential diagnosis

The virus can usually be grown from the mouth or saliva. Serology is rarely necessary but distinguishes primary infection from reactivation which rarely causes oral lesions. Other causes of stomatitis include *Candida albicans*; herpangina, and hand, foot, and mouth disease both due to coxsackieviruses, cause oral vesicles which rapidly progress to ulcers. Shingles can cause unilateral vesicles on the palate but associated skin rash is usually present (Chapter 3).

Management and prognosis

Herpetic disease resolves completely after about 7–10 days but requires careful and sympathetic nursing with attention to oral hygiene. Analgesics can be helpful. Acyclovir can speed recovery given early in the illness. It is also helpful in the immunocompromised where infection can be low-grade and chronic. Attention to fluid intake is necessary and occasionally parenteral fluids are

(a)

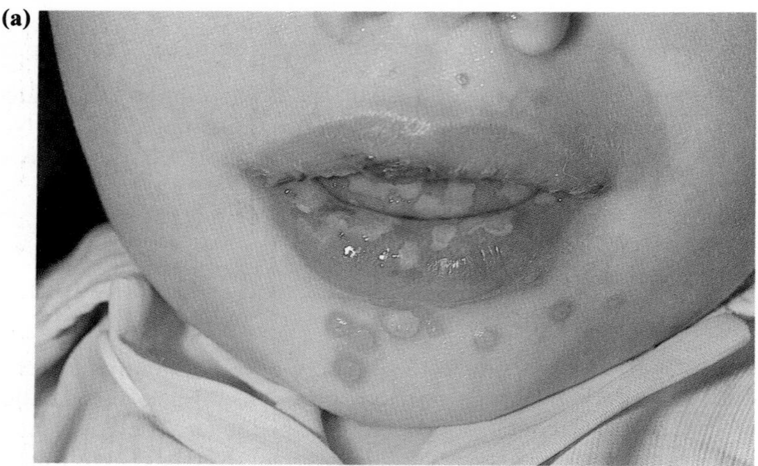

(b)

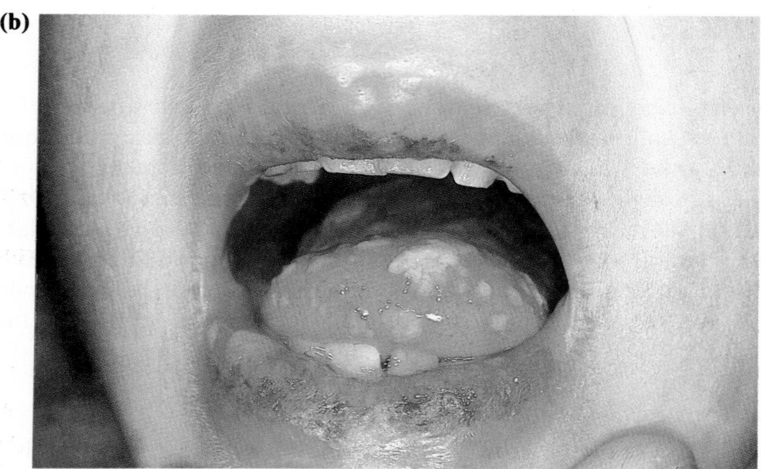

Fig. 4.11 Primary herpes simplex infection: **(a)** stomatitis with satellite vesicles over the chin; **(b)** vesicles in the mouth rapidly rupture, leaving shallow ulcers.

required. Attendants should avoid infection from saliva into cuts or nailfolds: gloves are advisable (Fig. 4.12).

Prevention

Apart from careful hygiene when clinical illness or vesicles are present, there are no practical means of prevention. Vaccine may be available in the future.

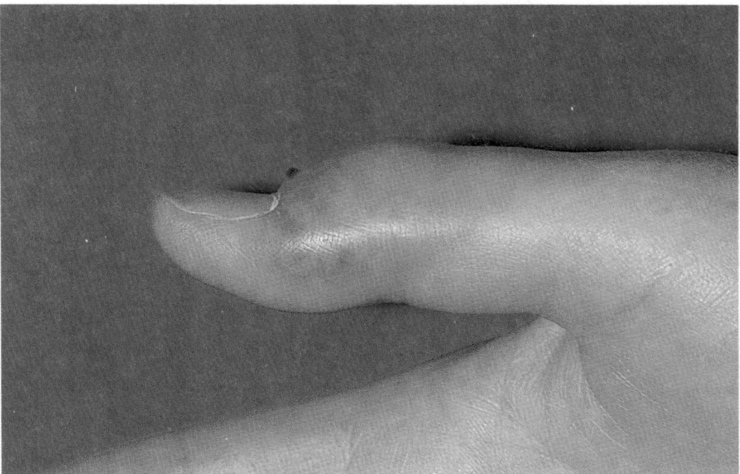

Fig. 4.12 Herpes simplex paronychia: incision and drainage are contraindicated.

Enteroviral infections

The alimentary tract is the main site of entry for enteroviruses which spread by faecal–oral and sometimes respiratory droplet routes. It is doubtful whether they cause diarrhoea, and their main effects are upon non-alimentary organs (Chapters 3, 6, and 7). Some can cause oral lesions: diagnosis can be confirmed by isolating virus from mouth lesions or faeces.

Herpangina

This acute, febrile but self-limiting illness is usually caused by coxsackieviruses, especially of group A. Commonest in autumn and winter months, it affects all age groups but mostly children. After a short incubation period of about 3–5 days, onset is acute, with fever and sore throat. Discrete vesicles, which soon rupture leaving shallow ulcers, develop around the soft palate, often along the junction with the hard palate and also on the uvula and tonsillar fauces. There is no specific treatment but analgesics may be required to relieve pain which can be severe.

Hand, foot, and mouth disease

This condition is also usually caused by group A coxsackieviruses with outbreaks in Britain every three or four years. Young children are mainly involved. It appears to be more common in rural of semi-rural areas but must not be confused with foot and mouth disease of cattle, sheep, and pigs, which very rarely affects humans. The incubation period is about 5 days. It is differentiated from herpangina in that vesicles in the mouth are commonly seen inside the cheek and

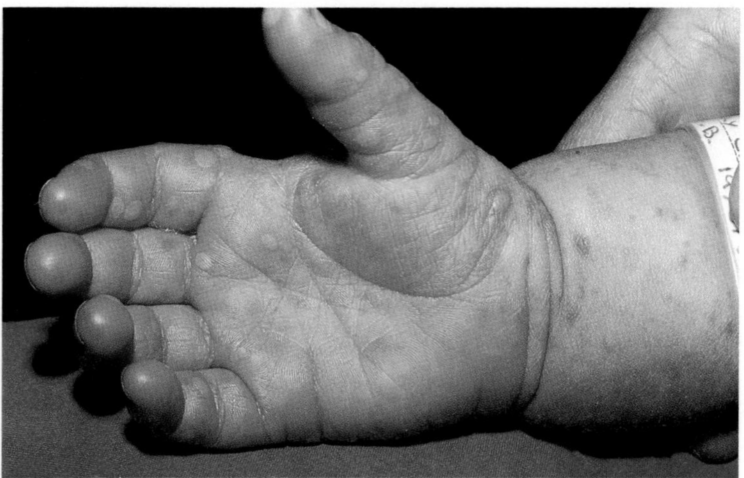

Fig. 4.13 Unusually abundant vesicles on the hand in hand, foot, and mouth disease.

along the gums and tongue as well as in the posterior half of the oral cavity. They also affect other sites, especially the hands and feet (Fig. 4.13). Constitutional upset and pain are generally less severe than with herpangina, and no treatment is usually required.

Mumps

Aetiology and epidemioloy

This systemic paramyxovirus infection is spread through saliva and respiratory droplets. Maximum infectivity is around the onset of illness but up to 60 per cent of infection is subclinical which may also be infectious. Symptomatic illness is commoner in family contacts than, for example, in school contacts. Infections are commonest in winter but sporadic cases occur throughout the year. Immunity after infection is lifelong.

Clinical features and differential diagnosis

The typical illness begins with fever, coryza, and sometimes conjunctivitis. The parotid duct openings into the mouth may be inflamed and swollen. Pain quickly develops over one or more of the parotid or submandibular salivary glands which become enlarged although much of the visible swelling is of surrounding soft tissues (Fig. 4.14). Pain and swelling can cause difficulty is opening the mouth (trismus). Fever commonly lasts four or five days but swelling takes up to two weeks to resolve. Mumps must be differentiated from acute tonsillitis with associated lymphadenopathy, where tenderness is localized to the upper anterior cervical lymph nodes and examination of the pharynx shows the source of the

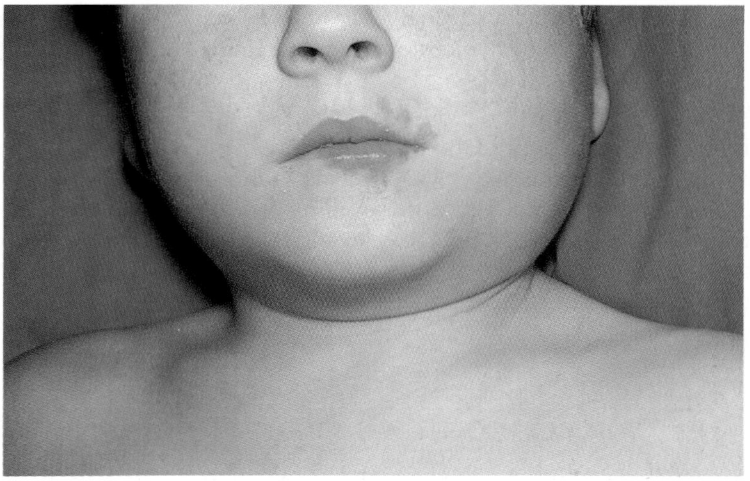

Fig. 4.14 Parotid and soft tissue swelling in mumps (the febrile illness has conincidentally caused 'cold sores').

infection. Acute parotitis associated with calculi or mucus plugging of the salivary duct is usually unilateral with rapid onset, severe pain and minimal fever unless there is secondary infection.

Aseptic meningitis is common. Mumps meningitis may precede, follow or present without obvious salivary gland involvement. The illness is usually mild especially in children. When there is confusion or altered conscious level encephalitis, which is rare, should be suspected. Especially in adults persistent headaches can be aggravated by exercise or premature return to normal activities. Deafness is another rare complication.

Complications

Orchitis affects only 10–15 per cent of post-pubertal males, often beginning during the second week of illness with swelling, pain and return of fever. It is usually unilateral but even when bilateral sterility rarely ensues. Oophoritis and mastitis occasionally occur, also pancreatitis which can cause abdominal pain but is not often severe.

Investigation and management

The diagnosis is confirmed serologically but virus may be isolated from saliva or swabs from around the opening of the parotid duct. When the central nervous system is involved virus can often be isolated from cerebrospinal fluid. Lumbar puncture may be necessary to exclude bacterial meningitis or subarachnoid haemorrhage when meningism is present.

Symptoms can be relieved with analgesics and bed rest during the acute illness. A short course of systemic coriticosteroids may relieve the pain of orchitis.

Prevention

There is little evidence that isolation prevents spread of infection, partly because the infectious period begins before the onset of symptoms. However, it is usual for children to be kept off school until the swelling has subsided. A live attenuated vaccine is available and, since 1990, has been given to children in Britain during the second year of life together with measles and rubella vaccines (MMR: Chapter 16). It can be given separately to older persons.

MYCOLOGICAL INFECTIONS

Alimentary candidiasis

Aetiology and epidemiology

Candida albicans is often present in the alimentary tract as a commensal. Symptoms are especially likely when its growth is encouraged by suppression of competing organisms by broad-spectrum antibodies, in diabetes mellitus and during immunosuppression, especially by AIDS, when generalized and systemic infections may occur.

Clinical features and diagnosis

Pain, discomfort, or irritation affect the mouth and pharynx; dysphagia can result from involvement of the oesophagus. The yeast can contribute to pruritus ani and nappy rash. White plaques of yeast and debris, usually multiple and adherent, may be seen on the affected surface (Fig. 4.15). Especially in adults, plaques may be inconspicuous with little to be seen except redness and a dry appearnce of the buccal mucosa. Yeasts may be seen on direct microscopy and confirmed as *C. albicans* by culture.

Treatment and prevention

Topical nystatin or other antifungal drugs are usually effective. Adults can slowly suck tablets or lozenges; gels are available to rub around the mouth of babies. Fluconazole orally or if necessary parenterally can help in more severe infections especially in immmunocompromise. Parenteral amphoteracin is occasionally required. Treating vaginal infection in pregnancy, good hygiene especially between mother and baby, good control of diabetes mellitus, and avoiding unnecessary or prolonged use of broad-spectrum antibiotics help in prevention. Early suspicion of infection and prompt treatment are important for those at risk of systemic spread.

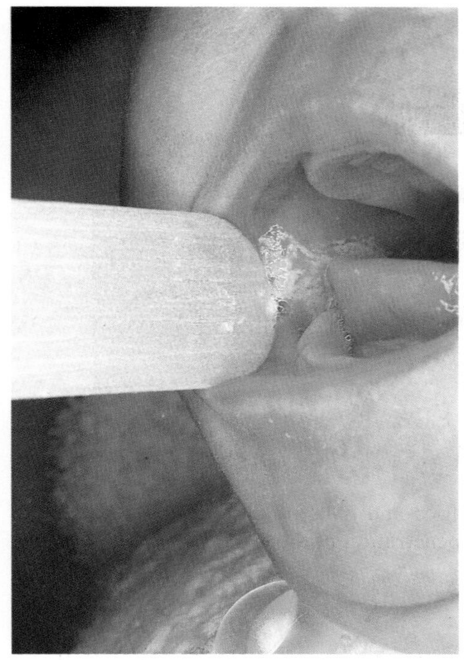

Fig. 4.15 The white plaques of oral candidiasis.

HELMINTHIC INFECTIONS

The commonest helminthic infection seen in Britain is due to the thread-worm. Tapeworms, hookworms, whipworms, strongyloides, and other roundworms are occasionally encountered. Infections are commonly asymptomatic.

Threadworms often pass from one individual to another by personal contact rather than through food. Several members of a family may be affected. Eggs laid around the anus are picked up on fingers through scratching induced by irritation from the worms. Roundworm eggs laid in the intestine are passed in stools. Careful hand-washing after defecation effectively interrupts transmission where sewage disposal is efficient.

Hookworms are common in warm climates where sewage disposal is inadequate and barefoot walking allows larvae from contaminated soil to penetrate the foot (Fig. 4.16). These then travel via lymph to blood to lungs, penetrate alveolar walls and travel up airways to be swallowed and attach themselves to the upper intestinal wall where they feed on blood, mature and lay eggs which are excreteted in faeces.

Tapeworms are contracted from infected meat, usually pork or beef, and control depends upon examination of carcasses and adequate cooking. Whipworms are spread through faecal contamination of food and water.

Fig. 4.16 Many agricultural workers in Africa and Asia are chronically infected with hookworms by larval penetration of the feet.

Clinical features and investigations

Threadworms (*Enterobius vermicularis*). Pruritus ani is common especially at night when worms are most active. Sometimes worms are seen in stool or on toilet paper, but syptoms other than the pruritus are unusual. If worms are not seen on direct examination of stool, ova may be demonstrated by pressing the sticky side of transparent adhesive tape onto the anal skin after awakening and transferring it to a microscope slide for examination.

Roundworms (*Ascaris lumbricoides*). The usual complaint is of seeing worms in faeces. They may pass spontaneously, occasionally through the mouth or nose. Heavy infestation can cause abdominal pain or even intestinal obstruction by a ball of worms. If worms are not seen, typical ova can be seen by microscopy of stool samples, if necessary using concentration methods.

Hookworms (*Ancylostoma duodenale; Necator americanus*). 'Ground itch' is a dermatitis caused by larval penetration, usually on feet. Allergic manifestations occur during migration within the body. Anaemia from blood loss may be severe with heavy intestinal parasitization. Diagnosis is by finding ova in smears of faecal concentrates.

Strongyloides. These worms are common in warm, wet regions where sanitation is poor, and have assumed importance elsewhere because of their propensity to cause serious disease in the immunocompromised, especially AIDS patients. Chronic diarrhoea, abdominal pain, and urticaria can occur. Eosinophilia, unusual in other intestinal worm infections, may be marked. Diagnosis is by stool microscopy, sometimes using concentration techniques, or the 'string test' described above (p. 117); serology is available.

Tapeworms (*Taenia* spp.). The head of the tapeworm adheres to the wall of the upper small intestine. Proglottid segments are continuously produced by *T. saginata* and *T. solium* which after fertilization break off and are passed in the stools. Proglottids of *T. saginata* (beef tapeworm) are mobile and may emerge from the anus spontaneously. *Taenia saginata* is harmless. *Taenia solium* (pork tapeworm) eggs, contained in disintegrating segments, if ingested can develop into larvae, invade tissues, and encyst causing *cysticercosis*. According to the sites of the cysts, symptoms may develop several years later—cerebral disease is particularly serious. Adult worm infection is diagnosed by detecting segments or ova in stools. Cysticercosis is diagnosed by biopsy of lesions or suspected by seeing calcified cysticerci, usually in muscle, by X-ray.

Whipworms (*trichuriasis*)

Mild infections with these nematodes are usually asymptomatic but may cause epigastric or right iliac fossa pain, vomiting, flatulence, and weight loss. Diarrhoea may be profuse, especially with associated *Entamoeba histolytica* or pathogenic *E. coli* infection, often with blood and mucus in the stools. Eggs are seen by stool microscopy. In severe cases endoscopy shows worms clinging to the rectal mucosa, though usually they reside mostly in the caecum.

Treatment

Threadworms are effectively destroyed by piperazine salts or thiabendazole; re-infection of patient and family members is difficult to prevent. Piperazine salts paralyse roundworms which are then passed in the stool; if numerous, the worms may form a ball and cause obstruction. Mebendazole or bephenium hydrocynaphthoate kill round-worms which are then commonly digested. Mebendazole or albendazole are used to treat trichuriasis and strongyloides infections. Mepacrine hydrochloride causes expulsion of intact tapeworm and for *T. solium* is usually preferred to alternatives which may cause release of eggs in the upper intestine and risk cysticercosis. Treatment of cysticercosis is symptomatic only and prevention therefore is important. For other tapeworms, niclosamide and praziquantel are effective.

Hydatid diesase

Man may act as accidental intermediate host to *Echinococcus granulosum* (the dog tapeworm) when eggs are swallowed through contact with faeces of a dog which has been infected by ingesting infected viscera of sheep, the normal intermediate host. Farming families are especially at risk and the disease is endemic in the north and west of Scotland and central Wales. Symptoms may arise from formation of cysts in vital organs, often liver or lung, causing pressure effects or secondary infection. Eosinophilia is common and serological tests are useful in diagnosis. Radiography may reveal calcification. Treatment is usually surgical, with great care to avoid spillage of cyst contents with can cause anaphylaxis. However, treatment with mebendazole or albendazole can be tried initially and, even if unsuccessful in reducing the size of the cysts, can reduce the risk of 'seedling' spread at operation.

Flukes

Fasciola hepatica, the sheep liver fluke, is the indigenous trematode in Britain which occasionally infects man after consumption of raw watercress contaminated by the intermediate host, a snail. Intermittent fever may be associated with vomiting, diarrhoea, myalgia, hepatic pain and tenderness, urticaria and migratory subcutaneous nodules. Eosinophilia is usual and ova may be found in faeces or duodenal aspirate. Bithional provides effective treatment. Other types of fluke infections occur in tropical areas, especially schistosomiasis (Chapter 14).

5

Infections of abdominal organs

INTRODUCTION

This chapter deals with infections of the liver, pancreas, renal tract, and abdominal contents other than the gastrointestinal tract itself (see Chapter 4). Infections of the genital tract are dealt with in Chapters 10 and 11. Symptoms from the abdominal organs are common and pose three important questions. First, what organ is affected? Second, is infection the cause of the symptoms? Third, is the problem best treated by antibiotics? Negative answers to either of the last two questions mean that the patient may require a surgical opinion. Many patients with infections of abdominal organs have non-specific systemic symptoms such as fever and malaise but the two commonest localizing features are pain and jaundice. The sequence of events can help in assessing pain which, for example, in acute appendicitis often starts around the umbilicus and then becomes localized in the right iliac fossa. Jaundice can be easily missed in artificial light but can usually be detected in the sclerae (Fig. 5.1) before it becomes obvious in the skin. Some foci of infection—for example, subphrenic or pelvic—may require examination by ultrasound, X-ray, or isotope scan.

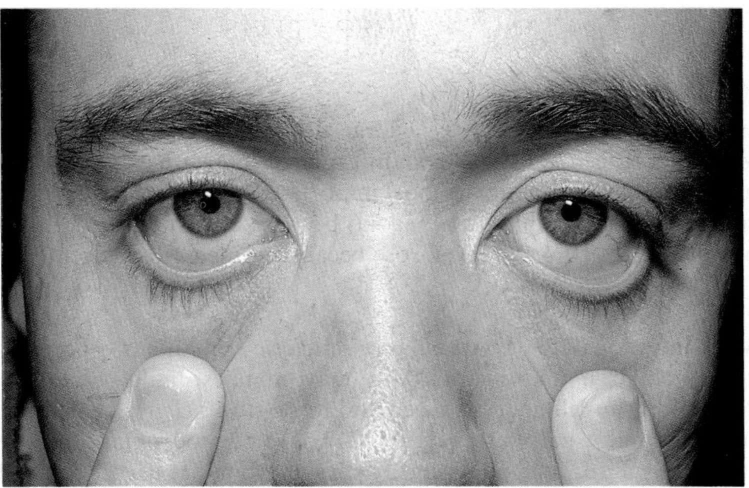

Fig. 5.1 Jaundice of the sclerae.

It is usually possible to identify the affected organ, but sometimes this may be difficult; for example, differentiating a huge left kidney from an enlarged spleen. In such cases history can help—thus a patient with fever and rigors who has just returned from a malaria-endemic area is more likely to have an enlarged spleen. Abdominal symptoms may also be caused by disease outside the abdomen; for example, referred pain from pleurisy, or vomiting from raised intracranial pressure.

The order in which infections are dealt with in this chapter is listed in Table 5.1.

Table 5.1. *Infections of the abdominal organs*

Hepatitis
Cholecystitis and cholangitis
Liver and subphrenic abscesses
Splenic infection
Mesenteric adenitis
Pancreatitis
Peritonitis
Other intra-abdominal abscesses
Urinary tract infection
Prostatitis

HEPATITIS

Aetiology and epidemiology

Hepatitis may be caused by a large number of infectious agents (Table 5.2). Viruses are the major cause, hepatitis A virus (HAV), hepatitis B virus (HBV) and hepatitis C (HCV) being the most important. HAV is no longer endemic in some countries (for example Scandinavia) because of better standards of living. The prevalence of HBV and HCV rose because of increasing parenteral drug abuse.

Clinical presentation

Most infections are asymptomatic. *Anicteric* hepatitis is especially common in young children and infants who may manifest mild vomiting, abdominal discomfort, unwillingness to feed and loose stools. *Icteric* hepatitis is much less common and usually preceded by 2–7 days of prodromal illness with malaise, anorexia, nausea, and vomiting. These symptoms are usually exacerbated by the smell of food or tobacco smoke. Smokers often stop. Myalgia, urticarial rash, and arthralgia suggest hepatitis B. Fever is usually absent but is commoner in early hepatitis A. Dark urine and pale stools precede onset of scleral icterus and more widespread jaundice. Tender hepatomegaly is common, occasionally with splenomegaly. Most patients begin to feel better 7–10 days after the onset of

Table 5.2. *Common infectious causes of hepatitis*

Viral	Hepatitis A
	Hepatitis B
	Hepatitis C
	Hepatitis D (delta agent)
	Hepatitis E
	non-A non-B hepatitis
	Cytomegalovirus
	Epstein–Barr virus
Others	*Coxiella burnetii*
	Toxoplasma gondii
Tropical	Yellow fever virus

symptoms, but the jaundice may take several weeks to clear. One cannot usually distinguish the various causes of hepatitis on clinical grounds, although the history can be suggestive.

Non-infective causes of hepatitis such as alcohol, drugs, chemical, or other diseases (systemic lupus erythematosus, primary biliary cirrhosis) must also be considered. Differential diagnosis of jaundice includes biliary obstruction without infection, and haemolysis. In leptospirosis and malaria, jaundice may be prominent but inflammation of the liver is unusual. Leptospirosis is considered here and malaria in Chapter 14.

Investigation

In viral hepatitis, hepatocellular damage (Fig. 5.2) usually raises the serum alanine and aspartate aminotransferase levels markedly. Intrahepatic cholestasis, if present, raises alkaline phosphatase also. High levels of alkaline phosphatase and cholesterol can result from extrahepatic biliary obstruction and be useful in distinguishing obstructive from hepatocellular jaundice. Transaminases are increased in icteric, anicteric, and even asymptomatic hepatitis.

Early in symptomatic illness, urobilinogen and bilirubin are usually present in the urine. Urobilinogen then disappears, to reappear in the recovery phase when bile again reaches the gut. In jaundice resulting from haemolysis, urobilinogen is increased but bilirubin is usually absent. In severe or prolonged disease, hypoprothrombinaemia and hypoalbuminaemia (Fig 5.3) are the best guides to liver function. The former may also be due to poor intake and absorption of vitamin K. Hypoglycaemia, hyponatraemia, and low serum urea are common.

The cause of viral hepatitis is identified by serological tests for specific antigens and antibodies. Ultrasound of liver and gall-bladder may detect gallstones or abscesses. Liver biopsy is unhelpful in acute viral hepatitis as histology does not usually distinguish the different causal viruses.

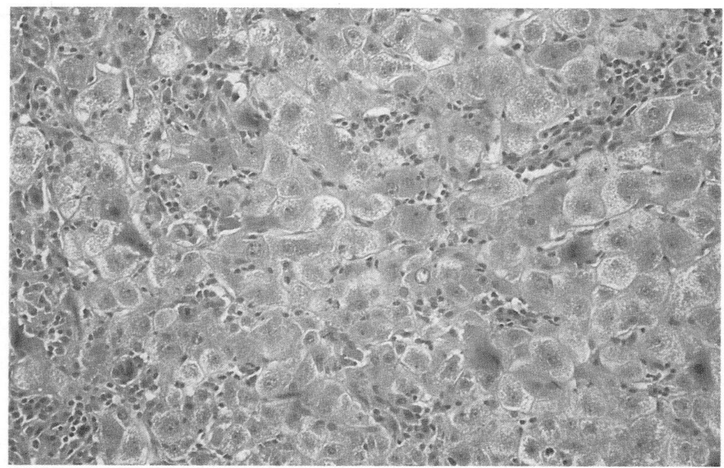

Fig. 5.2 Viral hepatitis showing ballooning and acidophilic degeneration of cells.

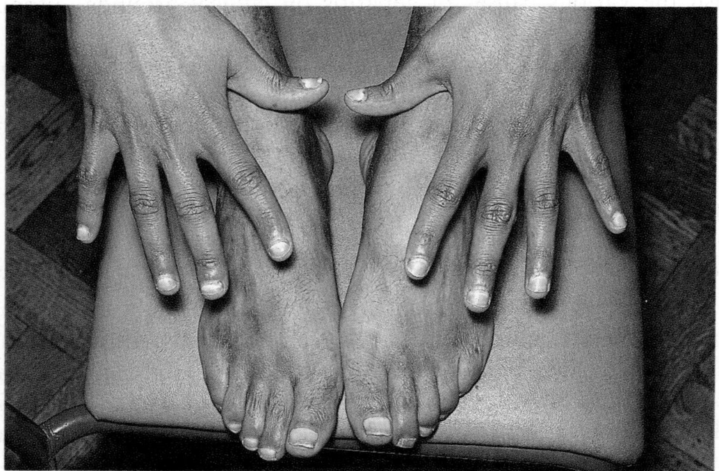

Fig. 5.3 White nails of hypoalbuminaemia.

Management

No specific treatment is yet available for acute viral hepatitis. Convalescence should not be hurried. Dietary restrictions and strict bed rest are unnecessary, but hepatotoxic agents such as oral contraceptives and alcohol should be avoided. Exercise should be taken up only gradually during convalescence. In some patients, especially those with prolonged cholestasis, itching may be relieved by calamine lotion, oral cholestyramine or aluminium hydroxide which bind bile salts within the gut. Three days of parenteral vitamin K will correct a prolonged prothrombin time if due to inadequate intake.

Liver failure

This can develop early during the illness, obscuring other signs of hepatitis, or as a late complication often following prolonged cholestasis. Characteristically there is irritability, confusion, forgetfulness progressing to coma, abnormal neurological signs (dysdiadochokinesis; flapping tremor), hepatic foetor, and typical EEG findings (symmetrical, high-voltage slow waves). Serum transaminase levels are high at the start but fall with progressive liver destruction. Blood ammonia rises; glucose, urea, and clotting factors decline.

Treatment is supportive until liver function recovers. Corticosteroids can reduce cholestasis but probably do not speed up liver cell recovery. Glucose in high concentration is given. Ammonia production in the colon is decreased by low-protein diet, oral neomycin to suppress colonic bacteria, and lactulose orally to decrease transit time and minimize absorption. Cimetidine or ranitidine lower acidity and reduce the risk of gastrointestinal haemorrhage.

Hepatitis A (infectious hepatitis)

Aetiology

Hepatitis A virus (HAV) is an enterovirus (type 72) (Fig. 5.4) which cannot normally be cultivated in the laboratory.

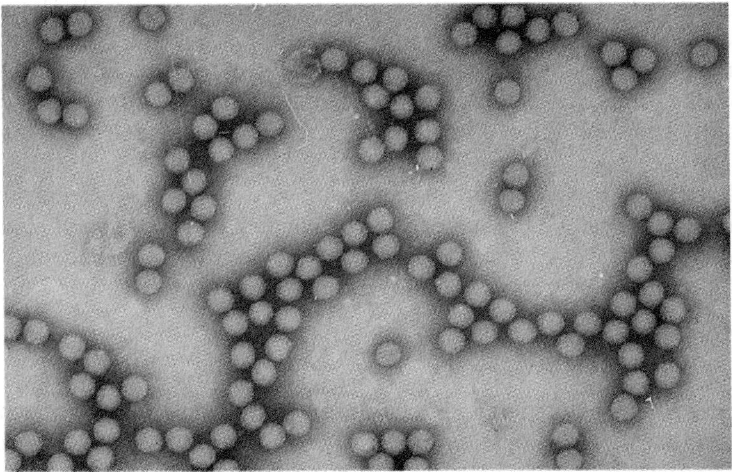

Fig. 5.4 Electronmicrograph of hepatitis A virus in faeces. Virus is usually undetectable in faeces if patient is asymptomatic.

Epidemiology

HAV infection is endemic in most countries, including the United Kingdom, but local outbreaks due to a common source are not uncommon. It is principally transmitted by the facecal–oral route, but is also excreted in urine. Infection often follows ingestion of contaminated water or food but can also spread from person to person, especially in children. Shellfish from polluted waters have caused outbreaks. Virus is present in faeces for about two weeks before onset of jaundice. By the time the patient is jaundiced or in hospital little or no virus is excreted. The exact prevalence is difficult to assess because of the large number of asymptomatic cases, but many adults in the United Kingdom have antibodies to HAV, especially over 50 years old and those in lower social classes. In Africa and Asia most of the population is infected during childhood whereas in Scandinavia (where the disease is no longer endemic) only 1 per cent of the population under 16 years old now has antibody.

Clinical presentation and investigation

The incubation period is 15–40 days. The clinical course is as described for hepatitis above. In anicteric cases, the diagnosis may be suggested by a history of jaundice in a contact. In icteric cases, other causes of heptatitis (see Table 5.2) should be considered. The diagnosis is made by finding IgM antbodies to HAV in the patient's serum; IgG antibodies indicate past infection and immunity.

Management and prophylaxis

When the diagnosis is unclear, the course prolonged, or complications arise, admission to hospital may be necessary. As hepatitis A is most infectious in the late incubation period, transmission of the HAV to contacts has mainly taken place by the time the patient has symptoms; hospital admission does not significantly reduce the spread of infection. General management is as stated above.

Prophylaxis with normal immunoglobulin is useful for some non-immune travellers to tropical areas (Chapter 16) who can first be tested for IgG antibody. Vaccine is given to frequent or long-term travellers to endemic areas. Immunoglobulin given during the incubation period may attenuate symptoms.

Prognosis

The disease is usually benign with complete recovery in several weeks, but may last some months. The mortality of icteric cases is 0.1 per cent or less, usually from acute liver failure, the most important complication, which mainly affects older adults. Marrow aplasia is a rare and usually fatal complication. Hepatitis A virus does not cause chronic hepatitis.

Hepatitis B (serum hepatitis)

Aetiology

Hepatitis B virus (HBV) is made up of a central core containing a core antigen (HBcAg) and *e*-antigen (HBeAg), and a surrounding envelope with surface antigen (HBsAg) (Fig. 5.5).

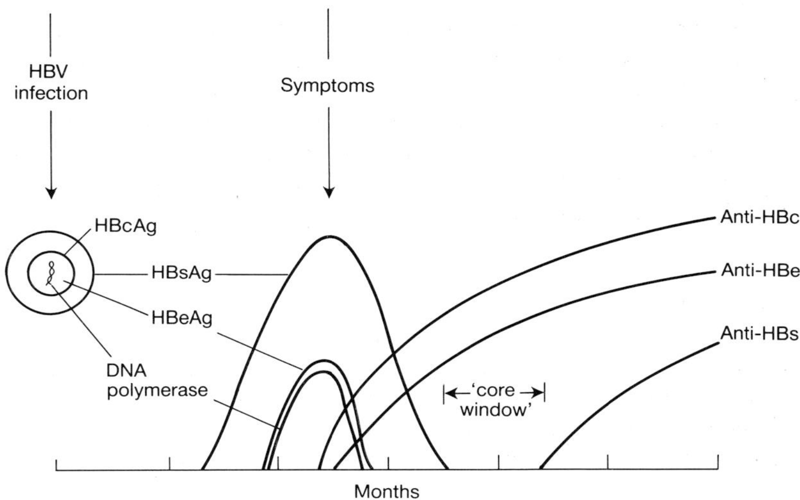

Fig. 5.5 Hepatitis B antigens and antibodies.

Epidemiology

Infection from HBV is endemic world-wide. It is spread by the parenteral route and blood is the main source of infection, although HBsAg has been found in other body fluids (for example, semen). In the developed world the risk of infection is increased in drug abusers who share needles (Fig. 5.6); where unsterile instruments are used as often in ear-piercing (Fig. 5.7), tattooing, and acupuncture; and potentially after blood transfusion and in haemodialysis units. Sexual transmission occurs and is common in promiscuous and homosexual groups. In institutions for the mentally handicapped prevalence is usually increased through transfer of blood from bites, scratches and sores, poor hygiene, and the defective immunity of many of these patients. In the developing world, perinatal and horizontal transmission in young children is particularly important. The mechanism of transfer of the virus horizontally is unclear but may include blood in saliva, cuts, sores, and biting insects.

In most patients (>90 per cent), HBsAg disappears from the blood in 4–6 months. It persists longer than six months in a few who are termed 'carriers'.

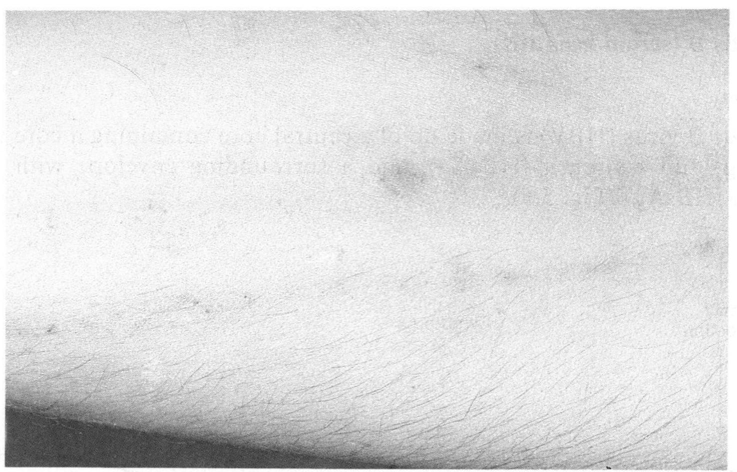

Fig. 5.6 Thrombophlebitis and hepatitis B infection are common in drug addicts sharing needles.

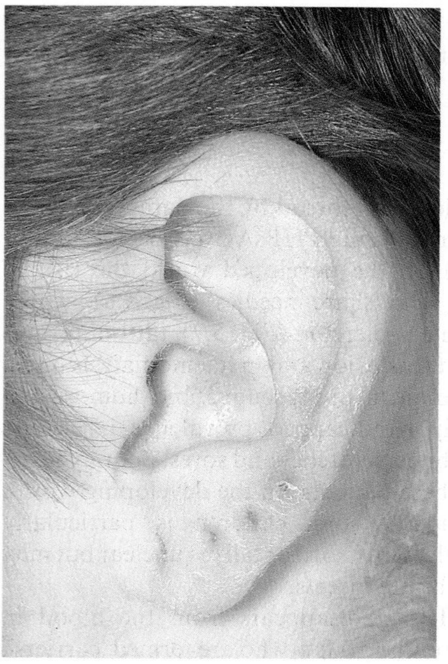

Fig. 5.7 Hepatitis B may be inoculated by unsterilized instruments used for ear-piercing.

The carrier state usually results from asymptomatic infections without jaundice. It can persist for many years but is not usually lifelong except after perinatally acquired infections. In Northern Europe and North America the prevalence of carriers is about 0.1 per cent, whereas in parts of Africa and Asia it is as high as 20 per cent. Carriers usually show serological evidence of HBe antigen or its antibody (anti-HBe), the former group being very much more infectious than the latter. In Japan, most asymptomatic carriers have HBe antigen, whereas in the UK 80 per cent have anti-HBe. HBe is also more often found in younger age groups. In populations with high carriage rates infants commonly acquire infection from their mother before or during birth or in the next few months. Since the infant's immune system cannot deal with the infection efficiently this is usually asymptomatic but with subsequent carriage, the main mechanism for maintaining high infection rates in these populations.

Clinical presentation

In the UK, most patients come from high-risk groups, especially intravenous drug abusers, homosexuals and some immigrant groups. After an incubation period of between 30 and 180 days, illness resembling hepatitis A may follow. The onset of HBV is usually insidious, and up to a quarter of the patients develop arthralgia or skin rashes during the prolonged incubation period. These prodromal symptoms are probably due to circulating immune complexes and sometimes present as a more florid illness like serum sickness. Clinically, hepatitis B cannot easily be distinguished from other forms of viral hepatitis.

Investigation

Diagnosis is based on distinctive serological findings (Fig. 5.5): HBsAg is detectable first, closely followed by the appearance of specific DNA, DNA polymerase, and HBeAg. These markers may precede the onset of symptoms. The presence of HBsAg typifies current infection and infectivity, but should be interpreted with caution since it is also present in carriers whose infectivity may be minimal yet who may develop hepatitis from some other cause. Definitive diagnosis can be made by finding IgM anti-HBc; this is particularly useful in the 'core window', the period when markers for HBsAg and anti-HBs are undetectable (Fig. 5.5). Carriers do not usually have IgM anti-HBc but instead have IgG anti-HBc. Rising levels of antibody (anti-HBs or anti-HBc) may confirm recent infection. Some permutations of results are shown in Table 5.3: sometimes the only way to distinguish acute infection from carriage is to repeat tests at 2–4-week intervals. In some cases of severe liver disease, chronic active hepatitis, hepatocellular carcinoma, and in Far Eastern carriers, HBeAg is present without DNA polymerase and these patients are less infectious.

Prophylaxis

Hepatitis B virus immunoglobulin given promptly after exposure is effective, but expensive. It is indicated where there is serious risk of contracting hepatitis B,

Table 5.3. *Some of the permutations of serological markers of hepatitis B*

Antigen		Antibodies				Comments
HBs	HBe	anti-HBs	anti-HBe	anti-HBc		
				IgG	IgM	
+	−	−	−	−	−	current infection
+	+	−	−	−	−	current infection
+	+	−	−	+	+	current infection
+	+	−	−	+	−	carrier
+	−	−	+	+	+	currrent infection
+	−	−	+	+	−	carrier
+	−	−	−	+	+	current infection
+	−	−	−	+	−	past infection
−	−	−	+	+	+	recent infection*
−	−	+	+	+	+	recent infection
−	−	−	+	+	−	past infection
−	−	+	+	+	−	past infection
−	−	+	−	−	−	past infection*
−	−	−	−	+	−	past infection
+	+	+	+	+	+	current*/past

*Rare, usually occur in drug addicts and others with repeated exposure to HBV.

either because of the type of exposure or where the suspected source has HBsAg. Immunoglobulin is most effective if given within 48 hours of exposure. Before giving immunoglobulin blood should be taken to establish the patient's HBV status—perhaps already carrier or immune. An accelerated vaccine course may provide a useful option after exposure. The neonate of an infected mother requires hepatitis B immunoglobulin and vaccine as soon as possible (Chapter 11).

Hepatitis B vaccine is safe and effective in prevention. It is offered to groups at high risk of infection (Chapter 16). Those vaccinated must be checked for development of antibodies four weeks after the last dose. The level of antibody allows prediction of the time for the next booster, 6 months to 5 years later. Vaccinated individuals should be given a booster after exposure if a recent antibody level is not known. In countries with high carriage rates, vaccine for all neonates (Chapter 11) may reduce the incidence of hepatomas in later life.

Carriers should be told their status and counselled on transmission of infection. They should not share toothbrushes or razors and should use a barrier method of contraception unless their partners are immune, They should not donate blood and should be aware of its infectious potential; they should also inform their medical and dental attendants of their condition. It is important to emphasize that ordinary domestic, social, and working activities carry negligible risk of transmitting infection. Carriers are characterized into 'high risk' (HBe antigen) or 'low risk' (HBe antibody) of transmitting infection. Periodic checks (6–12 months) are usual to detect change in risk category (for example the patient may develop antibody or HBsAg may disappear in a few spontaneously). Treatment with alpha interferon for 3 months has cleared carriage in up to 60 per cent of such persons. A combination of β- and γ-interferon may be more effective.

Prognosis

Of patients with acute hepatitis B fewer than 1 per cent die of liver failure and over 90 per cent recover completely. In about 5 per cent of infections HBsAg persists more than six months, but usually clears over the next year, although a few develop persistent heptatitis and 1 per cent chronic active heptatitis. Hepatoma is an important late sequel in many chronic carriers after a lapse of many years, often complicating cirrhosis.

Hepatitis C

Aetiology and epidemiology

This recently identified agent (HCV) probably causes three-quarters or more of what was formerly called 'non-A, non-B' hepatitis. It is a small RNA virus transmitted parenterally by blood like HBV, and it is common in intravenous drug users. About 1 per cent of healthy persons with no risk factor for HBV show evidence of past HCV infection.

Clinical presentation and investigation

After an incubation period intermediate between HAV and HBV, hepatitis develops, often without symptoms. Infection is detected by finding anti-HCV but antibodies take some time (average 9 months) to become detectable by current tests—more sensitive and specific tests are likely to be developed soon.

Prognosis

This is good, but patients seem more likely to develop chronic liver disease. Patients with primary hepatocellular carcinoma also show a high prevalence of anti-HCV. HCV and HBV coinfection may cause more severe disease.

Delta hepatitis (hepatitis D)

Aetiology and epidemioloy

The delta agent is a defective RNA hepatitis virus which requires the presence of HBV as a 'helper'. Delta infection can therefore only occur in the presence of concurrent acute or chronic HBV infection. Delta infection has been described world-wide, especially among some groups at high risk of HBsAg carriage (in

UK, intravenous drug abusers and patients who have had multiple transfusions) among whom there may outbreaks. Infection spreads by blood-to-blood contact but may be sexual, though it is uncommon among homosexuals.

Clinical presentation and investigation

Infection usually results in clinically apparent disease—for example, fulminant hepatitis when concurrent with acute hepatitis B, or biphasic illness 3–10 weeks apart when sequential, or in HBV carriers an acute attack of hepatitis. The diagnosis is made by detecting delta antigen after treatment of serum with detergent to remove the HBsAg coat, or by finding IgM anti-delta. The latter is usually of short duration and its clearance marks recovery from infection.

Prognosis

Illness is more severe in patients with delta infection and acute hepatitis B than in those with hepatitis B alone. Chronic active hepatitis and cirrhosis appear to also be commoner sequelae in hepatitis B carriers when coinfected with delta agent.

Hepatitis E

This small RNA virus (HEV), recognized only recently, resembles a calicivirus and is transmitted by the faecal–oral route like HAV. It is an important cause of water-borne epidemics in developing countries with high (over 20 per cent) mortality in pregnant women. Sporadic infections also occur, and before the virus was identified they were included within the enterically-transmitted form of 'non-A non-B' hepatitis. The virus is detectable in stool by electron microscopy but no serological test is yet available.

Non-A non-B hepatitis

This obsolescent term was introduced for cases of viral hepatitis in which tests excluded hepatitis A and B. Causal agents recognized include HCV, the common cause of post-transfusion hepatitis not due to HBV, and HEV, the main cause of enterically-transmitted hepatitis not due to HAV. Other agents may be identified in future.

Chronic hepatitis

Chronic persistent hepatitis

This is characterized by persistent raised transaminases and sometimes symptoms (malaise, mild abdominal discomfort) lasting more than six months. It may be caused by HAV, HBV, HCV, non-A non-B hepatitis, drugs, alcohol or chronic inflammatory bowel disease. Liver biopsy shows inflammation of the portal tracts but no necrosis. It resolves usually within 12 months; no treatment is required.

Chronic active (aggressive) hepatitis

This is more serious, with more severe clinical and biochemical abnormalities persisting at least six months. Liver biopsy shows inflammation beyond the portal tracts and 'piece-meal' necrosis. Causes include HBV (Fig. 5.8), HCV, non-A non-B hepatitis, drugs, alcohol, autoimmunity, and Wilson's disease (but *not* HAV). If due to HBV infection, mainly males are affected and serological markers for both HBsAg and HBeAg are usually found.

Corticosteroids and azathioprine may be of value in treatment of chronic active hepatitis, but where the cause is HBV the benefit is less clear. As mentioned above, 3 months' interferon treatment may clear HBsAg and HBeAg carriage and give biochemical and histological improvement of chronic active hepatitis. The effect of interferon on chronic non-A non-B hepatitis is being tested.

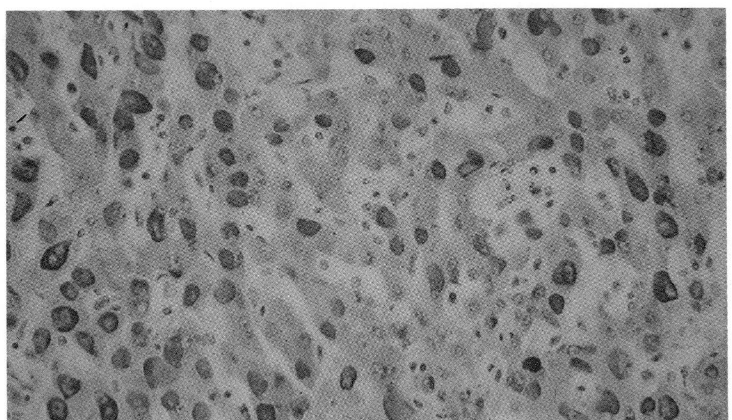

Fig. 5.8 Chronic hepatitis with hepatitis B surface antigen stained dark brown.

Epstein–Barr virus

This virus which causes infectious mononucleosis is more fully discussed in Chapter 2. Although sore throat, pyrexia, and lymphadenopathy are the commonest presenting features, 85 per cent of cases in hospital (usually the most severe) have abnormal liver function tests and 15 per cent have jaundice. Hepatic dysfunction is usually transient returning to normal within six weeks.

Cytomegalovirus

Other aspects of cytomegalovirus infection are considered in Chapters 11 and 12. Hepatitis is a common clinical feature seen in congenitally infected infants or young adults with infectious mononucleosis-like illness. Infection may also be transmitted in the leucocytes of donor blood; especially after massive transfusion, post-tranfusion hepatitis is common in susceptible individuals. After renal transplantation the rate of infection may be up to 60 per cent, mostly reactivated infections. The hepatitis is usually self-limiting, but about 2 per cent of renal transplant patients die of disseminated disease with liver and kidney failure. It is best to use CMV–antibody negative blood and organs in patients who are susceptible to CMV. Prophylaxis of transplant patients with ganciclovir is effective in reducing disease.

Toxoplasmosis

This disease caused by the protozoon *Toxoplasma gondii* is more fully discussed in Chapters 11 and 12. Most cases are asymptomatic. The commonest clinical manifestation of primary infection is malaise, fever, and lymphadenopathy 10–25 days after ingestion of the parasite from sources contaminated with cat faeces (unwashed vegetables, hands after gardening) or consumption of undercooked meat or unpasterurized milk. Hepatitis is a rare complication. Congenital toxoplasmosis is rare but jaundice at birth is common in those affected. In immunocompromised patients, infection is often disseminated and can be fatal.

Leptospirosis

Aetiology

Disease is caused by the pathogenic species, *Leptospira interrogans*, of which there are 202 serovars in 23 serogroups. In Britain, important serovars are *icterohaemorrhagiae* and the cattle-associated serovar *hardjo*.

Epidemiology

Leptospirosis affects wild and domestic animals world-wide, especially rodents, but is not often transferred to man. Infection enters through skin abrasions or mucous membranes from infected animal urine or other excretions. Farmers, veterinary surgeons, sewer workers, fish-farm workers and those bathing in contaminated water are particularly at risk. Many recreational waters are contaminated and users must be aware of possible infection.

Clinical presentation

After an incubation period of 2–17 days, there is often biphasic illness, initially 'flu-like' with pyrexia, headache, myalgia, and arthralgia. A wide range of more specific systemic features may follow, such as cough, diarrhoea, or myopericarditis; syndromes discussed below may be identifiable. There is considerable overlap between the clinical pictures caused by different serovars, although some are usually, but not always, associated with particular presentations. Many asymptomatic or mild infections occur.

Weil's disease. The major cause is *icterohaemorrhagiae* serovar carried by rats. After initial short illness, there is persistent pyrexia, jaundice, and renal failure with haemorrhagic rash due to disseminated intravascular coagulation. The illness is at its worst in the third week when death can result especially from renal or cardiac failure. About 10 per cent of the reported illnesses have this severe character.

Aseptic meningitis (Canicola fever). This is less severe and is usually caused by serovar *canicola* (carried by dogs and pigs). It usually presents as lymphocytic meningitis (Chapter 6).

Pretibial (Fort Bragg) fever. After initial illness, an erythematous maculopapular rash appears on the pretibial areas. The rash 'vaguely resembled erythema nodosum' and splenomegaly was common when the syndrome was first described. The causative agent is serovar *autumnalis*.

Investigation and management

In the first week, leptospires can be cultured or identified by dark ground microscopy in blood, cerebrospinal fluid, or urine in which they may be detectable for several months. After the first week, agglutination or complement fixation tests can detect rising titres of antibody. Benzylpenicillin, erythromycin, and tetracycline may be effective early in the illness but not often after the first week. Patients with serious complications require supportive care. Mild forms of the disease require no treatment. In some countries, but not in UK, vaccines are available for workers at high risk. In patients with Weil's disease, Hantavirus infection should be excluded serologically; patients are managed as for leptospirosis.

Prognosis

Recovery is usually complete. The mortality rate among jaundiced patients is around 10 per cent but considerably higher in older age groups. Aseptic meningitis usually resolves without sequelae.

Hantavirus

These viruses are named after the Hantaan river in Korea where they cause Korean Haemorrhagic Fever. Infection is usually spread by the urine and respiratory secretions of infected rodents, especially field mice. Several types of virus exist. Apart from asymptomatic infections there are two main clinical syndromes:

Nephropathia epidemica. This is commoner in Scandinavia and Finland, but also found elsewhere in Europe. After a flu-like illness, renal failure leads to oliguria. Raised liver enzymes often reveal liver involvement. Mortality is less than 0.5 per cent and the prognosis is good.

Haemorrhagic fever with renal syndrome. Severe cases often present with hypovolaemic shock carrying a mortality rate of up to 20 per cent. Less severe infections are probably more common. At the stage of renal failure, liver dysfunction is frequent. Diagnosis is made by demonstrating specific hantaviral antibodies; treatment is supportive.

OTHER ABDOMINAL INFECTIONS

Acute cholecystitis

Aetiology and epidemiology

This is primarily an infection of the gall-bladder often associated with obstruction by calculi. The infecting organisms are predominantly bowel flora, especially *E. coli.*

Clinical presentation and investigation

Sudden onset of pyrexia often with rigors and right upper quadrant pain with tenderness are characteristic, often leading to jaundice. Tenderness is more pronounced on deep inspiration (Murphy's sign). The patient may notice pale stools, dark urine, jaundice, and pruritus. Liver function tests show the pattern of cholestasis. Plain abdominal X-ray pictures reveal gallstones in 10 per cent (most stones are radiolucent). Ultrasound may show gallstones or dilated ducts. There is neutrophil leucocytosis, and blood cultures may be positive.

Management and prognosis

Over 90 per cent settle with bed rest, analgesics, and antibiotics. Since recurrence is likely, elective cholecystectomy is usually performed 2–3 months later. The prognosis for patients so managed is excellent.

Cholangitis

Charcot's triad of jaundice, biliary colic, and spiking fevers with rigors is characteristic of this ascending infection of the biliary tree. If untreated, liver abscess, septicaemia, and shock can develop. Non-suppurative acute cholangitis is more common and may respond to conservative treatment with antibiotics. Suppurative acute cholangitis requires urgent surgical drainage and broad spectrum antibiotics. Mortality rates are high in the presence of septicaemia and shock (Chapter 8).

Liver abscess

Aetiology and epidemiology

This usually complicates some other intra-abdominal infection draining through the portal venous system, especially appendicitis, diverticulitis, or after bowel surgery when the organisms are usually faecal bacteria (Fig. 5.9). Infection may also result from direct extension or systemic spread. In Wales and Scotland, hydatid disease must be considered. In the tropics amoebic abscess of the liver (Fig. 5.10) often complicates amoebic infection (Chapter 10).

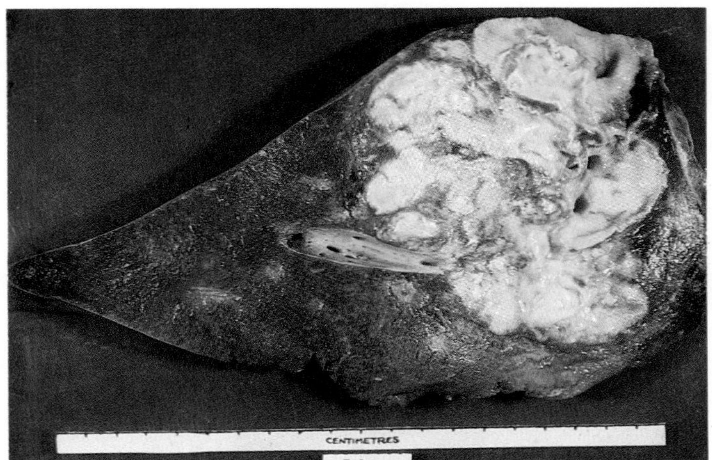

Fig. 5.9 Pyogenic liver abscesses due to faecal bacteria.

Clinical presentation and investigation

Pyrexia, right upper quadrant pain and tenderness and hepatomegaly are usual. Neutrophilia, raised ESR, and elevated alkaline phosphatase are common. Large abscesses may be diagnosed by ultrasound, computed tomography, or isotope scan; small abscesses may be undetectable. Most patients with amoebic abscesses have antibodies to *Entamoeba histolytica*. Amoebae are difficult to demonstrate in pus but may be found in biopsy material from the abscess wall.

Management and prognosis

Treatment of choice is drainage (Fig. 5.11), but the patient may be too ill for open surgery. Needle aspiration is indicated for single abscesses. Appropriate antibiotics or specific drug therapy are indicated, even after surgery. Amoebic abscesses do not usually require surgical drainage. The prognosis depends on the primary cause of the abscess.

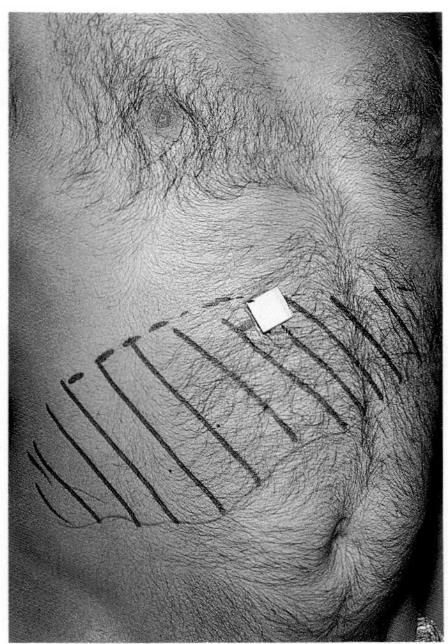

Fig. 5.10 Amoebic abscess of liver with gross hepatomegaly. Aspiration site shown.

Fig. 5.11 Pus from amoebic abscess.

Subphrenic abscess

Features resemble those of liver abscess, except that symptoms and signs are frequently obscure—often just unexplained fever. Investigation is as for liver abscess; treatment is by surgical drainage and appropriate antibiotics.

Splenomegaly in infections

The spleen may become enlarged as a reaction to generalized infections especially with bacteraemia, occasionally with focal infection or abscess. Non-infective causes of splenomegaly such as malignancy, haematological disorders, metabolic disease, connective tissue disease, and congestive splenomegaly predominate in Britain, but in warmer and tropical countries infection is the commonest cause.

Acute splenomegaly in Britain may be due to septicaemia, subacute bacterial endocarditis, typhoid, or other salmonella infection (possibly with splenic abscess), infectious mononucleosis or cytomegalovirus infection; the commonest chronic infections involved are syphilis and brucellosis. World-wide most splenic infections are caused by malaria, leishmaniasis, and schistosomiasis. Tropical splenomegaly may be an abnormal response to malaria.

Mesenteric lymphadenitis

This results from infection of the mesenteric lymph nodes, usually by an adenovirus but sometimes by *Yersinia pseudotuberculosis*. The infected nodes produce signs and symptoms like acute appendicitis. It is commoner in children, often with lymphocytosis in young children in whom intussusception can develop. Prognosis is good but surgery may be required for intussusception. Yersiniosis is usually self-limiting but the bacteria are generally sensitive to penicillin.

Pancreatitis

Most cases are associated with alcoholism, biliary disease, and trauma. Viruses (mumps and enteroviruses, especially coxsackie) do not usually cause severe acute pancreatitis but mild symptoms of abdominal tenderness, anorexia, or vomiting. There is leucocytosis and raised serum amylase. Most cases are self-limiting and treatment is supportive.

Peritonitis

Aetiology and epidemiology

Localized or generalized infection of the abdominal cavity may result from infection introduced from the exterior (for example, stab wounds), from an intra-abdominal organ (for example, appendicitis) or from septicaemia. In Britain, the main infective causes are perforated appendix or peptic ulcer, abdominal surgery, and peritoneal dialysis. Typically peritonitis is caused by the patient's own intestinal flora. Occasionally, especially in children, primary pneumococcal peritonitis can occur. Tuberculous peritonitis is discussed in Chapter 9.

Clinical presentation and investigation

Severe pain, vomiting, and pyrexia are common. Abdominal examination characteristically shows distension, rigidity, tenderness, rebound tenderness, and absent bowel sounds. However, abdominal findings may be minimal in the very young, the elderly, or if the patient is on high-dose steroids. Diagnosis is made on the clinical features. Investigations have limited value but there is usually neutrophilia. In most cases of perforation X-ray examination shows free gas under the diaphragm.

Management and prognosis

The cause should be found and dealt with. Supportive care with pain relief, nasogastric suction, and intravenous fluid is often required. Antibiotic therapy should be started early and reassessed when bacterial sensitivity results are known. The prognosis depends on the cause: overall mortality is about 10 per cent.

Other intra-abdominal abscesses

Aetiology and epidemiology

Intra-abdominal abscess can result from infection or perforation of an abdominal organ. A quarter are visceral abscesses, the rest intra- or retro-peritoneal. The infection is usually caused by mixed intestinal bacteria. Tuberculous abscesses can follow lymph node involvement or extend from the spine or other organs (Chapter 9).

Clinical presentation and investigation of acute abscesses

Classically, intermittent pyrexia, drenching sweats and rigors are present. Abdominal pain and tenderness may help to localize the abscess. Tuberculous abscesses present less dramatically. There is usually neutrophilia and a raised ESR. Radiographic studies, computed tomography, ultrasound and nuclear scans are useful in locating an abscess, but false negative results are not uncommon. Intraperitoneal abscesses (especially subphrenic or pelvic) are difficult to diagnose as pyrexia may be the only finding.

Management and prognosis

Surgical drainage is often required. Antibiotics are used to prevent further spread of infection and to treat areas around the abscess that cannot be drained. Without treatment, mortality is high. Tuberculous abscesses usually respond to antituberculous chemotherapy.

Urinary tract infections (UTI)

Aetiology and epidemiology

Among infections seen by the general practitioner, those affecting the urinary tract are second only to respiratory tract infections in frequency. The general term UTI usually connotes the presence of bacteria in the urine with or without symptoms. More specific terminology (Table 5.4) indicates the anatomical area involved but often there is overlap. Other pathogens such as viruses, chlamydias, and fungi, may colonize the urinary tract but do not usually cause symptoms other than described in Chapter 10. Infection is commoner in females, except for newborns and the elderly where the male incidence is similar. In hospitals 40 per cent of infections involve the urinary tract, most of them in catheterized patients.

Some vesico-ureteric reflux (retrograde movement of urine from bladder into ureter) is probably normal but not detected by radiological techniques. Excessive reflux is particularly common in children and pregnant women. In children, reflux can be caused by antomical abnormality of the urinary tract which can predispose to infection.

Table 5.4. *Specific urinary tract infections*

Upper urinary tract infections

Pyelonephritis (acute and chronic): inflammation of the renal pelvis, calyces, and renal parenchyma.

Lower urinary tract infections

Cystitis: inflammation of the bladder—commonly referred to by the unqualified use of the term UTI.

Urethritis: inflammation of the urethra, considered in Chapter 10.

Acute urethral syndrome: similar presentation to cystitis but urine often sterile; patient rarely helped by antibiotics.

Clinical presentation

Many infections remain asymptomatic. Acute upper urinary tract infection (acute pyelonephritis) causes pyrexia, rigors, and loin pain. In chronic infections of the upper tract (chronic pyelonephritis), pyrexia with hypertension in adults or enuresis in children may be the only findings. In lower urinary tract infection one cannot distinguish clinically between acute cystitis and acute urethral syndrome. Patients complain of urgency, dysuria, frequency, often with suprapubic tenderness. Children under two years may present with fever, irritability, diarrhoea, and vomiting.

Investigation

In symptomatic subjects, UTI is diagnosed if there are pus cells in the urine and culture produces a pure growth of bacteria, regardless of bacterial numbers. Asymptomatic infection is recognized by a positive midstream urine culture, formerly defined as $>10^5$ bacteria/ml but now considered significant if there are repeated counts of 10^4–10^5 bacteria/ml. If there are difficulties in interpretation, suprapubic aspiration in children or 'in-and-out' catherterization in adults showing 10^2–10^3 bacteria/ml is significant. There may be pyuria (>5 leucocytes/high power field), proteinuria and haematuria. A Gram stain of unspun urine correlates well with urine culture (any bacteria seen in each high-power field represent $>10^5$ bacteria/ml). In patients with pyrexia, and rigors, septicaemia may be present.

Management

Asymptomatic bacteriuria does not always require antibiotic therapy except in young children or pregnancy. Instead, one should seek predisposing factors such as diabetes, pregnancy, the presence of a catheter, or urinary tract obstruction (especially in males), and give appropriate treatment. Infection associated with an indwelling urinary catheter is generally treated by antiseptic bladder washouts and special attention to hygiene rather than antibiotics. Other patients with symptomatic infection require appropriate antibiotics, commonly trimethoprim, for 7–14 days. In women with uncomplicated lower UTI, antibiotics in a single dose may be equally effective. Increased fluid intake may be useful in UTI but has little advantage once appropriate antibiotics are used.

Prognosis

Although many patients with symptomatic lower UTI clear infection spontaneously from their urine, they are usually treated to prevent progression up the tract. Recurrent UTI is often due to reinfection by a different strain of the organism. Relapse due to the same bacterial strain usually indicates upper UTI and requires fuller urological investigation (for example, intravenous pyelogram, cystourethrogram, and cystoscopy). Women with recurrent UTI, usually of the lower tract, and no urological abnormality may require prophylactic antibiotic treatment. However, some of them may prevent symptoms by increasing fluid intake and voiding urine more often and after sexual intercourse. These measures may also help women with the acute urethral syndrome.

In young children, infection can damage the growing kidney. Every child with proven UTI should have a plain abdominal X-ray and abdominal ultrasound examination to detect renal abnormalities, obstruction, or calculi. Those with abnormal findings should have an intravenous pyelogram and micturating cystourethrogram to identify the cause. For vesiculo-ureteric reflux, long-term prophylactic antibiotics are given until the reflux disappears. If no reflux is detected or the child is over 5 years old, the child is followed up and investigations repeated if UTI recurs.

Prostatitis

The usual causes are bacteria (especially *Staphylococcus aureus* and *Staph. epidermidis*), *Chlamydia trachomatis*, and *Ureaplasma urealyticum* but often no causative organism is identified. Presentation is acute with fever, frequency, urgency, and dysuria. Perineal, scrotal, and back pain are common. Rectal examination reveals an enlarged, tender prostate. Mucous threads and pus cells are present in the urine and urine culture is usually positive. In some cases semen culture may be necessary. Most patients recover after appropriate antibiotic treatment. In a few, chronic prostatitis develops and is characterized by relapsing urinary tract infections in adult males.

6

Infections of the nervous system

The nervous system is not readily accessible to infectious agents, which mainly reach it indirectly by blood-borne routes. The infections can be quickly fatal or result in long-term functional impairment. The various syndromes can occur separately or in combination with each other. Similarly, the many pathogens of the nervous system may produce concurrent infection at other sites. Infections of the nervous system show seasonal and geographic variations. In the UK the summer and autumn months are times of maximal prevalence of enterovirus infections, most of them asymptomatic or trivial. However, echoviruses and coxsackieviruses are the commonest cause of meningitis at these seasons, different enteroviruses predominating in successive annual epidemics. Mumps virus, especially when epidemic out of the enterovirus season, causes a major proportion of CNS infections; mainly meningitis. Meningococcal meningitis can cause localized epidemics, usually during winter and spring in temperate climates, but in the sub-Saharan 'meningitis belt' of Africa (from 8 to 16 degrees north of the equator) widespread epidemics occur in the dry season, some with high mortality. Epidemics of viral meningoencephalitis, like Japanese B encephalitis, spread by arthropod vectors, especially mosquitos in warm climates, can incur serious mortality (up to 20–75 per cent), or disabling sequelae.

DEFINITIONS

Meningitis is inflammation of the meninges, clinically manifest as the meningeal triad: (i) constant and severe headache, aggravated by movement; (ii) photophobia; (iii) neck stiffness due to spasm of the spinal muscles.

Meningism is the occurrence of features of meningitis, especially neck stiffness, without infection of the meninges. It is particularly common in adults with subarachnoid haemorrhage or in children with acute febrile illness such as severe pharyngitis or urinary tract infection.

Encephalitis is inflammation of the brain which can present with symptoms varying from mild alterations of consciousness to coma. Emotional lability, irritability, and focal neurological signs are common. Encephalitis may be accompanied by meningitis (meningoencephalitis) or meningism, especially with viral infections. Likewise, meningitis is often accompanied by some inflammation of the brain and spinal cord (encephalomyelitis).

Brain abscess is a local collection of purulent material within or adjacent to the brain. Headache is the commonest complaint, although there is no distinctive clinical syndrome. It may complicate meningitis or result from spread of infection from other sites.

Neuritis is inflammation of a nerve. One nerve may be affected (mononeuropathy) as in tuberculoid leprosy, but more usually many nerves are involved at the same time and the terms polyneuritis or polyneuropathy are then more appropriate.

This chapter deals with meningitis, encephalitis, brain abscess, dementia, Guillain–Barré syndrome and finally with acute poliomyelitis.

MENINGITIS

Aetiology

Many agents may produce this condition: Table 6.1 summarizes the infective causes found in Britain. Viruses are commonest; but bacterial infections are usually more dangerous but treatable. The patient's age is a good indicator of the most likely bacterial cause (Table 6.2; Fig. 6.1). Sinusitis, otitis media, upper or

Table 6.1. *Cause of meningitis in Britain*

Common causes	
Viruses (60–70 per cent)	
enteroviruses (polio, echo, coxsackie)	
mumps	
Bacteria (30–40 per cent)	
Neisseria meningitidis	
Haemophilus influenzae	
Streptococcus pneumoniae	
Less common causes	
Mycobacterium tuberculosis	
Leptospira species	
Listeria monocytogenes	
Enterobacteriaciae, especially *E. coli*	usually in neonates
Group B streptococci	usually in neonates
Rare causes	
Toxoplasma gondii	usually in immunocompromised
Mycoplasma pneumoniae	usually in immunocompromised
Cryptococcus neoformans	usually in immunocompromised
Amoebae (*Naegleria* spp.)	usually in immunocompromised
Staphylococcus aureus	usually after CNS surgery
Coagulase negative streptococcus	usually after CNS surgery

Table 6.2. *Common causes of bacterial meningitis in Britain by age-group*

Aetiology	Neonates	Children	Adults
N. meningitidis	+	+++	+
H. influenzae	+	+++	+
Strep. pneumoniae	+	++	+++
Group B streptococci	+++	−	−
Gram-negative bacteria, especially *E. coli*	+++	−	−

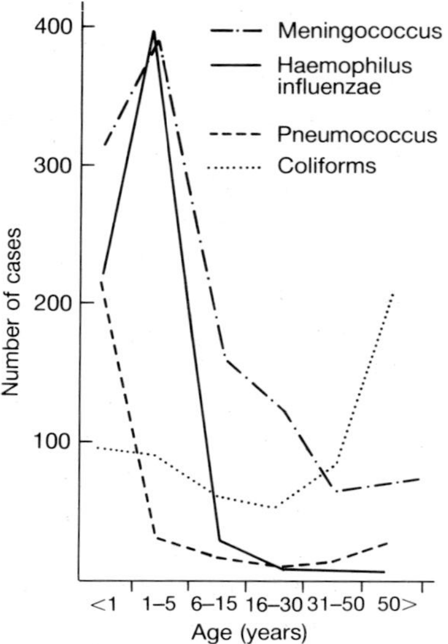

Fig. 6.1 Age distribution of bacterial meningitis in Scotland 1970–84.

lower respiratory tract infection, recent head injury or surgery, alcoholism, the presence of a ventricular shunt, immunosuppression or a previous history of meningitis, can all predispose to the condition.

Epidemiology

In Britain, viruses account for three-quarters of the cases of meningitis, with the highest incidence in children. *Neisseria meningitidis*, *H. influenzae*, and *Strep. pneumoniae* account for three-quarters of all bacterial meningitis.

Clinical presentation

Characteristically, there are symptoms of the meningeal triad. Neck and back stiffness can be detected by the patient's inability to put the chin on the chest, the head between the knees, or by a positive Kernig's sign (with a flexed hip, extension of the knee produces reflex spasm of hamstrings).

Fever, nausea and vomiting are usually present and drowsiness suggests cerebral oedema and perhaps raised intracranial pressure. Vomiting is especially common in children, and in all patients is accompanied by intensification of headache. Convulsions are common in young children but unusual in adults. Cranial nerves, especially the third, fourth, sixth, and seventh, are transiently involved in about 10 per cent of cases. A petechial or haemorrhagic rash suggests meningococcal meningitis, but other bacterial infections very rarely produce similar rashes. Viruses, especially echoviruses, sometimes cause a petechial rash in babies and children with meningitis. In the absence of a rash, *H. influenzae* is the most likely cause of bacterial meningitis in young children, whereas in all patients a head injury or otitis media suggest *Strep. pneumoniae*.

In young children and the elderly, the features of the meningeal triad may be absent and fever and irritability the only constant findings. Occasionally an infant's fontanelle may bulge, but it may be depressed with dehydration. In the immunosuppressed, signs may be even fewer and many patients have no fever. Those who have been partially treated with antimicrobial drugs may also lack typical clinical features. Thus, in all these patient groups, despite scanty clinical evidence, meningitis should be considered.

Several conditions are important in the differential diagnosis. Onset of subarachoid haemorrhage is usually abrupt, but in cerebral abscess and malignant infiltration the onset is insidious. Since meningism may accompany any acute febrile illness the possibility of infection at other sites, especially streptococcal tonsillitis and urinary tract infection, should be sought. Toxins such as lead or glue solvent may produce confusion and drowsiness. The many causes of coma may have to be considered.

Investigation

The diagnosis can be confirmed only by examination of the cerebrospinal fluid (CSF). Papilloedema or focal neurological defects require emergency computed tomography (CT) scan before lumbar puncture which carries a risk of medullary herniation (coning). Lumbar puncture findings, normal and in patients with different types of meningitis, are shown in Table 6.3. Culture of CSF allows microbiological confirmation of the cause. In very early bacterial meningitis, especially meningococcal, the CSF may appear normal yet the organism may grow in culture. Prior antibiotic treatment reduces the chance of organisms growing, especially if the drug readily enters the CSF.

In three-quarters of patients a Gram-stained film of CSF indicates the cause of bacterial meningitis. Other methods available in some laboratories for detecting

Table 6.3 *CSF findings in meningitis*

Characteristic		Type of meningitis		
	Normal	Bacterial (pyogenic)	Viral and leptospiral	Tuberculous
Appearance	Clear	Turbid/purulent	Clear/opalescent	Clear/opalescent
Leucocyte count ($\times 10^6$/litre)				
Range	<10	10–2000	10–500	10–1000
Usual count	0–5	>1000	<200	<200
Type	Lymphocytes	Neutrophils	Lymphocytes (neutrophils initially)	Lymphocytes (neutrophils initially)
Protein (g/litre)	0.15–0.4	0.5–5.0+	0.5–1.0	1.0–6.0+
Glucose (mmol/litre)	2.5–5.5	Very low	Normal (low in mumps)	Low
Gram stain	No organisms	Usually present	No organisms	Organisms usually stained by Ziehl–Neelsen

antigen in CSF, are by counterimmunoelectrophoresis (CIE), latex agglutination or enzyme-linked immunosorbent assay (ELISA). Antigen may be detected in small amounts even when culture is negative but the sensitivity of the test varies with different organisms. In one study, CIE was 100 per cent sensitive for *H. influenzae* meningitis but only 50 per cent for *N. meningitidis* meningitis. The limulus assay can be used to detect endotoxin-producing Gram-negative bacteria, but is negative in meningitis due to Gram-positive bacteria. Lactic acid levels are raised in bacterial meningitis but also in some non-infectious CNS disorders. Increased C-reaction protein is a sensitive indicator of bacterial aetiology.

In addition to the CSF, two sets of blood cultures should be taken and cultures of pharyngeal secretions, sputum, faeces, or urine where infection is a possibility. Serological tests are useful for detecting either antigen or antibody evidence of recent infection. Other laboratory tests provide nonspecific information. Neutrophilia in peripheral blood is usual in bacterial meningitis, whereas a normal white cell count is usual in viral meningitis.

Management

Treatment of most viral infections is symptomatic. Specific antimicrobial therapy is discussed below. Direct, intrathecal therapy should not be given because of possible toxicity and unpredictable distribution in the CSF. If no specific agent has been identified, therapy should be directed at all causes considered likely in view of the patient's age and clinical features. Those with meningism (with normal CSF findings) can be closely observed and the CSF re-examined in 12–24 hours, or earlier if there is clinical deterioration.

Treated patients who recover rapidly within 72 hours, becoming afebrile and mentally alert, require no further evaluation of CSF. However, in those who have persistent pyrexia, remain comatose or are difficult to evaluate, repeated CSF examination may help to document progress and establish whether antibiotic levels in the CSF are adequate. Development of focal neurological signs may indicate abscess, subdural effusion or vascular thrombosis. Recurrence of pyrexia after a week of treatment suggests drug allergy or relapse. Therapy should normally continue for 7–10 days after the patient is afebrile.

Complications

Long-term complications are rare in viral meningitis and affect under 10 per cent of those with bacterial meningitis. Abscess formation (subdural or cerebral) and mechanical damage (hydrocephalus) may cause epilepsy or focal neurological signs. Paralysis of cranial nerves is not uncommon, especially in tuberculous and pneumococcal meningitis. Partial or complete nerve deafness used to be a common complication of meningococcal meningitis. Long-term less serious complications, such as intellectual under-performance, behavioural problems, and neuromuscular abnormalities, may affect a quarter of children. Recurrent bacterial meningitis is generally associated with an abnormal communication between the meninges and the exterior, usually basal skull fracture or congenital defect. It is usually caused by pneumococci, occasionally by meningococci.

Prognosis

The prognosis for viral meningitis is excellent and the patient usually makes a spontaneous, complete recovery. For bacterial meningitis the outcome depends on:

Age: In neonates and the elderly the disease is more severe with a higher fatality rate.

Type of organism involved: Strep. pneumoniae, H. influenzae and *M. tuberculosis* are less responsive to therapy.

Early treatment: Delay in starting therapy can be fatal especially with *N. meningitidis*.

Severity of the disease: Convulsions, a deteriorating mental state or mechanical complications carry a poorer prognosis.

Viral meningitis

Aetiology and epidemiology

Viruses are the commonest causes of meningitis (Table 6.4). There is variation from year to year depending whether there is a mumps epidemic (every 3–4 years) and whether the current strain of echovirus or coxsackievirus is one markedly associated with meningitis. In some countries, poliovirus and arboviruses (alphaviruses and flaviviruses) are common causes of meningitis.

Enteroviruses spread by the faecal–oral route with highest incidence during the summer months and in the first 10 years of life. Mumps virus is spread by droplet infection and infected fomites, with highest incidence in the 5–12 year age-group.

Table 6.4. *Viral causes of meningitis in Britain*

Virus	Proportion (per cent)
Echo viruses	40 or more
Coxsackieviruses	30 or less
Mumps virus	20
Others*	10 or less

*Others include varicella-zoster, herpes simplex, influenza, adenovirus, arbovirus.

Clinical presentation and investigation

Clinical features in viral meningitis are similar irrespective of the virus. There is usually prodromal illness (fever with upper respiratory or gastrointestinal symptoms) followed by classical signs of meningitis. A maculopapular rash can occur in echo and coxsackie infections, especially in those under three years old. Mumps meningitis in half to two-thirds of cases follows salivary gland involvement, usually after about a week, sometimes before and often without salivary gland involvement. Disorientation is uncommon and, if present, encephalitis must be considered.

Lumbar puncture reveals CSF lymphocytosis and raised protein (Table 6.3) but early in the illness up to a quarter of patients may have an excess of neutrophils. The most important differential diagnoses are tuberculous meningitis and brain abscess. Other causes of lymphocytosis in the CSF include toxic agents, toxoplasmosis, brucellosis, borreliosis, and leptospirosis. These may be identified by serological tests. CSF should be cultured for viruses, though not all are detectable in CSF. From samples of stool, echo and coxsackie viruses are relatively easily grown (Fig. 6.2a and b) though their presence there may be coincidental and not necessarily the cause of the meningitis. Mumps, adenovirus, and herpesvirus isolated from throat swabs may also be coincidental. Evidence of infection by these and by leptospires is more often obtained by serological examination of paired sera or demonstration of IgM antibodies.

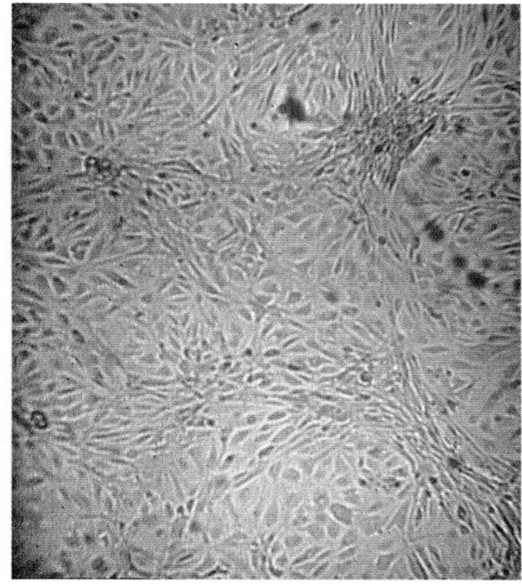

(a)

Fig. 6.2 (a) Normal monkey kidney tissue culture without any viral cytopathogenic effect.

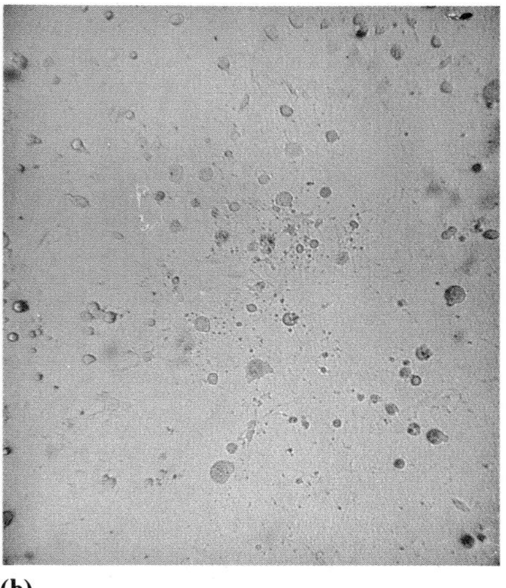

(b)

Fig. 6.2 (b) Enterovirus cytopathogenic effect from a sample of stool in monkey kidney tissue culture.

Management and prognosis

Treatment is symptomatic. Bed rest in a quiet, darkened room, adequate analgesics and antiemetics if indicated are usually sufficient. Cases with severe vomiting may need parenteral fluids. Most patients recover completely within a week. In a few, fatigue and headaches may persist for several months—they require adequate analgesia, rest, and reassurance.

Meningococcal meningitis (Fig. 6.3)

Aetiology

The causative organism, *N. meningitidis*, has several serogroups (most commonly groups A, B, and C) that produce similar disease. In Britain most cases have been due to group B.

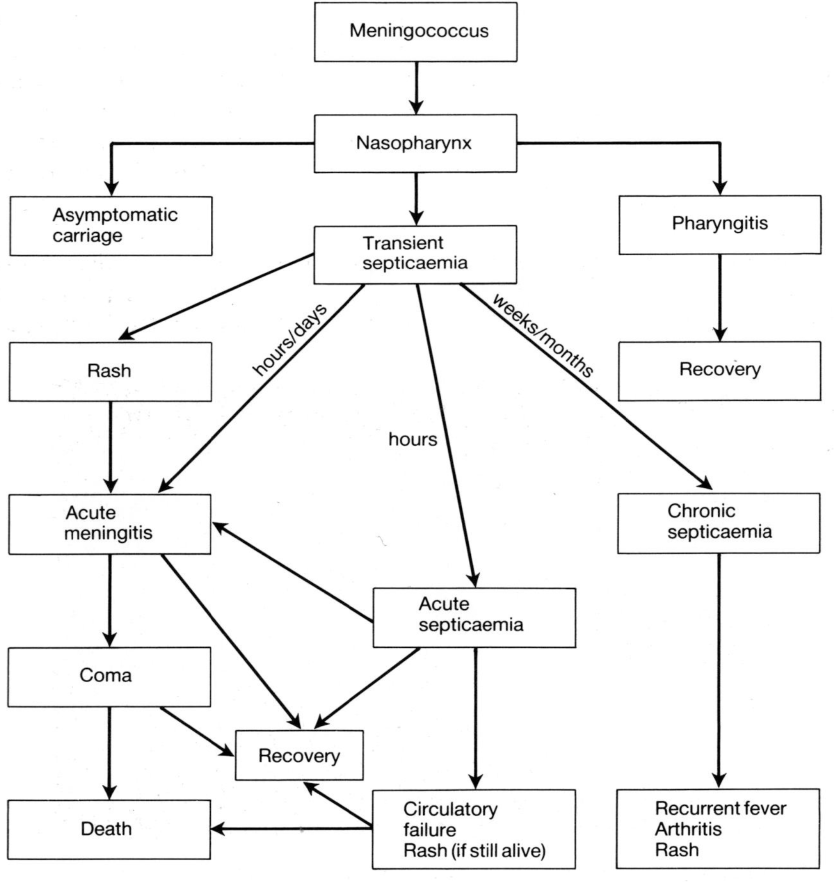

Fig. 6.3 Course of meningococcal infection.

Epidemiology

The disease is endemic world-wide with regular and irregular epidemics. In Britain it is the commonest cause of bacterial meningitis. Infection spreads by droplet transmission, usually from asymptomatic carriers. Carriage rates are variable but may be 25 per cent or more of the contacts of a case as compared with 10 per cent in the normal population. Carriage is transient with the carrier developing immunity. The incidence of disease is increased when conditions favour droplet spread of infection, for instance, during winter or with overcrowded living conditions, or when mucosae are damaged by dry heat as in sub-Saharan Africa. Children under five years of age are the most susceptible although persons of any age may be affected.

Clinical presentation

Infection produces a wide spectrum of clinical illness from mild to severe, with meningitis in 90 per cent of illnesses. After an incubation period of a few days (occasionally longer) symptoms of meningitis develop. Occasionally, severe disease can develop over a few hours. Most patients have a rash. This may consist of small discrete macules, especially over the buttocks and thighs, but more usually it is petechial (Fig. 6.4). The spots may be sparse or cover the entire body with coarse purpura in severe infections (Figs. 6.5 and 6.6). Subconjunctival haemorrhage may occur (Fig. 6.7).

In severe fulminating disease the patient usually dies within 24 hours. Fulminating septicaemia is not preceded by meningitis as these patients usually die before signs of meningitis can develop. They rapidly develop circulatory failure with widespread bleeding into the skin and disseminated intravascular

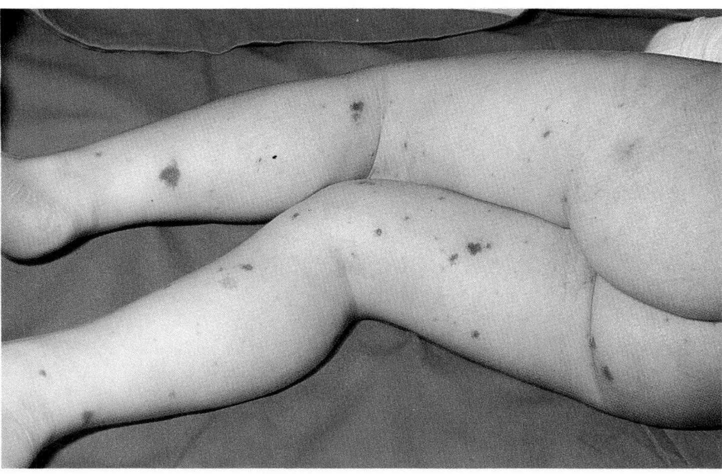

Fig. 6.4 Meningococcal meningitis with petechial and purpuric rash on buttocks and thighs.

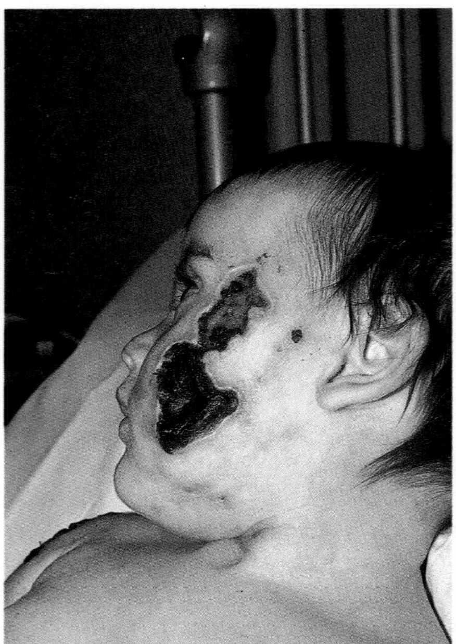

Fig. 6.5 Convalescent meningococcal meningitis with severe necrotic rash of face.

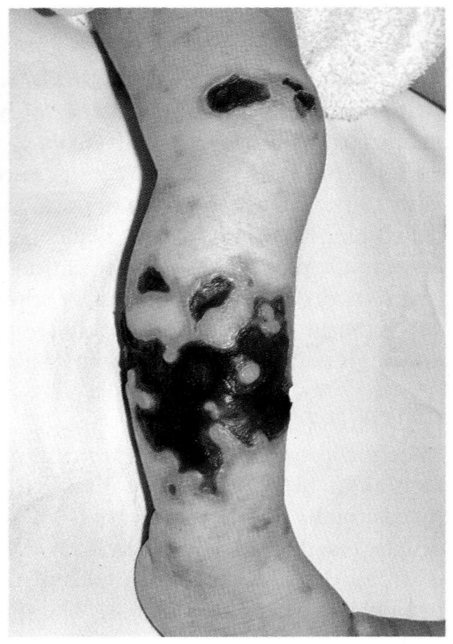

Fig. 6.6 Convalescent meningococcal meningitis: crusts and epithelial necrosis from severe rash.

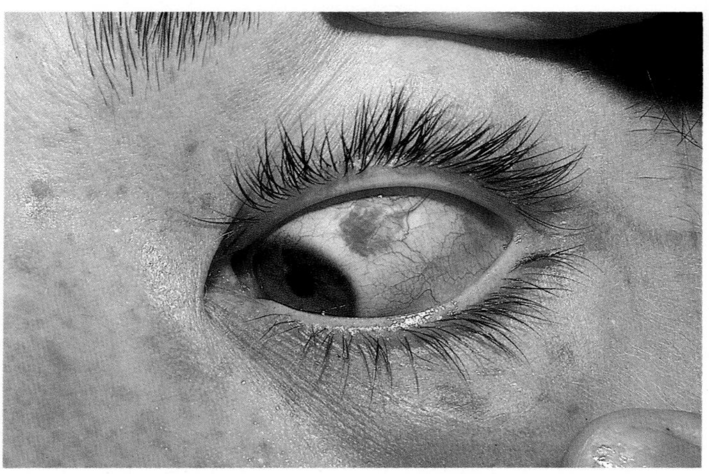

Fig. 6.7 Meningococcal meningitis with subconjunctival haemorrhage.

coagulation (Waterhouse–Friedrichsen syndrome). Chronic or intermittent septicaemia with recurrent fever and arthritis can also occur and is discussed in Chapter 8.

Investigations

The lumbar-puncture findings are those of bacterial meningitis (Table 6.3). Gram-negative diplococci, both free and within leucocytes, can usually be seen in a film of CSF (Fig. 6.8). Antigen can also be demonstrated in blood and CSF, and this is particularly useful if the patient has been given antibiotics. Meningococci may also be cultured from blood, CSF, and sometimes from skin lesions (Fig. 6.9). Finding them in a throat or nasal swab is only suggestive of causation.

Management

Start treatment immediately in severely ill patients before referral or starting investigations. The drug of choice is benzylpenicillin, in high dose parenterally. Meningococci are very sensitive and the dose may be reduced if the condition improves over the next few days. Continue treatment for several days after fever subsides and recovery has taken place. Chloramphenicol or cefotaxime may be used, especially if organisms are not seen and other bacterial causes are possible. Sulphonamides are not often used in Britain as resistant strains are now common. When disseminated intravascular coagulation is severe, blood products may be required. Adequate antibiotic treatment usually corrects mild coagulation defects quickly. A short course of dexamethasone may decrease the frequency of deafness.

Immediate antibiotic treatment is indicated to eliminate carriage in close family members or saliva contacts in nursery and day-schools. Since benzylpenicillin does not affect the carrier state, rifampicin is given for 48 hours or

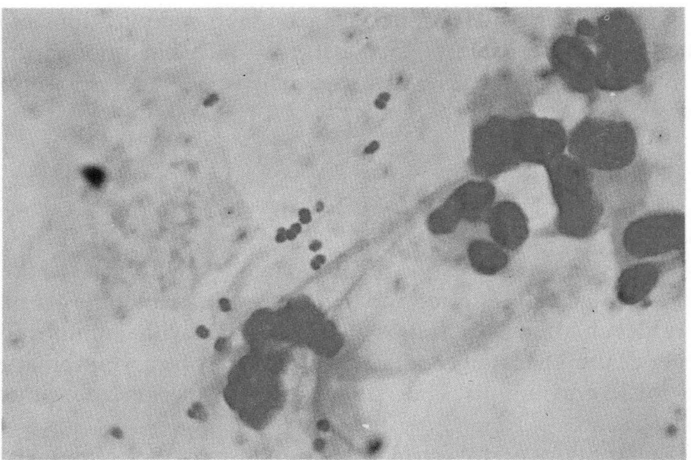

Fig. 6.8 Gram-negative diplococci in CSF of a case of meningococcal meningitis.

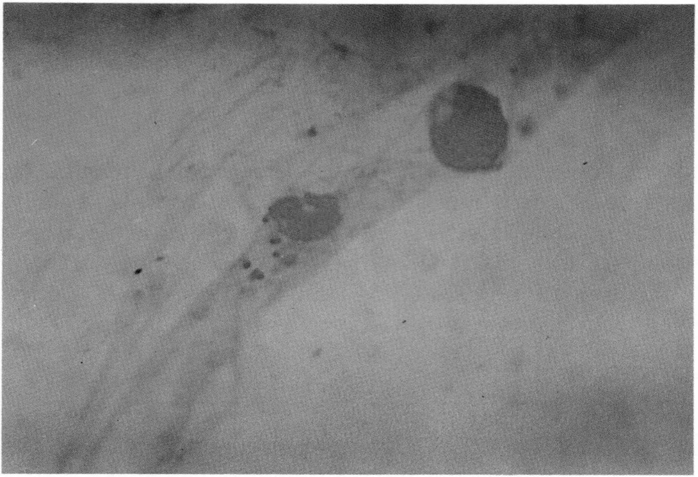

Fig. 6.9 Gram-negative diplococci in a skin lesion of a case of meningococcal meningitis.

single-dose ciprofloxacin. Sulphonamides are effective in parts of the world where the organism is sensitive. Contacts need not be quarantined. Patients also need treatment to eliminate persistent nasopharyngeal carriage. Vaccines against Group A and C strains have been licensed but as yet there is no Group B vaccine.

Prognosis

Untreated, mortality is high. With current treatment overall mortality is 5–10 per cent but higher in the young and elderly and in epidemics. It is lower in young adults whose mortality is about 1 per cent. Mortality is high in fulminating disease. Poor prognostic indicators are shock, fits, hypothermia, heavy purpuric rash, low white cell or platelet counts. Uncommon complications of meningococcal meningitis are reactive polyarthritis, purulent monoarthritis, and pericarditis. Complement deficiencies, especially C5–C9 predispose to recurrent attacks.

Haemophilus meningitis

Aetiology and epidemiology

The causative organism is *H. influenzae*, capsulated type B strain. Children under three years are principally affected and the disease is rare after the age of six years. Disease in older children and adults is often associated with immunosuppression. Up to 50 per cent of the family contacts of a case may be asymptomatic carriers, sometimes for several months. Thus, siblings may be infected from either a carrier or from the case.

In the United States and Australia this organism is the commonest cause of bacterial meningitis.

Clinical presentation and investigation

The illness commonly has an insidious onset over several days to a week. It often follows a respiratory or ear infection when, even though receiving antibiotic treatment, the child becomes drowsy, irritable, or pyrexial before developing signs of meningitis.

Examination of the CSF shows evidence of bacterial meningitis (Table 6.3). The Gram film usually shows pleomorphic, Gram-negative coccobacilli and often both CSF and blood culture are positive. If the Gram film is negative the diagnosis may be confirmed by detecting antigen by immunoelectrophoresis. Complications, especially sterile subdural effusion, are commoner than in meningococcal meningitis.

Management and prognosis

Chloramphenicol or cefotaxime are the drugs of choice. Complications are more likely when treatment is delayed. Nevertheless, antibiotic treatment has reduced the mortality from 90 per cent to 5–10 per cent. Prophylaxis for the whole family with rifampicin is indicated. Vaccines are now available for protection of children (Chapter 16).

Pneumococcal meningitis

Aetiology and epidemiology

The causative organism is *Strep. pneumoniae*, a Gram-positive diplococcus with many serotypes only a few of which cause meningitis. Meningeal infection is usually acquired by spread from another site of infection; for example, sinuses, middle ear, or the site of skull fracture. Immunodeficient patients such as those with diabetes, lymphoma, or splenectomy have a greater incidence of pneumococcal disease.

The disease is not as common as meningococcal or haemophilus meningitis, but has a higher mortality. It affects all age-groups but is the commonest cause of bacterial meningitis in the elderly.

Clinical presentation and investigation

Onset is acute with symptoms and signs of meningitis developing over 1–2 days. Sources of infection such as the ears, sinuses, lungs and cranium should be sought. Lumbar puncture reveals evidence of bacterial meningitis (Table 6.3) and a Gram stain demonstrates the organism. Blood and CSF culture and serological evidence of pneumococcal antigen may confirm the diagnosis.

Management and prognosis

Treat with high dose intravenous benzylpenicillin, or chloramphenicol if there is penicillin allergy. In South Africa and some other countries where there are penicillin-resistant strains, chloramphenicol, or cefotaxime may be initial drugs of choice until sensitivities are known. Disease is severe with overall mortality of 20 per cent, increasing with age. Complications, especially abscesses, cranial nerve defects, and mental impairment, are commoner than in other forms of meningitis. Most recurrent meningitis is due to the pneumococcus.

Neonatal meningitis

This is a world-wide problem. In the West, its incidence is 1:2500 live births, but it is more common in low-birthweight infants and its incidence is much higher in poorer countries. Infection probably occurs around the time of delivery from contact with mother's faeces or urine. The range of organisms is therefore wide, especially *E. coli* and Group B streptococci, but also *Staph. aureus*, *Listeria monocytogenes*, and other Gram-negative bacilli. Classical features of meningitis are absent and the neonate shows nonspecific features such as poor feeding and irritability. A bulging fontanelle is present in only 15 per cent of cases. Diagnosis is made by lumbar puncture. There is no ideal drug combination. Cefotaxime or ampicillin and gentamicin are widely used but the latter penetrates CSF poorly. Chloramphenicol can be toxic as the neonate is unable to metabolize or excrete the drug quickly. Close liaison with the laboratory may determine the best antibiotic. Mortality is high (50 per cent) but is much lower in specialized centres. Neurological sequelae are common.

Tuberculous meningitis

Other manifestations and treatment of *Mycobacterium tuberculosis* are dealt with in Chapter 9. Meningitis may be part of miliary disease or may be the only manifestation. Onset is insidious. After a few weeks of vague illness progressing to anorexia, irritability and vomiting, more definite mental and neurological changes appear, with severe headache and signs of meningitis. In the absence of treatment, spastic paraplegia or cranial nerve palsies may develop and coma which heralds death.

Ophthalmoscopy can show choroidal tubercles in disseminated disease and papilloedema is more common than in other bacterial forms of meningitis. When miliary disease is not present, laboratory diagnosis depends upon the CSF findings (Table 6.3). Acid-fast bacilli are seen in at least a quarter of cases, but experienced laboratories may visualize the organisms in 80 per cent of cases. Repeated lumbar punctures may be necessary before the diagnosis can be made. When antibiotics have been given before CSF examination, differentiation of bacterial meningitis from tuberculous meningitis or viral meningoencephalitis can be especially difficult. Trial of anti-tuberculous therapy may be required in cases of uncertainty. The use of corticosteroids may reduce adhesions and subsequent CSF obstruction. Early diagnosis and treatment (see Chapter 9) are essential for a good prognosis in this formerly fatal disease.

Staphylococcal meningitis

Staphylococcus aureus meningitis usually involves postoperative neurological patients, patients who have undergone a ventriculogram or myelogram, and those with staphylococcal endocarditis, or septicaemia. *Staphylococcus epidermidis* and *Staph. aureus* are the commonest organisms to infect patients with CSF shunts, causing half the ventriculo-atrial and a quarter of the ventriculo-peritoneal cases. Since resistance to benzylpenicillin is common, flucloxacillin with fusidic acid or rifampicin with co-trimoxazole should be used but the shunt may need replacing. Intrashunt vancomycin has also been successful.

Leptospiral meningitis

Meningitis may be the sole manifestation of leptospiral infection or form part of more generalized illness. Other aspects of leptospirosis are dealt with in Chapter 5. The frequency and severity of meningitis depends on the serovar involved. Infection usually follows exposure to infected water or animals, generally during summer or autumn. The CSF findings resemble those of viral meningitis (Table 6.3) and diagnosis is usually made by serological examination of blood. Treatment is by large doses of penicillin, or tetracycline in patients allergic to penicillin.

Listeria monocytogenes meningitis

This disease is more common in neonates, pregnant women and immunocompromised patients (for instance with malignancy). The CSF features are those of bacterial meningitis (Table 6.3), Gram-positive rods are seen in the CSF and the diagnosis is confirmed by isolation of the organism from CSF or blood. In fresh samples the organism exhibits characteristic 'tumbling' motility. Treat with intravenous ampicillin or amoxycillin. In the presence of underlying conditions mortality is high (25 per cent).

Cryptococcal meningitis

This is usually part of a disseminated infection in patients with lymphoma, diabetes, AIDS or on steroid therapy. Symptoms usually develop insidiously and focal neurological signs are commoner than in other forms of meningitis. In 50 per cent of cases lumbar puncture shows cells of the fungus *Cryptococcus neoformans* in the CSF. These can be mistaken for lymphocytes unless India ink is used to reveal the capsule. Treatment is with fluconazole or intravenous amphotericin B. Recurrences are usual unless the underlying condition is controlled.

Meningitis of unknown aetiology

Some patients show evidence of bacterial meningitis in the CSF (Table 6.3) but with no organism recognized by microscopy or culture. Often the reason for inability to see or grow the organism is prior antibiotic treatment which, particularly in children, is often given orally before meningitis is suspected. Sterility of CSF is evidence that this treatment has been at least partially successful. Despite negative bacteriological findings, such patients probably have bacterial meningitis, and treatment is directed against the common bacterial pathogens. Cefotaxime or chloramphenicol are the drugs of choice.

Serological studies and examination of the CSF and serum by counter-immunoelectrophoresis may provide the diagnosis in partially treated bacterial meningitis. In patients whose CSF findings are minimally abnormal and who appear well, there is a case for withholding antibiotics but maintaining close observation and repeating the lumbar puncture in 6–24 hours. If the CSF findings are still abnormal, the patient requires a full course of antibiotics.

ENCEPHALITIS

Aetiology and epidemiology

This disease is rare in Britain and can have several presentations. Approximate frequencies of the main aetiological agents encountered in Britain are listed in Table 6.5 and encephalitis can complicate severe cases of meningitis (for example, tuberculous, pneumococcal).

Table 6.5. *Infective causes of encephalitis in Britain*

Cause	Proportion (per cent)
Unknown	40
Herpes simplex	25
Measles	10
Varicella zoster	10
Mumps	5
Influenza	5
Others*	5

*Others include retrovirus, rabies, polyomavirus, Creutzfeldt–Jakob disease agent, tuberculosis, pneumococci.

Sporadic encephalitis. This is the commonest presentation in Britain. Most cases are of unknown aetiology and viruses are the usual aetiological agents of the others. Herpes simplex is the commonest cause of severe cases, but mumps, enteroviruses, adenoviruses, and also cytomegalovirus, Epstein–Barr virus, polyomavirus, and human immunodeficiency virus (HIV) can cause the disease. In Britain, louping ill (a flavivirus transmitted by ticks from infected sheep) is a rare cause of sporadic cases, more so in the north-west of Scotland. Lyme disease (Chapter 3) is increasingly recognized as a cause of encephalitis. Rabies (Chapter 14) is a rare importation.

Epidemic encephalitis. In many countries, seasonal outbreaks of encephalitis, usually due to arboviruses, are common. Mosquito-transmitted disease is important in the United States (for example Eastern and Western equine, St Louis and La Crosse encephalitis) and E. and S. Asia (Japanese B encephalitis), whereas tick-borne disease is commoner in USSR (Russian Spring–Summer encephalitis) and Europe (Central European Encephalitis) where it may also be acquired by drinking infected milk from goats or cows.

Post-infectious meningoencephalitis. This rare complication of measles, chickenpox, influenza, mumps, rubella, and vaccinia develops some 10 days after the start of the viral illness.

Subacute sclerosing panencephalitis (SSPE). The characteristic pattern of illness is for a child to have a primary measles infection in the first two years of life. On average seven years later, an inexorably progressive encephalitis slowly develops. More rarely it is due to rubella.

Clinical presentation and investigations

The onset is usually acute with pyrexia, headache, vomiting, and photophobia. There may be mild confusion, behavioural abnormalities, and a depressed level of consciousness. Convulsions are common, especially in infants where other specific signs (for example, neck stiffness, pyrexia) are often absent, but they have little prognostic significance. Diagnostic clues may include mucocutaneous herpes simplex, or cerebellar ataxia which is commoner in chickenpox and louping ill. A history of recent illness, travel, or occupation may also give important pointers to the diagnoasis.

The differential diagnosis includes tuberculous meningitis, a space-occupying lesion (such as abscess or haematoma), cerebral malaria and demyelinating disorders. Lumbar puncture is probably the most useful investigation: it shows a raised lymphocyte count and protein content. Comparison of serum and CSF antibody levels to particular viruses can be diagnostic, but titres are not usually raised early in the illness. Recently, the polymerase chain reaction (PCR) has been used to diagnose herpes simplex encephalitis from CSF, and PCR will probably become more widely used to diagnose other infections such as tuberculosis. A computed tomography scan is mainly of use in herpes simplex encephalitis where there is usually localized necrosis of temporal lobes. EEG abnormalities (paroxysmal slow and sharp complexes) are found in herpes encephalitis and typical bursts of waves in SSPE. Neurosurgical opinion is often required. Brain biopsy with characteristic histology and the demonstration of virus by immunofluorescence or electron microscopy was needed in the past to diagnose herpes encephalitis; now with the advent of low-toxicity antiviral agents, such as acyclovir, this is rarely justifiable.

Management and prognosis

Specific treatment with acyclovir is available for herpes simplex encephalitis in which the mortality without treatment is 70 per cent. With acyclovir, mortality is 20 per cent. The prognosis is improved by treatment before coma develops.

For other types of encephalitis, only symptomatic and supportive treatment are available. Mannitol and/or dexamethasone or surgery may be used to control intracranial pressure. Mortality in epidemic encephalitis varies with the arbovirus between 10 and 70 per cent. In sporadic encephalitis, apart from herpes infection, mortality is about 15 per cent. Post-infectious encephalitis has a better prognosis. SSPE is uniformly fatal. In all forms of encephalitis, mortality is higher if the patient presents with widespread disease (convulsions, coma). Survivors of severe disease often have neurological sequelae such as paresis, intellectual and behavioural abnormalities.

Brain abscess

Brain abscess usually follows an identifiable source of infection: extension of infection of ears or sinuses; metastatic, especially after pulmonary infection; head injury or neurosurgery; or complicating meningitis. Clinical features are not distinctive and often reflect the underlying disorder. Patients complain of headache, lethargy, and vomiting. Seizures affect 25 per cent and, dependent on the site of the abscess, specific neurological signs may be detected.

If abscess is suspected, lumbar puncture is contraindicated, even in the absence of papilloedema, as it may cause neurological deterioration and the findings are not diagnostic. A computed tomography scan is the investigation of choice. This may be necessary as an emergency before CSF examination when meningitis is a possibility. The results of bacteriological culture of abscess contents vary but usually reveal mixed aerobes and anaerobes. The treatment of choice is probably surgical drainage of pus under antimicrobial cover such as chloramphenicol plus metronidazole. High dose, intravenous antibiotics without surgery are usually successful only if given early, before encapsulation of the abscess. Antibiotics should be continued for 4–8 weeks and progress monitored by computed tomography scans. It may be necessary to re-operate if there is recurrence. Half of the survivors have neurological impairment such as speech defects or seizures.

Subdural empyema (pus in the subdural space) usually results from infection of the sinuses or mastoid, but occasionally follows extension of brain abscess. The clinical features resemble brain abscess for which it is commonly mistaken. The computed tomography scan usually confirms the diagnosis and treatment is by early drainage of pus and high dose, intravenous antibiotic therapy. Mortality is greatly increased by delay in treatment, such as waiting for a brain abscess to localize.

Extradural abscess is rare and associated with osteomyelitis of a cranial bone (usually sinuses or mastoid). Treatment is with appropriate antibiotic and the site of primary infection may require surgical drainage.

DEMENTIA

Dementia is global impairment of higher intellectual function in an alert patient. Since the prevalence of dementia increases with age, it is more of a problem in developed countries. There is some degree of dementia in 10 per cent of the population aged over 65 years and 20 per cent over the age of 80 years. Dementia is arbitrarily divided into presenile (before 65 years) and senile (after 65 years).

Infectious agents as a cause of dementia are uncommon. More than half of the cases in Britain are due to Alzheimer's disease. The main infective causes are: syphilis (Chapter 10), Creutzfeldt–Jakob disease, HIV (Chapter 12), neuroborreliosis (Lyme disease, Chapter 3), and subacute sclerosing panencephalitis.

Creutzfeldt–Jakob disease can present in the late teens but usually between 50 and 60 years of age. The illness begins with minor signs such as gait problems, blurred vision or impaired memory, and progresses rapidly to dementia. There is no treatment and 80 per cent die within one year. The cause is an unconventional 'virus' (prion?) not yet fully identified. There is no evidence of non-invasive transmission and barrier nursing is not required. Instruments or other material in contact with blood, CSF, or viscera should be autoclaved or treated with 10 000 parts per million available chlorine.

GUILLAIN-BARRÉ SYNDROME

This acute polyneuritis of uncertain aetiology affects adults and older children, mostly with a previous history of respiratory or gastrointestinal illness. Some show evidence of *Mycoplasma pneumoniae*, cytomegalovirus, *Borrelia burgdorferi*, *Campylobacter jejuni*, Epstein–Barr or human immunodeficiency virus infection. It may also be a rare complication of chickenpox, measles, influenza A vaccine, or Semple rabies vaccine. Usually 1–2 weeks after upper respiratory tract illness, there is symmetrical sensory (all modalities) and motor loss beginning in distal parts of the limbs ('glove and stocking'). This slowly ascends, involving trunk and arms in flaccid paralysis. In severe cases, muscles of respiration are impaired and there are facial and bulbar palsies. The differential diagnosis includes poliomyelitis (typically asymmetrical), diphtheritic polyneuritis, and transverse myelitis (sensorimotor paralysis below a given spinal level).

Lumbar puncture shows the CSF to have normal white cell count and glucose concentration but a grossly elevated protein content. Plasmapheresis aids recovery in acute cases, otherwise treatment is supportive: recovery is usually complete within two months but in severe cases may occasionally take a couple of years. Many require tracheostomy, assisted ventilation and bladder catheterization. Respiratory and urinary infections are common and need prompt treatment. Steroids are of doubtful value. About 15 per cent of patients have residual neurological deficits, usually facial palsies. Fatalities (5 per cent) result from respiratory failure, cardiac dysrythmia, hypotension, or secondary infections. The outcome is poor if there is no improvement within three weeks of peak deficit.

POLIOMYELITIS

Paralytic poliomyelitis can cause considerable mortality, morbidity, and disability but is largely preventable by the use of vaccine (Fig. 6.10). In countries where vaccine is not yet widely used, such as India, scores of children become paralyzed every day. Nevertheless, countries with established polio vaccination

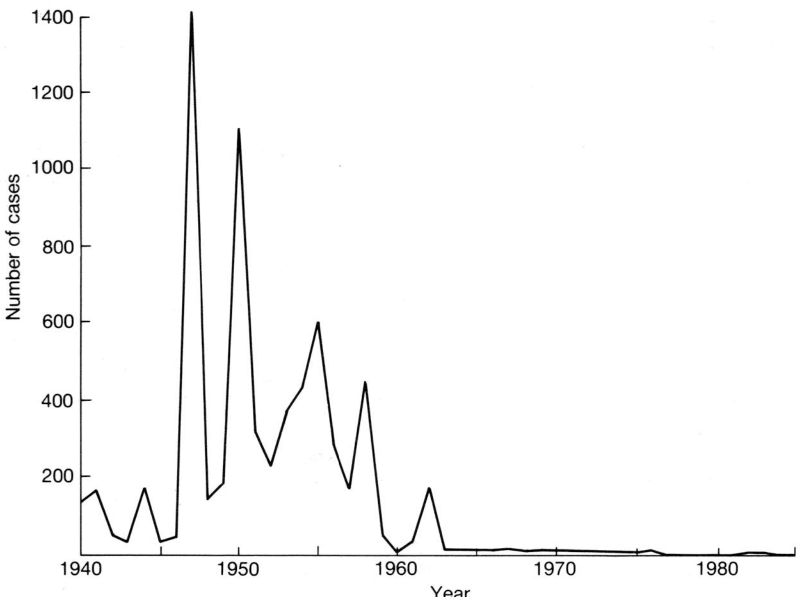

Fig. 6.10 Acute poliomyelitis notifications by year in Scotland (*Scottish Health Statistics*); inactivated polio vaccine introduced in 1956 and oral polio vaccine in 1962.

programmes may still have problems; for example, the 1031 paralytic cases in Taiwan in 1982, and the outbreak of paralytic cases in Finland in 1984 due to an antigenic variant.

Aetiology and epidemiology

The main causative agent is the poliovirus which has three serotypes. Infection is usually spread by the faecal–oral route, but in the acute phase may also be spread by droplets from nasopharynx. The pattern of disease depends on the degree of development of the country because of factors such as hygiene and vaccination programmes. In underdeveloped countries, the virus can infect virtually all children in the community, causing much crippling unless controlled by vaccination. The local adult population is largely immune and unaffected but the disease can affect non-immune visitors. Improvements in living standards allow many children to escape immunizing infection so that periodic epidemics appear involving older persons also (e.g. UK after 1947: Fig. 6.10). In developed countries, widespread use of vaccine makes disease rare, but the few cases may not be recognized because they are not expected.

Clinical presentation and investigation

The incubation period is on average 7–14 days (range 3–35 days). Infection produces three different results (Fig. 6.11):

(i) Asymptomatic illness in 90 per cent of cases with lifelong immunity to the same type of poliovirus;

(ii) Mild illness in up to 8 per cent. Viraemia is accompanied by pyrexia and flu-like illness, with rapid, complete recovery;

(iii) Major CNS illness in 1–2 per cent. There is usually a prodromal pyrexial illness followed by a few days of apparent well-being before classical signs of meningitis appear, or the patient may present with meningeal irritation. These patients may also have pain and spasm in the limb muscles. Within a week many recover completely, but two-thirds become paralyzed, with a range of effects varying from one muscle group (for example, dorsiflexors of the foot) to involvement of almost all muscle groups. Destruction of anterior horn cells (spinal poliomyelitis) is common, whereas medulla and brain-stem damage (bulbar poliomyelitis) is uncommon and destruction of the motor cortex (encephalitic poliomyelitis) is rare.

Flaccid, asymmetrical lower motor neurone paralysis can extend for several days until pyrexia abates. The upper limbs are most often affected, with no sensory abnormalities. Lower limb paralysis may be accompanied by bladder dysfunction which fortunately recovers in a few weeks. Bulbar poliomyelitis may cause life-threatening disorders of swallowing, breathing, or speech.

The main differential diagnosis of paralytic poliomyelitis is from Guillain–Barré syndrome and demyelinating disease. The former is distinguished by its symmetrical nature, sensory loss, and CSF findings (normal leucocyte count and

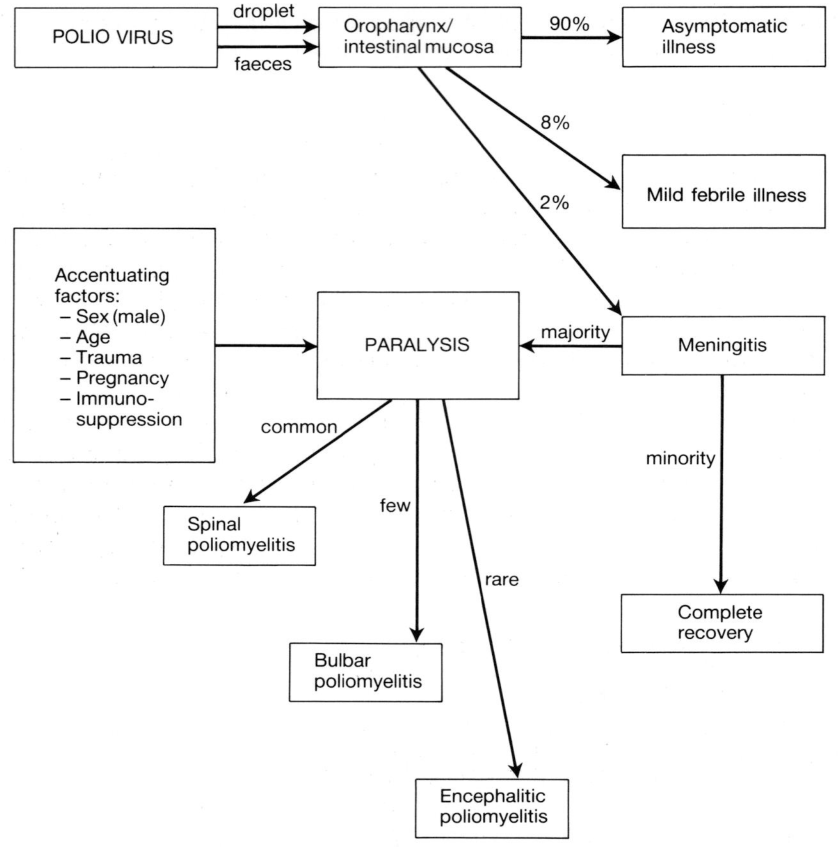

Fig. 6.11 Results of poliovirus infection.

high protein content). In the latter, pyrexia and meningeal symptoms are usually absent. Rarer causes of paralysis include snake bites, scorpion stings, tick-borne encephalitis, toxins (for example, from cassava or bites of certain ticks), chemical (for example, arsenic), and acute intermittent porphyria. In addition, other enteroviruses rarely cause paralysis (for example, coxsackie A7, enterovirus 70, 71).

Poliovirus is easily grown from stools (Fig. 6.2a and b) and in the first five days of illness from nasopharyngeal swabs. Stool excretion of virus may persist for a few weeks and up to three months in young children. Patients with polioviral meningitis show CSF features of viral meningitis (Table 6.3) but poliovirus is very rarely found in the CSF as the concentration of virus is low. Virus isolation from CSF is conclusive whereas an enterovirus found in faeces may represent coincidental carriage. The diagnosis may be confirmed by detecting high or rising titres of polioviral antibodies, but maximum titres have usually been attained by the time of the first blood sample.

Management and prognosis

Meningitis alone is treated as for viral meningitis. Paralytic cases require bed rest. If still pyrexial, the patient is observed closely for the development of difficulty in swallowing or respiration. Analgesia can be given for pain relief. Respiratory failure requires assisted ventilation usually with intubation to maintain a clear airway without obstruction by secretions. After the acute illness, graded physiotherapy is of great value. Patients in hospital should be isolated and barrier nursed, but quarantine measures at home are not justified. Food-handlers or those working with children should not return to work unless virus excretion has stopped.

Polio infection by wild virus is preventable. Sabin-type vaccine (an attenuated live virus) contains the three types of poliovirus and is given orally. It is remarkably free from complications, although paralytic illness is associated with about 1 in 4 million doses (usually within 28 days of administration of the vaccine). Household contacts (especially adults) may be secondarily infected by vaccine virus which has become less attenuated during passage. It is advisable to

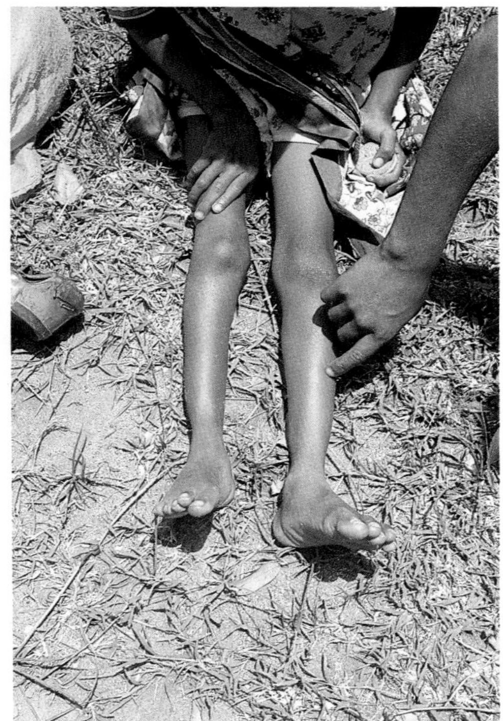

Fig. 6.12 The normal leg contrasts with the short wasted limb due to poliomyelitis.

give the vaccine itself to unvaccinated adults at the same time as the children of the household. Salk-type inactivated trivalent polio vaccine avoids this risk and is more stable but also more expensive and harder to administer requiring three injections. Improved inactivated vaccines of higher potency have been developed which require fewer injections but are even more expensive. They can be combined with bacterial antigens, have been used successfully in several countries and may become more widely used in future. Genetically-modified vaccines which have no tendency to revert to 'wildtype' virus are also under development.

In countries with epidemics (some Asiatic and African countries) the mortality of paralytic poliomyelitis is 20 per cent in adults and 5 per cent in children. The usual cause of death is respiratory failure from spinal or bulbar involvement. Bulbar palsies (dysfunction of speech, swallowing and breathing) usually recover well. Improvement in affected skeletal muscle groups may continue for up to two years. Experienced physiotherapy complemented by orthopaedic help can have dramatic effects but the legacy of crippling can be serious. (Fig. 6.12). Increasing fatigue and weakness of affected muscles may become manifest many years after the onset of poliomyelitis.

7

Infections with generalized rash

INTRODUCTION

Exanthems are acute febrile infections associated with the eruption of a skin rash which is often characteristic in appearance. Most result from blood-borne dissemination of an infectious agent, usually viral, or of microbial toxin in the case of scarlet fever. Similar lesions may be seen in the oral cavity, and are called enanthems.

Pathogenesis

Two main pathogenic mechanisms are concerned in producing rashes: in the first, *damage to skin capillaries* causes erythematous or haemorrhagic rashes (Fig. 7.1). If erythema is marked, red cells may leak out of dilated capillaries and, when erythema fades, may simulate a haemorrhagic rash. In the second, *damage to skin cells* (Table 7.1) usually produces vesicular lesions which typically evolve through macular, papular, vesicular, and pustular stages before crusting and healing. Whether a scar is left depends on the depth of skin involved. The two main pathogenic mechanisms overlap, and severe blood-vessel damage, as in Stevens–Johnson syndrome, can cause central vesiculation of the erythematous lesion. Likewise there is always some erythema around vesicular lesions from minor involvement of local capillaries.

Table 7.1. *Mechanism of rashes mediated through damage to skin cells (e.g. chickenpox)*

Pathological process	Lesions
Virus infects contiguous cells	Invisible
Local inflammatory capillary dilatation	Erythematous macule
Local inflammatory oedema	Erythematous papule
Increasing oedema separates cells; cell necrosis	Vesicle
Inflammatory cells enter fluid	Pustule
Fluid absorbed	Crust/scab

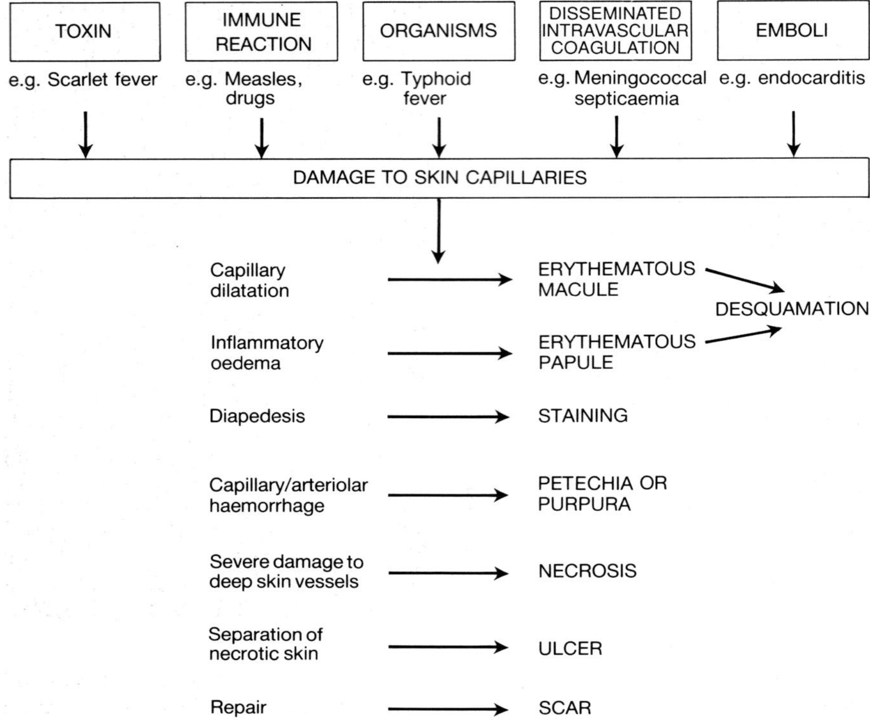

Fig. 7.1 Mechanisms and results of damage to skin capillaries.

Clinical approach

Although rash may be prominent in a patient's history and examination, one should avoid premature 'spot' diagnosis and determine the associated symptoms and signs and their order of development. In infections with rash, the *prodrome* refers to the interval between onset of symptoms and onset of rash as distinct from the *incubation period* (Chapter 1).

The rash is best defined by the nature of the individual lesions, their distribution, and how their nature and distribution vary with time—their evolution. Many rashes are widespread and consist of uniform small elements which are defined in Table 7.2. It is most important to recognize a haemorrhagic rash because possible causes (Table 7.3) include life-threatening meningococcal sepsis, which requires prompt antibiotic therapy. Erythematous (Tables 7.5 and 7.6), vesicular (Table 7.7) and urticarial (Table 7.4) lesions are considered below. If itchy, a rash may be excoriated or have superimposed scratch marks (Fig. 3.20). Scratching can also cause petechiae or purpura, for instance in thrombocytopenia.

The distribution of a rash may be diagnostic—for example, the usual

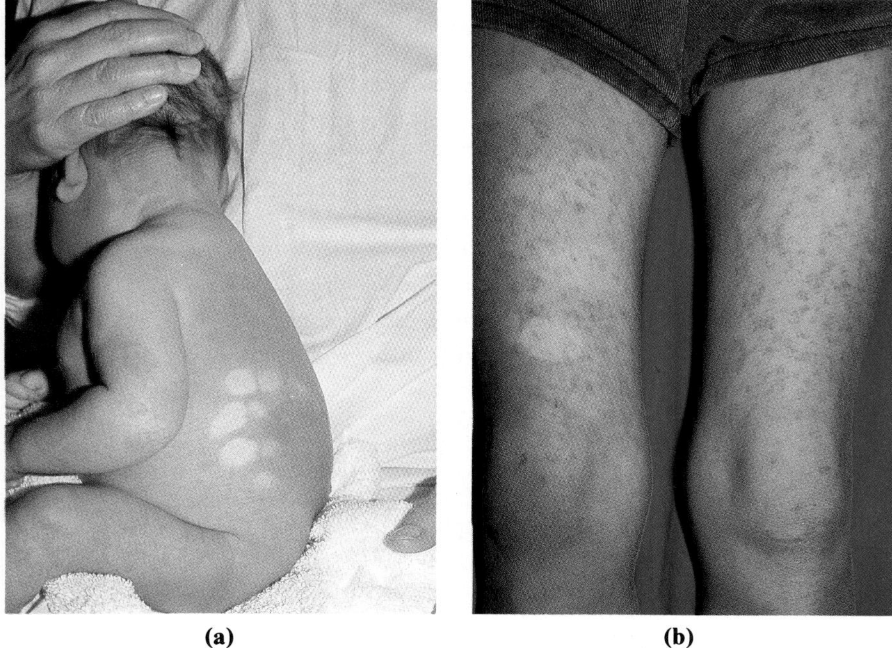

(a) (b)

Fig. 7.2 (**a**) Blanching from finger pressure on back of child with facial erysipelas and generalized scarlatiniform erythematous rash; (**b**) red cells in subcutaneous tissues whether from capillary haemorrhage or diapedesis will not blanch. Post-measles 'staining'.

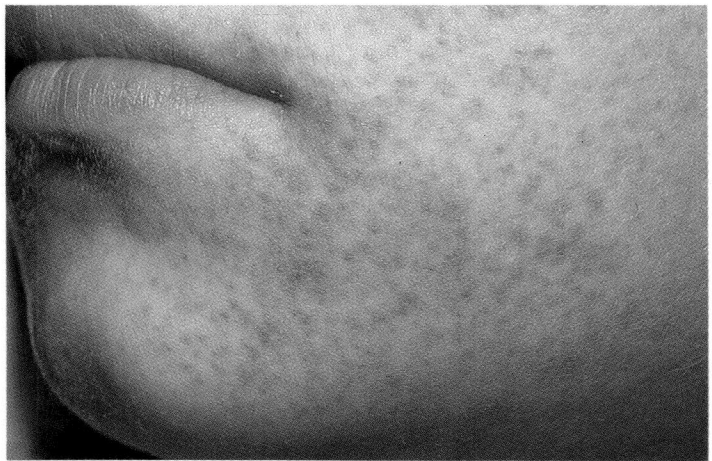

Fig. 7.3 Petechial rash from back pressure on capillaries during intense vomiting.

Table 7.2. *Definitions of common skin lesions*

Erythema	Area of red discoloration that blanches on pressure (Fig. 7.2a)
Petechia	Small (often punctate) haemorrhage (does *not* blanch on pressure) (Fig. 7.2b)
Ecchymosis	Larger haemorrhage within skin
Purpura	Numerous petechiae or ecchymosis
Macule	Small flat area of discoloration
Papule	Small raised area, usually discoloured
Vesicle	Small blister (less than 5 mm diameter) containing clear fluid
Pustule	Small blister containing cloudy fluid
Bulla	Large blister
Crust or scab	Dried exudate
Ulcer	Defect in an epithelial surface due to loss of tissue
Wheal/urticaria	Raised, often itchy lesion, with small surrounding zone of erythema and paler centre. Irregular, often up to 10 cm

localizations of shingles or 'cold sores'. Other distributions provide important diagnostic pointers such as the presence of lesions of secondary syphilis on palms and soles, or preferential spread of herpes simplex to eczematous areas (Kaposi's varicelliform eruption). Finally, evolution of a rash may be characteristic; for example, the steady progression of measles rash down the body over three or more days (Fig. 7.6) or the transition of varicella (or herpes simplex: Fig. 3.10) lesions from macular and papular stages to vesicular and pustular stages before drying and crusting.

On brown or black skins, erythematous lesions are less obvious. It may be impossible to see the macules of rubella, but more intense rashes, like measles, can usually be perceived, especially as these tend to be papular.

ERYTHEMATOUS RASHES

Measles (Figs 7.4–7.8)

Aetiology and epidemiology

Measles is caused by a paramyxovirus of which there is only one serotype. It has a world-wide distribution and is spread mainly by the airborne route from

Table 7.3 *Haemorrhagic rashes with fever*

Condition	Diagnostic pointers
Meningococcal sepsis	Irregular 'flea-bitten' rash, often sparse, may be petechial or ecchymotic. Occasionally erythematous initially. Sometimes subconjunctival haemorrhages. (Chapter 6)
Enteroviral infections (especially echo 9)	Uniform petechial rash, mainly over trunk, seen occasionally. Enteroviral meningitis may also be present. (Chapter 6)
Pressure effects leading to capillary rupture, sometimes seen in meningitis or fever of any cause	Fine petechical rash may appear over face and upper trunk after severe vomiting or retching (Fig. 7.3). Fine petechial rash may develop over the areas pressed when patient held firmly, e.g. during lumbar puncture.
Post-erythematous staining, e.g. after measles	Uniform, corresponding to distribution of previous erythema. (Fig. 7.8)
Drug reactions	Widespread distribution, generally uniform and purpuric (see Fig. 18.5b).
Henoch–Schönlein purpura	Distribution on peripheries and buttocks. Crops, usually petechial. Arthralgia. Haematuria. Previous upper respiratory infection. (Chapter 13)
Haemolytic–uraemic syndrome	Predisposing respiratory or diarrhoeal episode. Hypertension. Anaemia. Renal failure. Petechial rash. (Chapter 13)
Disseminated intravascular coagulation (DIC) severe septicaemia, rarely in measles or varicella	Shocked. Blotchy, widespread purpuric rash. Possible focus of infection. If haemorrhagic measles or varicella, purpuric rash superimposed on usual exanthem. (Chapters 7, 8)
Viral haemorrhagic fevers	Travel and occupational history. Symptoms within 21 days of leaving endemic area. Myalgia. Blotchy purpuric rash and shock. (Chapter 14)
Rickettsioses	See Chapter 14.

respiratory secretions of acute cases. It is highly infectious. Hence most children have normally been infected by five years of age, and almost all adults are immune. Immunity is lifelong. There is no long-term infectious state and no animal host. In large unvaccinated communities measles is endemic with epidemics every 2–5 years. Most cases present in colder months. In smaller and isolated communities measles circulation may cease altogether unless it is re-introduced. An acute case is infective from just before the onset of symptoms to

Table 7.4. *Common causes of urticarial rashes*

Agent/syndrome	Diagnostic pointers
Foods	Strawberries, shellfish
Drugs/chemicals (see Fig. 18.3)	Penicillin, salicylates, tartrazine dyes, plants
Bites/stings	Insects; mites or fleas from pets
Helminths	*Trichinella*, hookworm. Travel history
Prodromal hepatitis B (Chapter 5)	Also arthralgia. High-risk group or contact

the defervescence of the fever. Clinical infection under nine months of age is uncommon in developed countries because of passive protection by maternal antibody, but is common and often severe in malnourished infants, especially with vitamin A deficiency and in crowded conditions such as refugee camps. Measles is estimated to cause 1.5 million deaths world-wide each year.

Clinical

After an incubation period of 10–14 days a pronounced coryzal prodrome lasting 3–4 days develops. Fever up to 39 °C is present and the child appears miserable, with bleary eyes and runny nose. There is characteristic dry cough and sometimes croup. Occasionally, early in the prodrome, a transient macular rash is noted. Towards the end of the prodrome pathognomonic spots appear on the buccal mucosa in 80 per cent of patients. These small areas of white epithelial necrosis ('Koplik's spots') contrast with the background reddened mucosal surface (Fig. 7.4). They may also appear on other mucosal surfaces such as the conjunctivae or rectum. Within 24–48 hours of Koplik's spots appearing, the dusky red maculopapular rash of measles begins behind the ears (Fig. 7.5) and spreads rapidly to the face. Thereafter the rash spreads down the body, reaching the lower limbs in two to three days (Fig. 7.6); it usually stops just below the knees. In babies with nappies and those lying on their backs the rash may be accentuated in the buttocks and back respectively. Fever peaks at 39–40 °C as the rash appears on the upper part of the body (Fig. 7.7) and then subsides slowly, the patient being apyrexial by the time lesions stop appearing. About the time the rash appears on the chest, the dry cough moistens, and occasionally fine crackles heard throughout the lung fields indicate measles bronchiolitis. This resolves as the rash progresses. The mucosal surfaces of the conjunctivae and pharynx remain reddened but Koplik's spots disappear shortly after rash appears. Both tympanic membranes are injected peripherally. There is general, but slight, lymph-node enlargement. The rash fades from above downwards and commonly

Table 7.5. *Illnesses with generalized erythematous rashes consisting of mainly small elements*

Agent/syndrome	Diagnostic pointers
Measles	Marked coryzal prodrome with conjunctival injection and dry cough. Koplik's spots. Slow progression down body. Staining.
Rubella	Lymph node enlargement, especially occipital. Pink macules. Confluent on trunk on second day.
Enteroviruses	Commonly macular, or petechial. Uniform on trunk. No progression. No prominent lymph nodes. Fever, and often mild respiratory features.
Erythema infectiosum (Fifth disease)	'Slapped cheeks'. Lace-like lesions over trunk, evanescent. Can last one week. Arthralgia.
Exanthema subitum	Sharp febrile illness. Maculopapular rash only on fall of temperature.
Infectious mononucleosis	Prominent lymph-node enlargement. Tonsillar exudates. Palatal petechiae. Splenomegaly. Early macular rash occasionally. Dense maculopapular rash provoked by ampicillin-like drugs.
Enteric fever	Stepladder fever, headache, confusion, splenomegaly. May be 'rose spots' on trunk, usually sparse, may be profuse in paratyphoid B.
Pityriasis rosea	Herald patch. No systemic upset. 'Christmas tree' distribution. Peripheral scaling of lesions. Lasts weeks.
Scabies	Pruritus. Burrows. Scratchmarks. Distribution in axillae, groins, finger webs. Sometimes cheiropompholyx (Chapter 3).
Erythema multiforme/ Stevens–Johnson syndrome	Peripheral distribution. Target lesions. Mucocutaneous lesions. May be precipitating factor. (Chapter 13).
Drugs	Usually maculopapular on flanks and limbs, especially penicillins, sulphonamides, allopurinol. No progression.

leaves 'staining' from haemoglobin in subcutaneous tissue (Figs 7.2b and 7.8). Occasionally there is fine scaly desquamation after the rash has faded (see Fig. 2.5). Other causes of a similar rash, but without other features of measles, include rubella, enteroviruses, and drug reactions, especially to ampicillin (see Fig. 18.2).

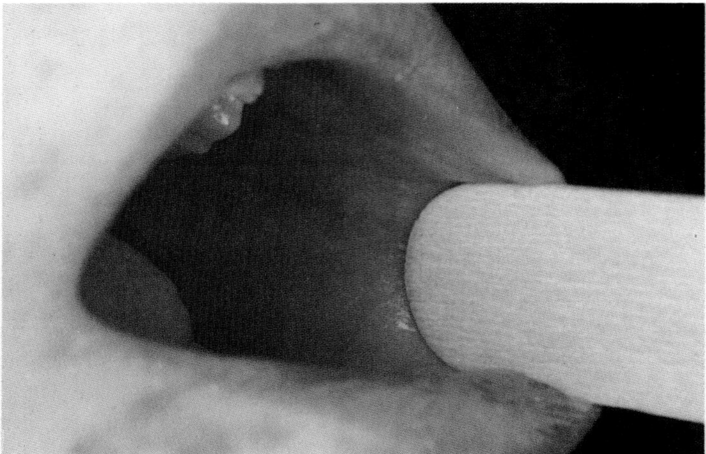

Fig. 7.4 Measles. Koplik's spots.

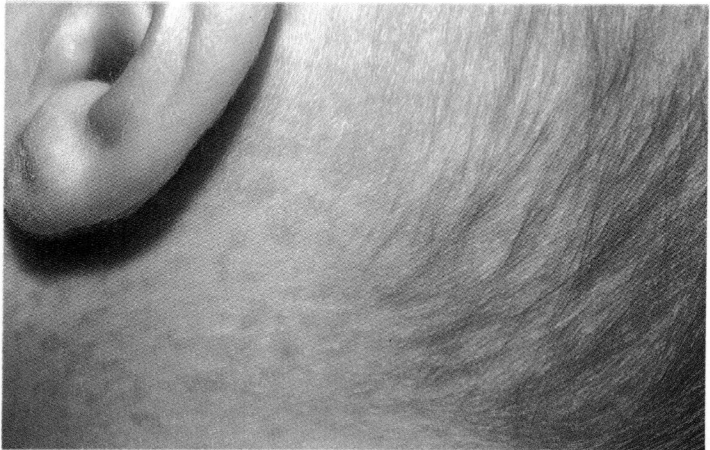

Fig. 7.5 Measles. Rash starts behind the ear.

Investigation

Measles can be diagnosed clinically by the marked catarrhal prodrome, with dry cough, and the progression of the rash. The usual pronounced rash indicates good cell-mediated immunity. However, in infants under nine months of age rash may be attenuated by protective maternal antibody. In the immunocompromised cell-mediated immunity may be depressed with only a sparse rash, if any, but there may be persistent Koplik's spots. In these cases serological tests are usually negative and diagnosis may depend on identification of measles virus in nasopharyngeal aspirate or throat swab. The prognosis is grave.

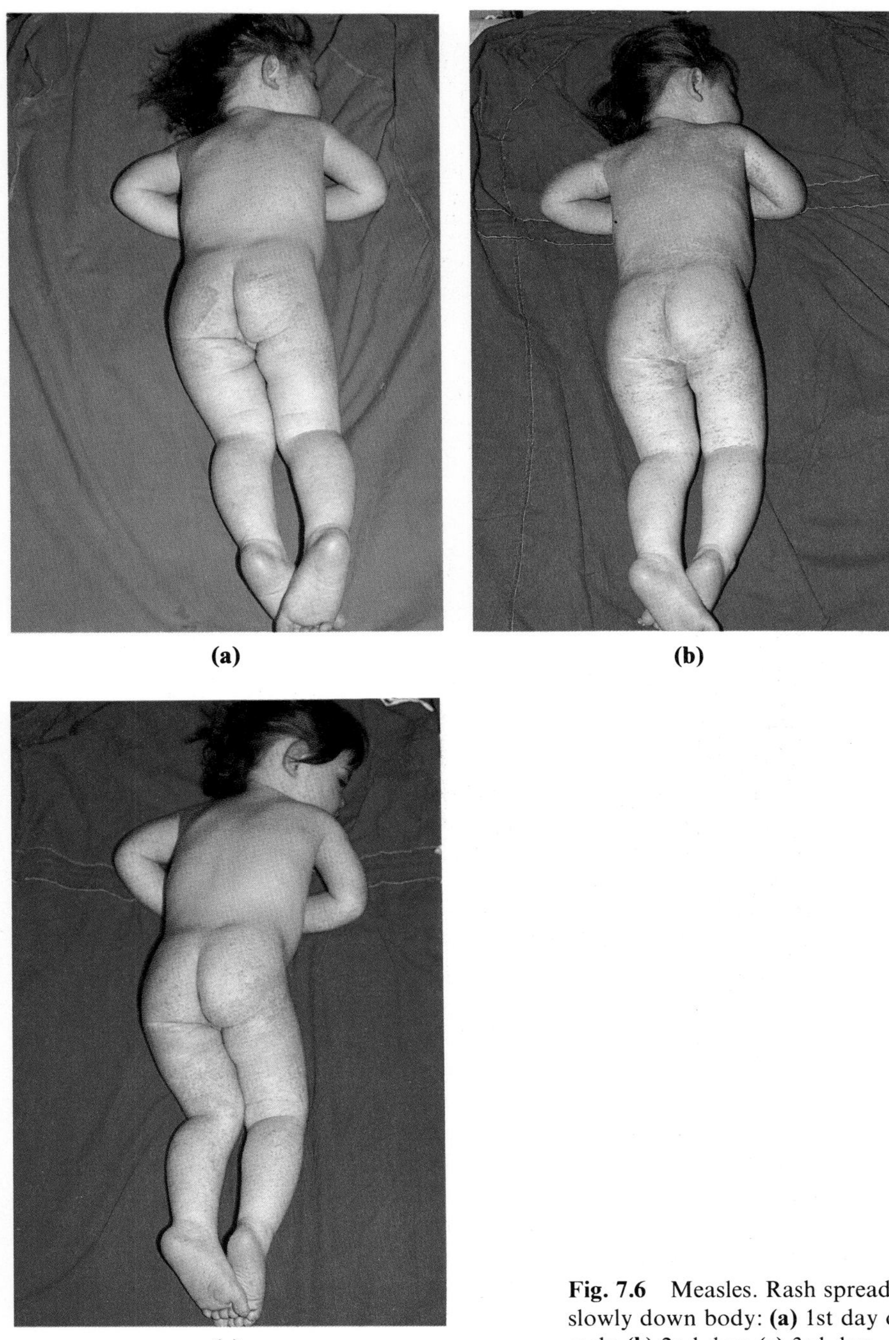

Fig. 7.6 Measles. Rash spreads slowly down body: **(a)** 1st day of rash; **(b)** 2nd day; **(c)** 3rd day.

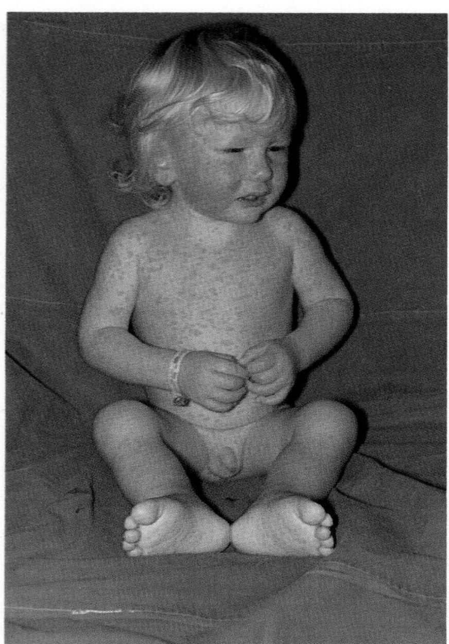

Fig. 7.7 Measles. Children do not enjoy measles.

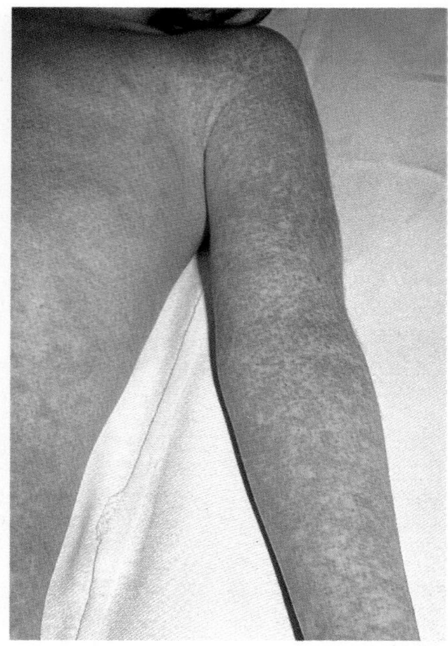

Fig. 7.8 Measles. Unusually heavy 'staining' referred to hospital as possible meningococcal sepsis.

Management and complications

Management is symptomatic except when complications occur, most commonly secondary bacterial bronchopneumonia and otitis media (Fig. 2.5). These affect up to 14 per cent of patients and should particularly be suspected if the temperature does not fall as the rash fades. The organisms responsible are usually *H. influenzae*, *Strep. pneumoniae*, or *Strep. pyogenes* and treatment should be by co-trimoxazole or amoxycillin–clavulanate. Amoxycillin alone is satisfactory where the incidence of penicillinase-producing *H. influenzae* is low.

Febrile convulsions develop in 1 in 700 patients either during the prodrome or at the onset of the rash. Post-infectious encephalitis (Chapter 6) affects 1 in 1000, usually 4–5 days after the onset of the rash, with alteration of consciousness, vomiting, or convulsions. Most of these patients recover without sequelae but of those that become comatose many die. In 1 in 1 000 000 the immune response does not clear all virus which persists in the brain, progressing as a slow virus infection to become symptomatic, on average, seven years later as sub-acute sclerosing panencephalitis [SSPE] (Chapter 6).

In many patients, scattered shadows on the chest X-ray indicate minor degrees of giant-cell pneumonia caused by the virus itself. Spontaneous resolution is usual. However, in the immunocompromised, impaired immune reponses may allow the virus to proliferate and cause extensive giant-cell pneumonia and interstitial pneumonitis (see Fig. 12.2). This often progresses to pulmonary fibrosis and death. Similarly, especially in leukaemic children proliferation within the brain is a rare cause of progressive encephalopathy.

In the tropics, measles is severe in the young malnourished child, with mortality rates as high as 10–20 per cent in parts of rural Africa. Since measles in well-nourished children resembles that seen in the developed world, it seems likely that the combination of protein–calorie malnutrition, vitamin A deficiency, concurrent infection with parasites such as malaria, and inadequate treatment of complications causes the high mortality. The rash may be atypical and giant cell pneumonia is common. Both protein–calorie malnutrition and vitamin A deficiency are exacerbated by measles causing kwashiorkor or keratomalacia, an important cause of blindness. Diarrhoea is common and may be severe and prolonged causing further protein loss. Cancrum oris (rapid necrosis of oral and facial tissues) may be seen. The Mantoux reaction may be suppressed during measles and tuberculosis exacerbated. Vitamin A given during acute illness reduces mortality and morbidity, even speeding resolution of pneumonia and diarrhoea, and a dose of 400 000 i.u. should be given to all with measles in countries where the measles mortality rate is 1% or more.

Prevention

Live attenuated measles vaccine is available, alone or combined with mumps and rubella vaccines (MMR) (Chapter 16). High uptake (over 95 per cent) in the USA resulted in a 99 per cent fall in incidence and most infections there now concern imported measles, campus outbreaks in groups of susceptible young adults or children in inner city areas with poor immunization uptake. Outbreaks have developed in schools with 98 per cent vaccination rates. Two to 10 per cent fail to seroconvert after vaccination, especially when given just after 12 months of age.

In contacts at risk of developing severe illness, like hospital in-patients with concurrent illnesses, human measles immunoglobulin can prevent or modify measles if given within three days of contact. Contacts with no contraindications to measles vaccine can be given vaccine promptly.

Rubella

Aetiology and epidemiology

Rubella is caused by a togavirus of which there is only one serotype. Rubella virus is spread mainly by the airborne route from acute cases whose infectivity usually lasts from about a week before symptoms to about a week after rash appears. It is highly transmissible and by adulthood about 80 per cent of the unvaccinated population have developed antibodies. There is no animal reservoir but there may be long-term excretion of virus from the throat and urine of those born with congenital rubella. Reinfections, demonstrated by antibody rise, are frequent, especially in those vaccinated, but are rarely viraemic and usually asymptomatic. Congenital transmission may follow viraemia in the pregnant woman but the probability of foetal infection and subsequent damage varies with the stage of the pregnancy when viraemia takes place (Chapter 11).

Clinical presentation

Postnatal rubella is generally mild, inapparent in at least 50 per cent, especially in children. The incubation period is 2–3 weeks and in adults there may be a short prodrome of mild fever, malaise, and upper respiratory symptoms. Rash appears first on the face but spreads rapidly to the rest of the body as discrete, mainly macular, erythematous lesions (Fig. 7.9) of lighter, rose-pink colour than the dusky red of measles. Lymph nodes are enlarged in all areas, especially occipital, and the spleen is often palpable. In severe cases conjunctival injection may be marked (Fig. 7.10). Fever is usually minimal and transient but may be up to 38 °C when rash appears. On the second day lesions on the face and trunk become confluent whilst those on limbs remain discrete. By the third day the rash has usually disappeared. Often there is no rash, though lymph nodes may still enlarge—usually this is recognized only in contacts.

Rubelliform rashes are seen in some enteroviral and parvoviral infections, as the occasional early rash of infectious mononucleosis, and in mild drug rashes. Occasionally, rubella rash on the first day looks rather like measles and on the second day like scarlet fever, but history and examination should distinguish these.

Investigation

Since rubella is not reliably diagnosed clinically, serological confirmation is essential to establish the diagnosis firmly when necessary. Thus in pregnancy, immediately possible rubella or exposure to rubella infection is suspected an initial blood sample should be collected. Isolation of virus from throat swabs is difficult and rarely necessary.

Management and complications

Management is symptomatic though further advice is urgently required in pregnancy (Chapter 11). Flitting arthritis of fingers, wrists, and knees may be seen particularly in the post-pubertal female, and occasionally thrombocytopenic purpura develops. Post-infectious encephalitis affects 1 in 5000 patients; SSPE is a very rare late complication.

Prevention

Live, attenuated vaccine is available, alone or combined in MMR, and its role in preventing congenital infection is discussed in Chapter 11.

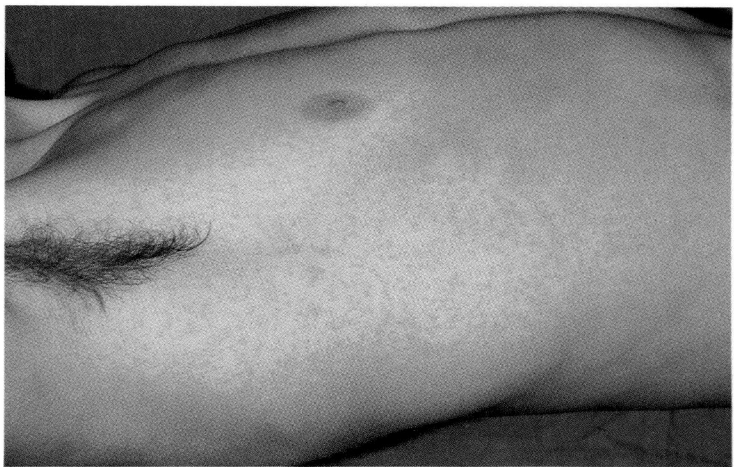

Fig. 7.9 Rubella. About 24 hours after first appearance, the rash is almost confluent on the trunk.

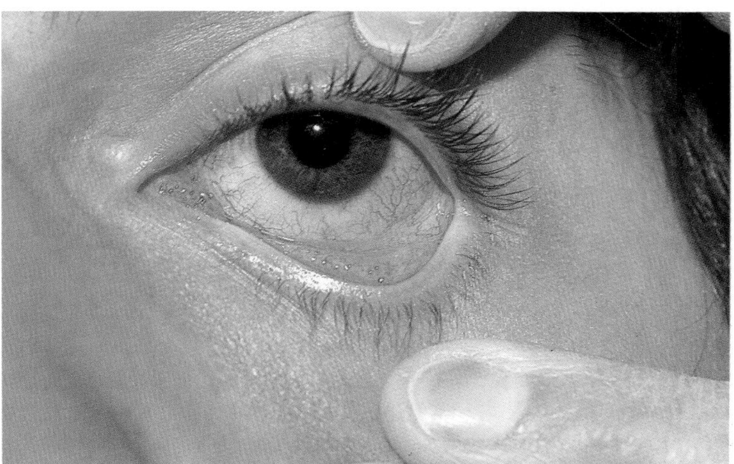

Fig. 7.10 Rubella. Conjunctival 'injection' in severe case.

Enteroviruses

Amongst the many clinical expressions of enteroviruses are rashes, particularly in young children. These are usually centrally distributed and macular or maculopapular (Fig. 7.11), although vesicular, petechial, and urticarial forms occur. As with most symptoms and signs of enteroviral infection, these rashes may be the major feature of the illness or may be combined with other features such as aseptic meningitis or upper respiratory infection. Some serotypes are

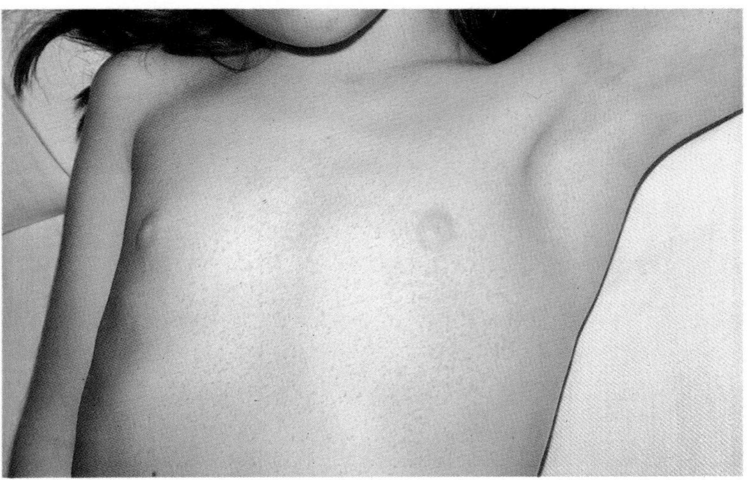

Fig. 7.11 Enteroviral exanthem.

more often associated with rashes, particularly coxsackievirus A9 and A16 and echovirus 4, 9, and 16. Particularly distinctive is the hand, foot, and mouth syndrome when vesicular lesions appear on hands and feet (see Fig. 4.13), with little systemic upset (Chapter 4). Management of enteroviral infection is symptomatic and diagnosis is by virus isolation. Occasionally serology is attempted when symptoms are severe or important diagnostic difficulty arises—as in distinguishing the rubelliform rash of echo 9 from rubella itself.

Erythema infectiosum (Fig. 7.12)

This illness, also known as slapped cheek disease and Fifth disease, is caused by human parvovirus B19, previously only associated with aplastic crises in patients with chronic haemolytic anaemias, especially sickle-cell disease. Outbreaks of erythema infectiosum usually affect young children, causing marked redness of cheeks with an irregular, slightly raised outline and circumoral pallor (Fig. 7.12). Occasionally there is a one- or two-day prodrome of mild flu-like symptoms. A maculopapular rash may later develop on the trunk and become confluent with fading centres in a lace-like pattern. This may fluctuate in intensity over the next week or two. Mild generalized lymph-node enlargement is present, and occasionally nonthrombocytopenic purpura. Symptomatic treatment is sufficient. Many adults suffer pharyngitis and joint pain and swelling that may persist for some months. Transient mild bone-marrow 'arrest' is demonstrable in normal persons but is more marked in those with congenital haemolytic anaemia. Chronic anaemia may develop in the immunodeficient as a result of persistent infection. The illness with its rash, lymph-node enlargement, and arthritis is similar to rubella and can likewise damage the fetus (Chapter 11). Lymph-node

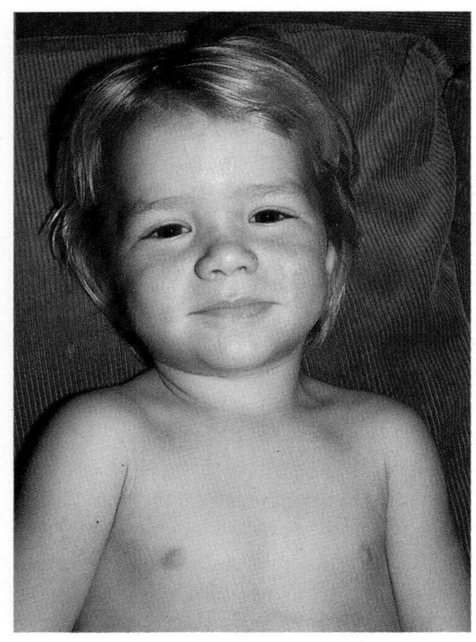

Fig. 7.12 Erythema infectiosum. 'Slapped cheeks'.

enlargement is more prominent in rubella, while arthritis and long duration of rash are more prevalent in parvovirus infection. Diagnosis is confirmed serologically.

Exanthem subitum (roseola infantum)

This infection is now attributed to human herpesvirus 6 (HHV6). Usually seen in children under the age of three years, its incidence tends to be sporadic. The source of infection is probably the saliva of adult cases. HHV6 specific DNA has been found in salivary glands suggesting viral persistence with intermittent excretion into saliva. Viral persistence has been detected in other organs. Initial febrile illness (up to 40 °C) is accompanied by moderate systemic upset and upper respiratory features, occasionally with febrile convulsions. After about three days the temperature abruptly returns to normal and the rash then appears. It consists of an erythematous maculopapular rash, predominantly on the trunk, lasting only 24–48 hours.

Pityriasis rosea

This is another infection of presumed viral aetiology. It mainly consists of the rash itself with no general upset. The first sign is usually the 'herald patch', an erythematous, slightly scaly lesion 2–3 cm in diameter (Fig. 7.13). This may be

Table 7.6. *Common causes of generalized erythematous rashes: consisting of mainly diffuse erythema*

Agent/syndrome	Diagnostic pointers
Scarlet fever	Punctate erythema. Pastia's sign (see text). Sandpaper feel. Enanthem—follicular tonsillitis, white strawberry tongue, later red.
Scalded skin syndromes Ritters disease Lyells disease (toxic epidermal necrolysis)	Nikolsky's sign (see text). Extensive separation of upper layers of skin leaving raw moist surfaces. Localized source of infection. Tender skin.
Staphylococcal scarlet fever	Primary focus of infection. Widespread erythema. Accentuation in skinfolds. No enanthem.
Toxic shock syndrome	Fever, vomiting, diarrhoea. Hypotension. Renal failure. Large irregular patches of erythema. Primary focus of infection. Myalgia. (Chapter 8.)
Kawasaki disease	Continuing fever. Conjunctival infection. Cracked lips. Red strawberry tongue. Indurated oedema/erythema of palms and soles. Blotchy erythema on trunk. (Chapter 13.)
Drugs (erythroderma)	Widespread erythema with fever. Drug administration especially sulphonamides, allopurinol, gold salts, phenylbutazone.

present on its own at any site on the body for up to two weeks before more generalized rash appears of similar, but smaller, lesions (Fig. 7.14). Sometimes the lesions simulate fungal lesions with their circular or ovoid shape, outer scaliness and erythema, and central pale areas, but ringworm does not have such a wide distribution. The rash of pityriasis rosea is heaviest on the trunk, particularly the back, and may have a 'Christmas tree' pattern with slightly oval lesions with their long axes sloping down to the sides. Lesions may continue to appear for 4–6 weeks without any upset except slight itch. There are no diagnostic tests and no known sequelae.

Epstein–Barr virus (see also Chapter 2)

In the early stages of infectious mononucleosis a generalized rash is occasionally seen. This is very similar to early rubella rash, consisting of pink macules with most lesions on the trunk. The presence of enlarged lymph-nodes add to diagnostic difficulty unless typical tonsillar appearances are present. This early rash disappears within 24 hours and should not be confused with the

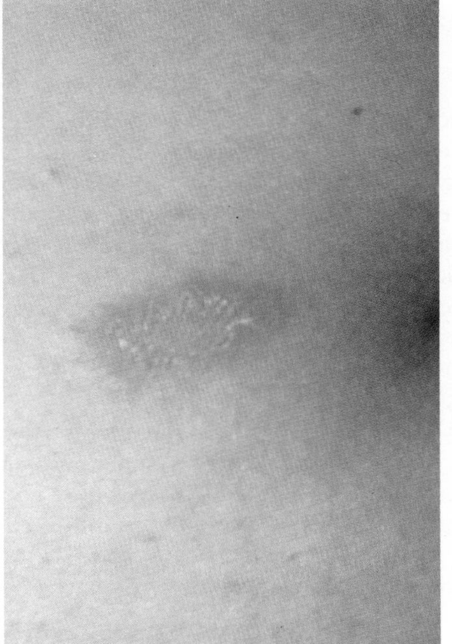

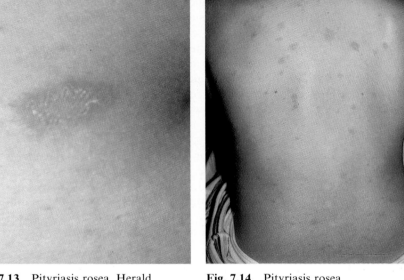

Fig. 7.13 Pityriasis rosea. Herald patch with active outer edge, inner scaling, and resolving centre.

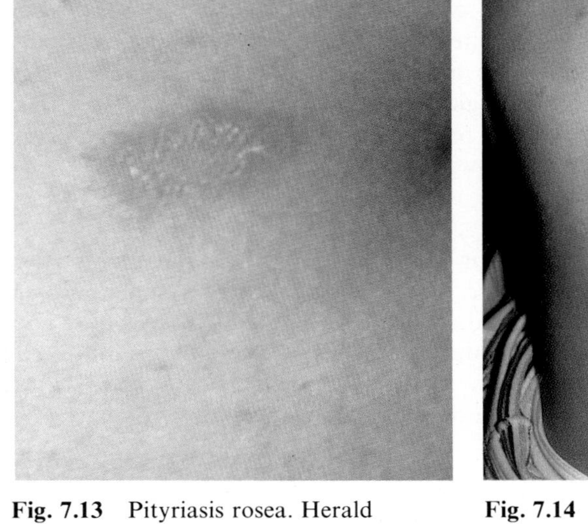

Fig. 7.14 Pityriasis rosea. Generalized rash.

hypersensitivity rash that develops in over 90 per cent of patients with infectious mononucleosis when given ampicillin or related drugs (see Fig. 18.2). This often appears about 10 days after starting the drug and is usually intense, erythematous, and maculopapular, heaviest on trunk but extending to limbs as well, resembling measles but without the prodrome and characteristic evolution. Although typical of a penicillin hypersensitivity rash, it does not indicate hypersensitivity to the penicillin nucleus and is not a contraindication to the use of ampicillin thereafter.

Enteric fever (see also Chapter 14)

The sparse 'rose spots' of typhoid fever are not prominent and are visible only on white skins. A few macular erythematous lesions are usually present on the lower thorax and upper abdomen, anteriorly and posteriorly, about 3 mm in diameter (see Fig. 14.9). They appear in crops for four or five days starting at the end of the first week of symptoms. *Salmonella typhi* or paratyphi can be isolated from the lesions. In paratyphoid fevers the rash tends to be heavier and in paratyphoid B can occasionally become profuse and generalized.

Scarlet fever (Figs 7.15 and 7.16)

Aetiology and epidemiology

Some strains of *Strep. pyogenes* produce an erythrogenic toxin. Scarlet fever develops when a patient infected with such a streptococcus is already sensitized but not immune to the toxin. There are several serotypes of toxin but scarlet fever is rarely suffered more than once. The site of the streptococcal infection is usually the tonsils (Chapter 2). Less often the site is elsewhere, usually cutaneous, for instance burns, varicella lesions or surgical wounds (surgical scarlet fever). The illness has become milder in Britain in recent decades. Rare similar illnesses are attributable to staphylococci (see below).

Clinical presentation

Besides local effects in the tonsils or skin, generalized erythematous rash develops, with circumoral pallor. The diffuse erythema has a superimposed punctuate appearance due to increased erythema around slightly elevated hair follicles (Fig. 7.15) giving a coarse, 'sandpaper-like' feel to the skin when stroked. Further accentuation of the rash is seen in skin creases (Pastia's sign), particularly at the elbow and axilla. Besides follicular tonsillitis, oral mucosa is

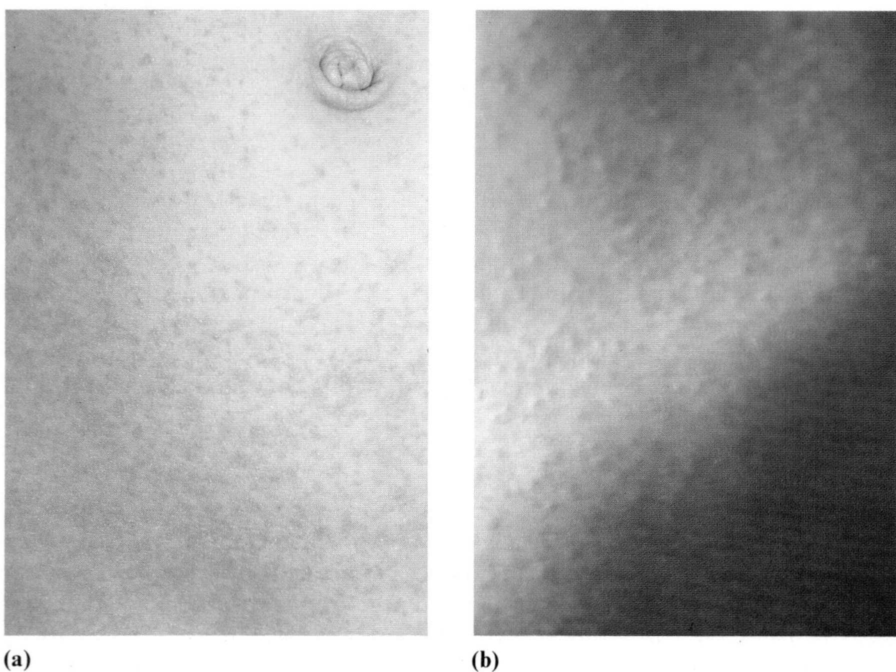

(a) **(b)**

Fig. 7.15 Scarlet fever: **(a)** punctate erythema; **(b)** side lighting reveals prominent hair follicles that give a coarse feeling to the skin.

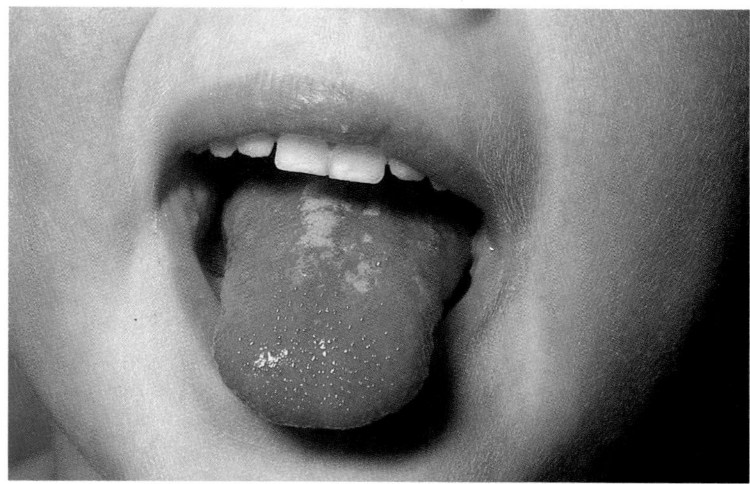

Fig. 7.16 Scarlet fever. The 'red strawberry' appearance left as the whitened filiform papillae peel off.

reddened and initially the filiform papillae of the tongue become white with prominent red fungiform papillae, the 'white strawberry' appearance. After further 24 hours the filiform papillae strip off leaving a 'red strawberry' appearance (Fig. 7.16). Although other diffuse erythemas do not usually have a punctate element, generalized erythroderma (exfoliative dermatitis), usually caused by drug hypersensitivity, and toxic epidermal necrolysis (scalded skin syndrome—see below) should be considered. The erythemas of toxic shock syndrome and Kawasaki disease (see Fig. 13.6e) are usually patchy in distribution.

Investigations

Neutrophil leucocytosis is present initially though eosinophilia is prominent in convalescence. *Strep. pyogenes* is usually cultured readily from the primary site of infection. Streptococcal antibodies subsequently increase.

Management

Penicillin is given, sometimes parenterally initially. Erythromycin is effective in those hypersensitive to penicillin.

Prognosis

Recovery is usually uneventful and the severe cardiotoxic effects seen earlier this century are now rare. Desquamation of the thicker epidermis of palms and soles is common 10–14 days after the initial erythema, but nowadays is unusual on other areas. As after other streptococcal infections, the varied hypersensitivity syndromes of rheumatic fever, glomerulonephritis, erythema nodosum, or Henoch–Schönlein purpura may follow.

Scalded skin syndromes

Phage group II *Staph. aureus* may produce an exotoxin causing exfoliation of outer layers of the epidermis. In the neonate this is known as Ritter's disease, usually with abrupt onset of generalized erythema in skinfolds and a coarse feel to the skin. Within two days Nikolsky's sign appears: wrinkling and separation of upper layers of skin caused by the trauma of light stroking. Large sheets of skin separated by fluid may peel off leaving a raw, tender surface. These areas dry, but show further desquamation as the rash fades over the next few days. Although the initial site of infection may have been minor and localized, staphylococci readily colonize affected skin and can easily be isolated. In the older child this syndrome is known as *toxic epidermal necrolysis* (Lyell's disease, Fig. 7.17). Some forms of staphylococcal impetigo (Chapter 3) are bullous and phage II types are often responsible.

Full-blown exfoliation may not follow the development of generalized erythema which is called *staphylococcal scarlet fever*. However, later there is marked desquamation over the *whole body*, which is unusual in streptococcal scarlet fever of which the typical enanthem is absent. The rash of toxic shock

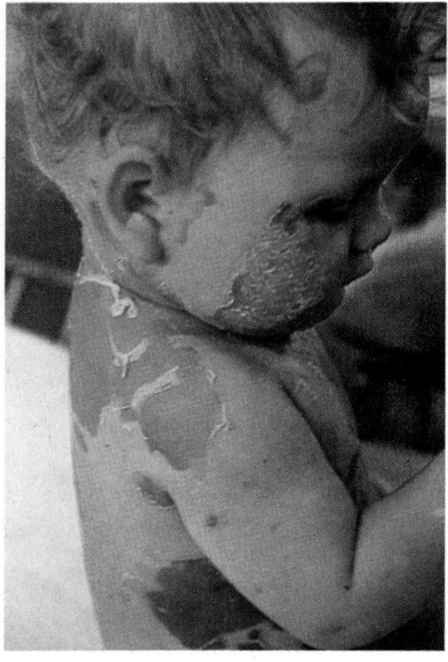

Fig. 7.17 Toxic epidermal necrolysis. Exfoliation results in large areas of denuded skin.

syndrome (Chapter 8) is similar but usually patchy in distribution a subsequently desquamates. A red strawberry tongue may be present.

VESICULAR AND PUSTULAR ERUPTIONS

Varicella

Aetiology and epidemiology

Varicella is caused by Herpes varicella-zoster virus (VZ). Highly infectious, it is transmitted by the airborne route, mainly from the few oral lesions of acute cases, but even more readily if pneumonitis develops. Although early skin lesions of varicella contain live virus, this is not readily disseminated. Like all herpes viruses, VZ can remain latent. It remains in sensory ganglia until reactivated as shingles (herpes zoster) (Chapter 3). Like varicella, the skin lesions of shingles are not very infectious. Varicella is mainly an infection of childhood although in adults it is relatively commoner in tropical countries. In developed countries, transmission persists throughout the year with peaks in spring and early summer.

Clinical features

Illness is usually mild but more severe in adults than children. Although there may be a short prodrome of fever and malaise, rash is often the first evidence of illness. It is pleomorphic with lesions at many stages of development from macule to papule to vesicle to pustule (Fig. 7.18). Most typical of the early lesions are the vesicles, which are very superficial and thin-walled, looking like water drops on the skin ('glass pox'). The distribution is centripetal, densest on trunk and face, decreasing peripherally along limbs. The rash may be sparse or dense. Fever accompanies development of the rash together with malaise and aches in the limbs in adults; illness may be trivial in young children. Fresh lesions may appear in crops for up to a week. Individual lesions usually crust within 3–4 days and are often itchy. Unusual distributions may develop, with more lesions in areas of skin inflamed during the incubation period from incidental causes like sunburn or near a subcutaneous abscess (Fig. 7.19). The crusts eventually drop off without scarring unless traumatized or secondarily infected. A few oral lesions are usually present on the hard palate (Fig. 7.20), tongue, or gums. These may be vesicular or have a raised white appearance, but often the roofs of vesicles are traumatized and shallow ulcers remain. In immunocompromised patients, skin lesions are larger, deeper, and often coalescent, evolve more slowly and may be accompanied by high, swinging fever for 5–10 days. Oral lesions too may be increased in number and size.

Investigation

Clinical diagnosis of varicella is usually easy and laboratory confirmation rarely required. Herpesvirus particles can rapidly be detected by electron microscopy of

Table 7.7. *Common causes of generalized vesicular/pustular/crusting rashes*

Agent/syndrome	Diagnostic pointers
Varicella	Pleomorphic and superficial. Crops. Central distribution. Some oral vesicles or ulcers on tongue or palate.
Kaposi's varicelliform eruption	Distribution that of eczematous skin. Primary source in patient or contact.
Hand, foot, and mouth disease	Lesions hands/feet and buttocks. Oral ulcers. Little systemic upset. (Chapter 4)
Stevens–Johnson syndrome	Vesicles develop in middle of target area in more severe cases with marked mucocutaneous involvement. (Chapter 13)
Gonococcal septicaemia	Sparse pustular rash on peripheries, with haemorrhages into pustule. Arthritis. Young adults. (Chapter 8)
Pseudomonal septicaemia (ecthyma gangrenosa) (see Fig. 12.4)	Immunocompromised. Lesions initially haemorrhagic and papular, later central necrosis, sometimes suppuration, then crusting.
Pseudomonal folliculitis	Use of communal tub or whirlpool bath 2–3 days earlier. Rash mainly on trunk. Mainly folliculitis but often small pustules.

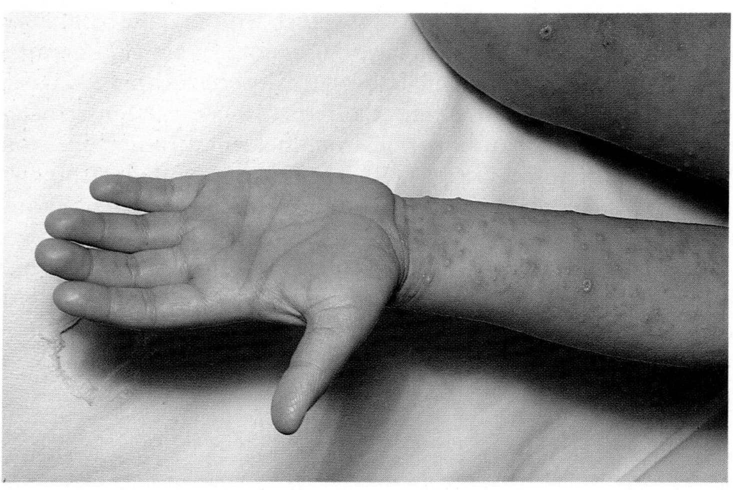

Fig. 7.18 Varicella. Pleomorphic rash.

(a)

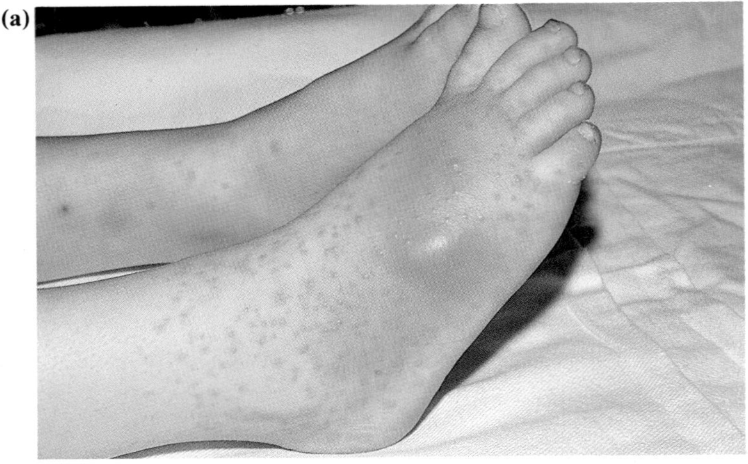

(b)

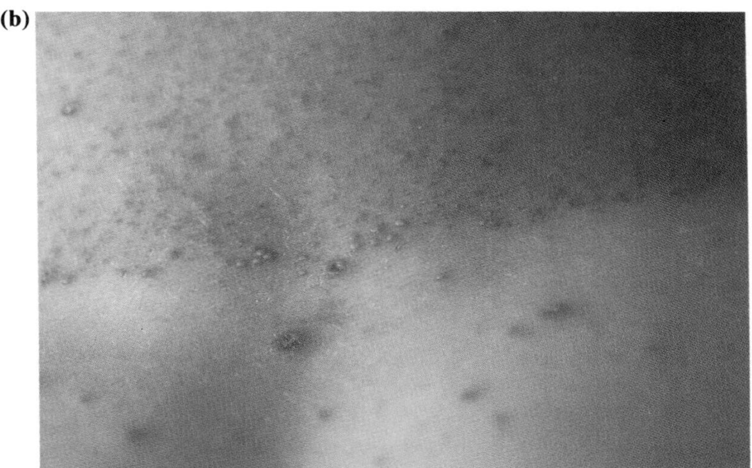

Fig. 7.19 Varicella. Dense rash: **(a)** around subcutaneous abscess that developed in the incubation period after standing on nail; **(b)** in sunburnt area.

vesicular fluid although this cannot distinguish VZ from herpes simplex. Antibodies may be measured, but a rise in titre may not be seen in immunocompromised patients. As there may be diagnostic difficulty in these patients, lesions should be cultured for virus.

Management and complications

Treatment is symptomatic, such as calamine lotion to relieve itchiness. Specific antiviral therapy with acyclovir is reserved for immunocompromised patents and severe viral complications such as pneumonitis. Given within five days of the

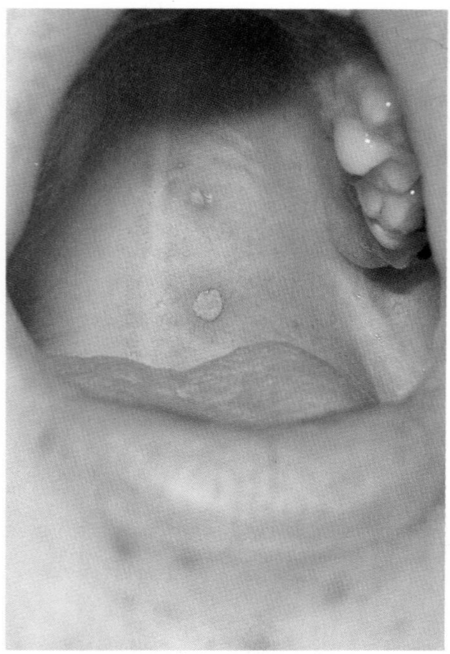

Fig. 7.20 Varicella. Oral lesions in patient with normal immune function.

rash, acyclovir speeds resolution and limits new lesions. Rarely, varicella interstitial pneumonitis develops, particularly in those immunocompromised, partially compromised (as in pregnancy), or occasionally those with apparently normal immunological defences. It is commoner in smokers. Chest auscultation may be normal but chest X-rays reveal soft nodular shadows (see Fig. 11.8). The respiratory rate increases and blood gases show deteriorating gas exchange. It may settle at any stage but can progress to require positive pressure ventilation. Widespread pulmonary fibrosis may still follow and death ensue.

Purpura due to thrombocytopenia is a rare complication, as is haemorrhagic varicella in which disseminated intravascular coagulation develops with high mortality. Varicella skin lesions provide a site of entry for the two common skin invaders, *Staph. aureus* and *Strep. pyogenes*. The former causes a small surrounding zone of erythema with some superficial peeling (Fig. 7.21a), the latter a rapidly spreading erysipelas or cellulitis, (Fig. 7.21b) depending on the depth of infection, scarlet fever or rarely septicaemia. Flucloxacillin or benzylpenicillin should then be prescribed. Post-infectious encephalitis may follow shortly after the rash has stopped appearing. In about half the cases this involves the cerebellum and may present with ataxia. The prognosis is good. Reye's syndrome (Chapter 13) is a rare complication.

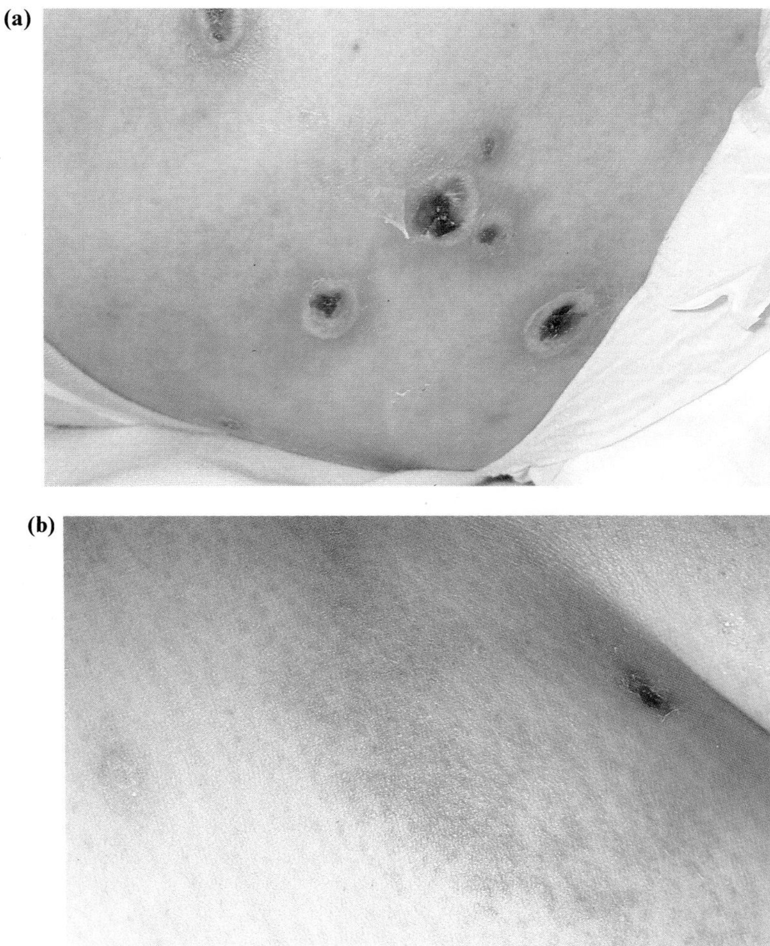

Fig. 7.21 Varicella. Secondary infection of skin lesions: **(a)** *Staph. aureus* in lesions; **(b)** *Strep. pyogenes* spreading from a lesion.

Prevention

Immunocompromised contacts should receive varicella–zoster immune globulin promptly. This can prevent or modify illness if given within four days. It may also be given to pregnant non-immune contacts and neonates (Chapter 11). A live attenuated varicella vaccine has been successfully developed. Its main role will probably be immunizing the immunocompromised; for example, leukaemic children in remission. It is not yet available in Britain.

(a)

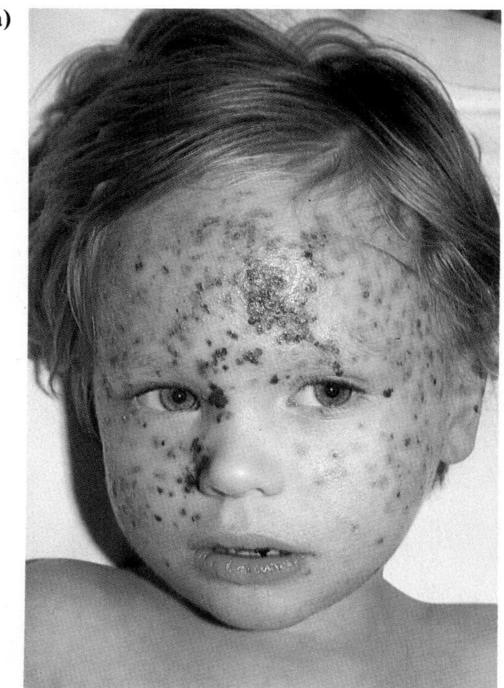

(b)

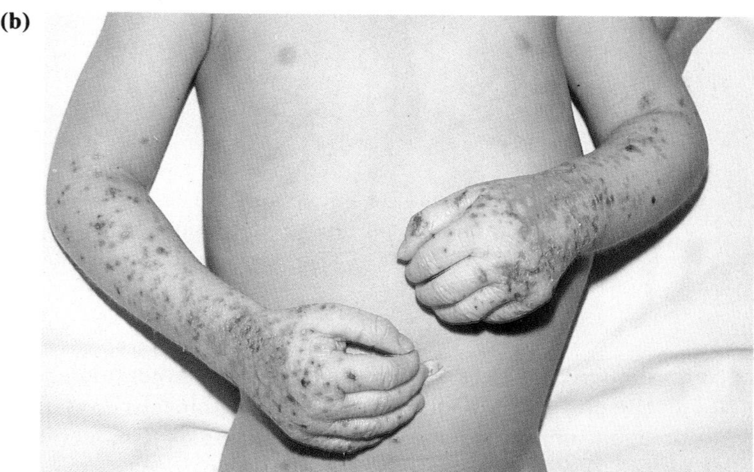

Fig. 7.22 Kaposi's varicelliform eruption. Herpes simplex and atopic eczema: **(a)** late lesions on face; **(b)** on lower arms.

Kaposi's varicelliform eruption

Some viruses like herpes simplex and vaccinia have a predilection for eczematous skin. Initially localized infections by these viruses in people with active eczematous lesions can spread rapidly in the distribution of the underlying eczema (Fig. 7.22). Illness may then be severe and febrile, even life-threatening in young children.

8

Septicaemia and shock

The incidence of septicaemia and shock among patients in hospital has been steadily rising for reasons which include increases in the numbers of susceptible patients (elderly or with chronic illnesses), immunosuppressive therapy, and invasive procedures. Infection in the bloodstream can rapidly spread through the body and, septicaemia in the community may complicate common conditions such as salmonella, gonococcal or pneumococcal infections. Fortunately, potential pathogens are normally removed quickly by the immune system. *Bacteraemia* (viable bacteria in the blood) is often transient (lasting only minutes) causing no symptoms as after tooth extraction or instrumentation (catheterization, endoscopy). *Septicaemia* is when clinical symptoms are caused by multiplication of bacteria in the blood, or by sustained entry of large numbers of organisms into the blood from a focus of infection. Inadequately treated patients with septicaemia often develop '*septic shock*', a condition of circulatory insufficiency and widespread damage to all body systems.

SEPTICAEMIA

Aetiology and epidemiology

Table 8.1 lists conditions associated with bacteraemia, which may progress to septicaemia. The routes of infection are mainly through skin or mucous membranes after surgery and invasive procedures or by extension of localized infections. Development of septicaemia depends on many host factors which influence immune status (Fig. 8.1). Septicaemia may result from conditions in Table 8.1. The most frequent causative organisms are shown in Table 8.2. Septicaemia can be an important feature of infections with *Strep. pneumoniae*, *N. meningitidis*, *S. typhi*, and other salmonellae, anthrax, plague and tularaemia. Unusual organisms and fungi may cause septicaemi in patients with immunosuppression (Chapter 12) or on parenteral nutrition.

Septicaemia is mainly diagnosed in hospital: about 1 per cent of in-patients are admitted with or develop it. Neonates, pregnant women, the elderly, and patients with underlying diseases such as diabetes, cirrhosis, and malignancy are particularly susceptible. A small proportion of febrile children with non-specific signs have bacteraemia, usually due to *H. influenzae*, *Strep. pneumoniae*, or *N.*

Table 8.1. *Conditions associated with bacteraemia*

Condition	Usual organism
Cuts, grazes, boils, chickenpox lesions	*Strep. pyogenes/Staph. aureus*
Dental extraction	*Strep. viridans*
Intravenous drug abuse	*Staph. aureus*
Intravascular catheters	*Staph. aureus*
After abdominal surgery or urological manipulation (e.g. urinary catheterization)	Enterobacteria
Extension of localized infection (e.g. pneumonia, cholecystitis, pyelonephritis)	Dependent on site of infection

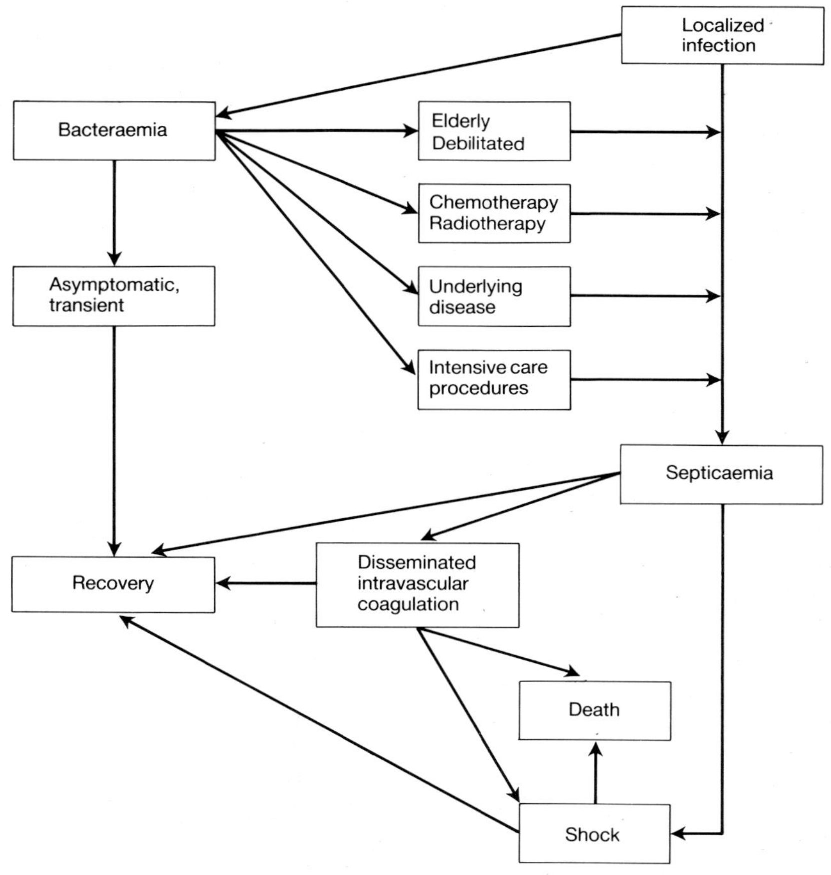

Fig. 8.1 Infection, septicaemia, and shock.

Table 8.2. *Commonest bacteria causing septicaemia in Britain*

1. *Escherichia coli*
2. *Streptococcus pneumoniae*
3. *Staphylococcus aureus*
4. *Staphylococcus epidermidis*

meningitidis. About a third resolve spontaneously, but whenever found, bacteraemia should be treated. Some rare but virulent infections often produce septicaemia.

Clinical presentation

Septicaemia is classically heralded by rigors, and intermittent fever which may be spiking. Malaise, mental confusion, nausea, vomiting, diarrhoea, and headache are common. At presentation, infection may already be widely disseminated and can provide clinical clues to the cause of septicaemia. A generalized haemorrhagic rash suggests *Neisseria meningitidis* (Chapter 6). Splinter haemorrhages and Osler's nodes suggest bacterial endocarditis. Abscesses and osteomyelitis are characteristic of *Staphylococcus aureus*, and abscesses should always be sought in *Strep. milleri* and *Fusobacterium necrophorum* septicaemias. Septicaemia commonly accompanies meningoccal meningitis and pneumococcal pneumonia. Jaundice may result from haemolysis, especially in clostridial septicaemias. Symptoms and signs of the initial, localized infection may also be present; for example, cholecystitis, pyelonephritis, arthritis, intravenous lines. A splenectomy scar should be sought as patients whose spleens have been removed are particularly prone to *Strep. pneumoniae* infection. However, in the neonate, elderly or severely ill patient, clinical findings may be minimal; for example, poor feeding in the neonate, confusion in the elderly, or lack of fever in severely ill patients.

Investigation

Blood cultures are essential. If possible, suspected primary sites of infection should also be sampled for bacterial culture. A Gram stain of buffy-coat preparations, samples from skin lesions or removed catheters may demonstrate the responsible organism. There is usually neutrophilia with left shift and toxic granulation. Detection of endotoxin (Limulus lysate test) is available in some laboratories.

Management

Choice of antibiotic depends on the clinical and available laboratory findings and is influenced by any concurrent liver and kidney impairment. If the likely causative organism is suggested by the clinical picture or recent laboratory reports, give specific antibiotics. Otherwise, treat with broad spectrum antibiotics in combination. Septic shock may need treatment (see below), and surgical drainage may be required for any abscess.

Prognosis

Outcome depends on the virulence of infection, the underlying disease and the immune competence of the patient. Response to appropriate early treatment is usually favourable, especially if there is no malignant disease. Nevertheless overall mortality is at least a third.

Streptococcus pneumoniae septicaemia. Pneumococcal septicaemia is common, usually complicating pneumonia (Chapter 2). Sinusitis and otitis media are also sources of infection. Alcoholics, those with AIDS and patients without a spleen are particularly prone to speticaemia and prophylactic pneumococcal vaccine should be given *before* splenectomy. Treat with benzylpenicillin or erythromycin. Penicillin resistance is rare at present.

Staphylococcal septicaemia. Staphylococci are normal commensals of skin and easily contaminate blood being withdrawn for culture. In genuine staphylococcal septicaemia the source of infection is usually obvious (Table 8.1; Fig. 8.2). Older patients, intravenous drug abusers and those with diabetes mellitus or skin diseases such as eczema are particularly prone to staphylococcal septicaemia. Complications include endocarditis, pneumonia, lung abscess, osteomyelitis, and meningitis. The commom organism is *Staph. aureus* and most British strains resist penicillin. Two antibiotics, usually including cloxacillin or flucloxacillin

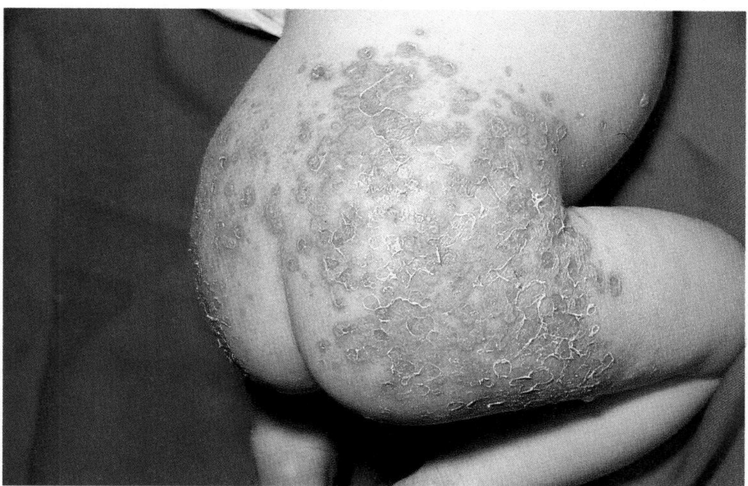

Fig. 8.2 Staphylococcal infection of the skin in a neonate.

should be given for at least six weeks in case of endocarditis. Combined fusidic acid and lincomycin give good bone penetration in osteomyelitis. Increasing prevalence of methicillin/flucloxacillin resistant strains of *Staph. aureus* may necessitate use of quinolones, vancomycin, or penems.

Streptococcus pyogenes septicaemia. Most patients have infection of the skin or subcutaneous tissues (Table 8.1; Figs 8.3 and 8.4) and some also underlying disease such as lymphoma. A scarlatiniform rash can occur. Treat with benzylpenicillin or erythromycin.

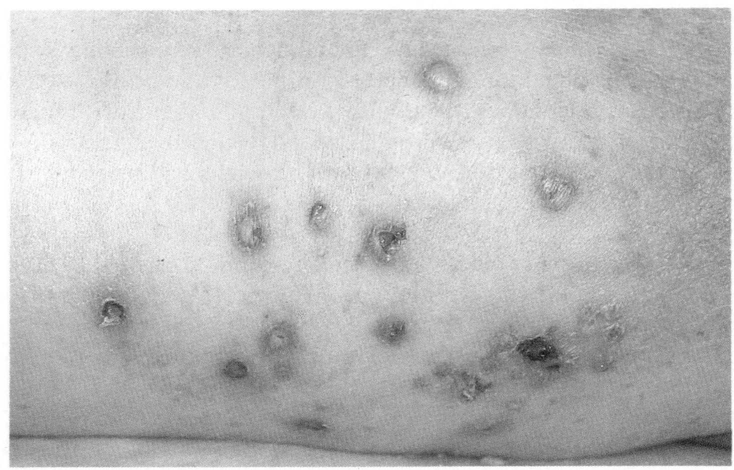

Fig. 8.3 Embolic staphylococcal skin lesions in septicaemia.

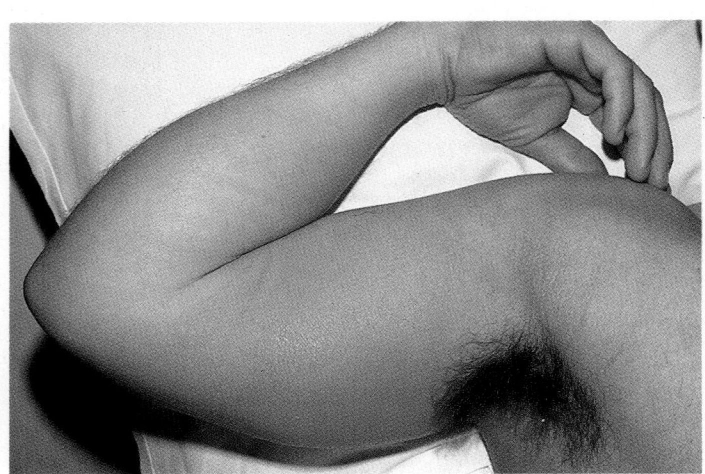

Fig. 8.4 Streptococcal cellulitis.

Enteric Gram-negative septicaemia. This important group of pathogens includes *Escherichia coli* (commonest) and *Klebsiella*, *Enterobacter*, *Bacteroides*, *Proteus*, and *Pseudomonas* spp. These commensals of the bowel frequently cause intra-abdominal infection, especially in the urinary tract and after surgery. Complications include abscesses and rarely osteomyelitis or endocarditis. These patients often present in septic shock.

Neisseria meningitidis septicaemia. Acute meningococcal septicaemia and meningitis (Chapter 6) are often associated, sometimes in epidemics. Fulminating septicaemia (Waterhouse–Friedrichsen syndrome) has a high mortality. Delay in treatment, usually with benzylpenicillin, may result in endocarditis, meningitis, disseminated intravascular coagulation, or nephritis. Chronic meningococcal septicaemia is rare, and may last many weeks. Symptoms are relapsing fever, a maculopapular or petechial rash and arthralgia. Splenomegaly is found in a fifth of patients.

Neisseria gonorrhoeae septicaemia. About 1 per cent of adults with gonococcal infection develop septicaemia, with fever, pustular (Fig. 8.5) or petechial skin lesions on extremities (especially fingers), and arthralgia especially in large joints.

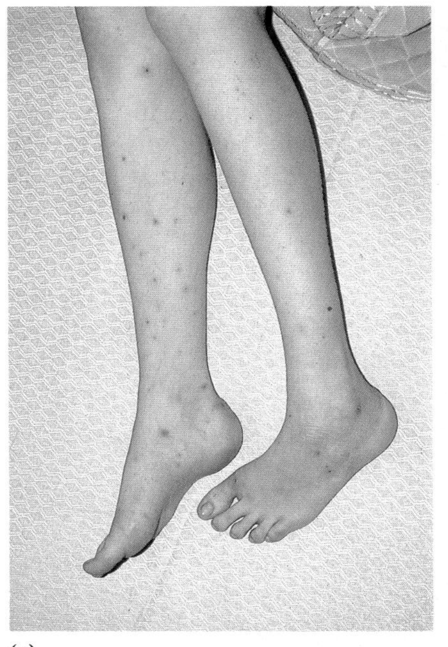

(a)

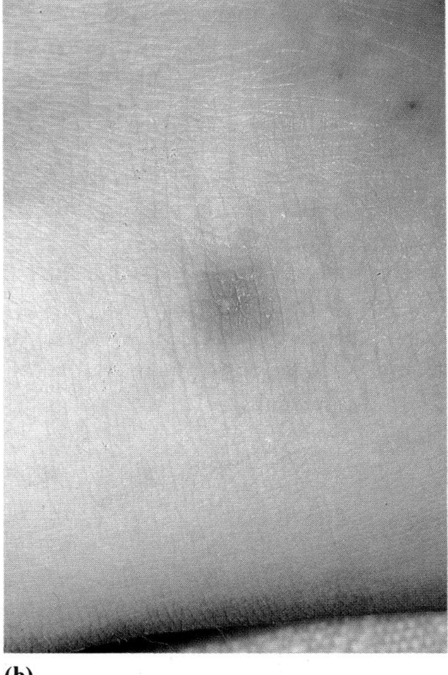

(b)

Fig. 8.5 Gonococcal skin lesions resulting from septicaemia: **(a)** scattered lesions on legs; **(b)** close-up of pustule.

Without treatment, septic arthritis or endocarditis may ensue. Perihepatitis (Fitz–Hugh–Curtis syndrome) can cause right upper quadrant pain. Treat with benzylpenicillin or a cephalosporin.

Salmonella (non-typhoid) septicaemia. These patients usually but not always have diarrhoea (Chapter 4). The illness may be prolonged and complicated by abscess and focal infections affecting heart, kidneys, bone, and joints. It usually responds to drainage of any abscess and cotrimoxazole, cefotaxime, or ciprofloxacin.

Haemophilus influenzae septicaemia. Children 6–24 months old are particularly liable to this illness, especially those with underlying haematological disease. Associated features include meningitis, pneumonia, epiglottitis, cellulitis (often facial), endocarditis and arthritis. Patients usually are toxic, have high fever and profound leucocytosis. Chloramphenicol or cefotaxime are the drugs of choice.

Plague

This zoonotic bacterial infection by *Yersinia pestis* is widespread mainly in ground rodents in endemic foci in much of the Americas, Africa, and central and south-east Asia. It is transmitted mainly by fleas and can spread to humans exposed to rodents or their fleas. The incubation period is 1–6 days. Cutaneous infections from the flea bites cause acute suppurative lymphadenitis ('bubonic plague') with high fever, often septicaemic and sometimes haemorrhagic ('Black Death'). Violent and rapidly fatal 'pneumonic plaque' is due to lung involvement from inhalation of airborne bacilli from infected or dead rodents, flea faeces or respiratory droplets from a pneumonic human case, or sometimes secondary to septicaemia. Human fleas also may transmit infection from person-to-person.

For treatment streptomycin is the drug of choice, alternatively chloramphenicol or tetracycline. Because of the high mortality and potential for epidemics plague is universally notifiable under International Health Regulations. Quarantine and rodent control measures are applied to prevent the spread of plague.

Tularaemia

This infection by *Franciscella tularensis* is also a zoonosis, mainly involving rabbits and hares in areas of North American and Eurasia. It can be acquired by the bite of certain ticks, deerfly, or commonly by hunters skinning or cutting up the animals and infected via cuts on hands, splashes or aerosol into the eyes, or inhalation. Infection commonly causes skin ulceration and involvement of regional lymph nodes ('ulceroglandular' form), or septicaemia, pneumonia or both—severe, dangerous fevers resembling plague. Streptomycin is the drug of choice.

Melioidosis

This is a rare infection by *Pseudomonas pseudomallei*, a saprophytic bacillus in certain soils and waters mainly in south-east Asia causing suppurative diseases which often resemble pulmonary tuberculosis. Co-trimoxazole is the drug of choice.

SEPTIC SHOCK

Aetiology

Septic shock results from inadequate tissue perfusion and is usually caused by Gram-negative septicaemia (especially with *E. coli* or *Pseudomonas*) but may follow Gram-positive septicaemia (pneumococcal, streptococcal). In healthy young people who have travelled abroad, causes of septicaemia rare in Britain such as anthrax (Chapter 3), plague, tularaemia, and melioidosis should be considered. Shock is precipitated by various factors, especially release of bacterial endotoxin into the bloodstream, and there need not be concomitant septicaemia.

Clinical presentation

In the early stages, vasodilation reduces peripheral vascular resistance. To compensate, cardiac output increases. The patient has warm extremities, a bounding pulse, tachycardia and hypotension; confusion is common, and there may be fever. In later stages vasoconstriction increases peripheral vascular resistance. Cardiac output and central venous pressure are reduced. The extremities become cold, moist, cyanotic, and urinary output is diminished. Tachycardia, severe hypoxia, and metabolic (lactic) acidosis develop, and stupor and coma quickly follow. Respiratory failure, cardiac failure, renal failure, and disseminated intravascular coagulation can all contribute to death.

Investigation

Results of investigations in patients with septic shock vary according to the cause and the stage of the shock, especially if complications have developed. Fortunately, diagnosis is not difficult in a patient with pyrexia, rigors, hypotension, and an obvious source of infection. However, fever may be absent in elderly and immunosuppressed patients with severe infection. Mental confusion and hyperpnoea with normal chest X-rays are useful indicators of septic shock in these patients. Conditions that can mimic it include myocardial infarction, pulmonary embolism, and silent haemorrhage. Laboratory investigations are less useful for diagnosis than for monitoring progress of shock, but they help to identify the cause.

Management

Septic shock is a medical emergency. Therapy aims first to maintain life with supportive care and then to deal with the cause of shock. For anoxia, respiratory support is best given early. Positive pressure ventilation treats both anoxia and pulmonary oedema. Blood volume replacement monitored by central venous pressure is carried out at the same time. Plasma protein fraction is the fluid of choice as it is not lost from circulation as quickly as electrolyte solutions, followed by whole blood if the patient is anaemic. Coagulation defects may require correction, and clotting factor supplements are usually given. Aggressive but monitored fluid replacement should not be withheld because of oliguria in a patient with hypotension. Vasoactive drugs are required for persistent hypotension: dopamine and dobutamine are drugs of choice as they increase cardiac output, blood pressure, and renal blood flow.

Antibiotics should be started as soon as blood cultures and other relevant samples have been taken. Choice of antibiotic at this stage depends on whether a possible site of infection is detected: if not, multiple wide-spectrum therapy is indicated. Patients with a focus of infection such as an abscess, necrotic bowel, obstructed gall-bladder or infected uterus, require surgical intervention under appropriate antibiotic cover.

The role of corticosteroids is controversial, but they may be helpful if given early. Anti-endotoxin therapy for Gram-negative septicaemia may also be of benefit.

Prognosis

The high mortality of septic shock (50 per cent) is mainly due to failure to recognize and treat the condition early enough. Since most patients are in hospital before onset of septic shock and since the sources of infection are so often venous or urinary catheters, there is scope for prevention. The prognosis is improved if the immune response is adequate, if antimicrobial treatment is effective, if there is no underlying disease and if complications are not severe.

Disseminated intravascular coagulation

Aetiology

Disseminated intravascular coagulation (DIC) results from activation of the haemostatic system in response to injury. Vascular injury can activate both coagulation and fibrinolytic systems simultaneously (Fig. 8.6). In severe cases, these processes result in consumption of coagulation factors ('defibrination' or 'consumption coagulopathy') producing clinical DIC. DIC can be caused by many unrelated illnesses (Table 8.3). Most cases are associated with sepsis, in

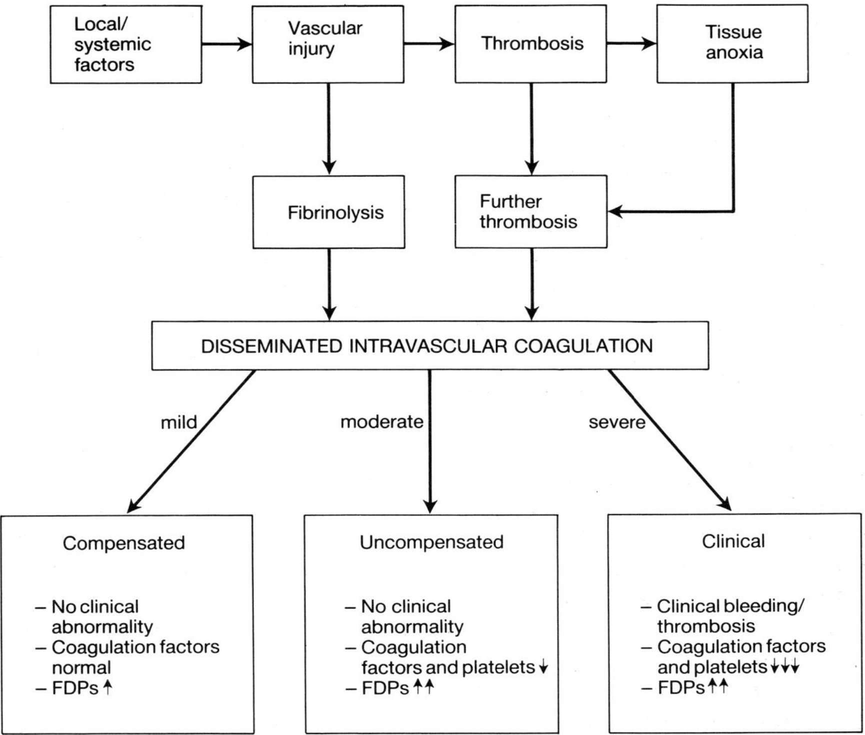

Fig. 8.6 Disseminated intravascular coagulation.

Table 8.3. *Disorders associated with disseminated intravascular coagulation*

Local factors
Vascular (massive thrombosis, pulmonary embolism)
Tissue damage (burns, obstetric complications)

Systemic factors
Shock (hypovolaemic, cardiogenic)
Infections (Gram-negative and Gram-positive septicaemia, some viraemias)
Immunological (anaphylaxis)
Malignancy
Liver disease
Miscellaneous (snake venoms, hypoxia, heat exhaustion)

particular Gram-negative (especially meningococcal) and staphylococcal septicaemia, obstetric complications and malignancy.

Epidemiology

Disseminated intravascular coagulation is not a disease itself but an exaggeration of the normal defence response to injury—most ill patients have mild degrees of DIC which are not detectable clinically. With severe underlying disease, DIC is part of the terminal process.

Clinical presentation

The paradox of simultaneous thrombosis and bleeding produces a clinical picture ranging from asymptomatic to fulminant disease. Bleeding is the commonest sign (90 per cent of severe cases) usually into skin and mucous membranes. Petechiae and purpura are common but there can also be frank bleeding with epistaxis, haematemesis, melaena, or vaginal haemorrhage. Oozing from venepuncture sites is an early sign. Major thrombosis affects about 25 per cent of cases, but there are probably microthrombi in the tissues of most patients. Renal lesions especially are common in DIC, reversible in the early stages. Gangrene of toes and fingers is rare except in the Moschowitz syndrome (thrombotic thrombocytopenic purpura).

Investigation

In severe cases diagnosis is easy since the blood does not clot in a container without anticoagulant. Milder cases can be difficult to recognize. The most sensitive tests are: elevated fibrinogen degradation products due to activation of the fibrinolytic system, positive ethanol gelation test (or protamine sulphate test), prolonged thrombin time, and reduced platelet count because of platelet consumption at sites of vascular injury. The kaolin cephalin time (partial thromboplastin time) may be prolonged and fragmented red cells (schistocytes, Fig. 8.7) may be seen. Buffy-coat preparations may reveal giant nuclear masses (probably endothelial cell nuclei).

Management

It is most important to deal with the underlying condition, usually septicaemia. Supportive measures are also required to reverse shock, acidosis, renal failure, and hypoxia. Many severely ill patients require artificial ventilation. Replacement of coagulation factors by fresh frozen plasma and platelets is necessary in severe bleeding. Antithrombin III concentrates have had encouraging results in severe cases, but antifibrinolytic agents are not indicated.

More specific therapy is rarely indicated. Heparin is used when peripheral gangrene, deep vein thrombosis, pulmonary emboli, or organ damage from thrombosis is suspected. It is given by intravenous infusion with regular testing of the thrombin time with and without protamine sulphate. Without these

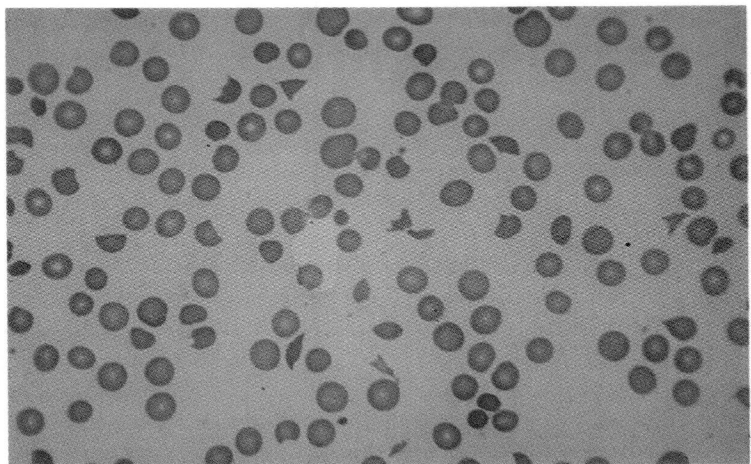

Fig. 8.7 Fragmented red cells in disseminated intravascular coagulation.

indications, heparin is probably of no benefit, but it may have a role in preparing a patient for surgery and when bleeding cannot be controlled by other means.

Prognosis

In severe disease, the mortality is 50 per cent. Mortality is lower with obstetric complications but high in septicaemia, especially with underlying malignancy. Mortality also depends on the expertise of supportive care.

Shock lung

Shock lung (adult respiratory distress syndrome) is a complication of septic shock and of malignant malaria characterized by pulmonary congestion and oedema in the presence of normal left atrial and pulmonary venous pressures. The oedema fluid has a high protein content and results from damage to pulmonary capillaries. The patient presents with progressive dyspnoea, cyanosis, and deepening levels of unconsciousness. Chest X-ray shows pulmonary oedema with normal size heart; blood gases show reduced oxygen with normal carbon dioxide. Treatment aims to remove the infection by antimicrobial therapy and reverse the hypoxaemia for which positive end-expiratory pressure ventilation may be necessary.

Toxic shock syndrome

Aetiology and Epidemiology

This syndrome was initially recognized in menstruating females who used tampons. *Staphylococcus aureus* toxins cause this illness. Exactly how tampons promote toxaemia is unclear but frequent changing or intermittent use of lower absorbency tampons reduces the risk. Most cases still involve women under 30 years of age but since 1981 the number of cases has significantly fallen. Non-menstrual toxic shock syndrome has proportionately increased, can develop in either sex at any age, and may result from staphylococcal sepsis at any site, for instance a surgical wound, burn, sinusitis, or pneumonia.

Clinical presentation

Patients present with protean signs and symptoms, some of them mild. For a diagnosis of toxic shock syndrome, certain criteria are necessary: patients should have pyrexia, a macular erythematous rash like sunburn which desquamates after 1–2 weeks and particularly affects the palms (Fig. 8.8) and soles (Fig. 8.9), and hypotension (systolic blood pressure <90 mmHg in adults) or orthostatic hypotension. Also, three or more organ systems should be involved: gastrointestinal (vomiting, diarrhoea); muscular (myalgia); mucous membranes (vaginal, conjunctival hyperaemia); renal in absence of urinary tract infection; hepatic; and central nervous system (disorientation or alterations in consciousness).

Investigation

Diagnosis depends on exclusion of similar presenting illnesses. Bacterial and virological cultures from throat, blood, and CSF should be negative (except for *Staph. aureus*), and tests should be negative for leptospirosis, measles, rubella, meningococcal infection, scarlet fever, and rickettsial spotted fevers. Kawasaki disease (Chapter 13) can have a similar presentation, but usually affects under-5-year-olds with prominent cervical lymphadenopathy in addition.

Patients should be examined for a focus of staphylococcal infection. Isolates

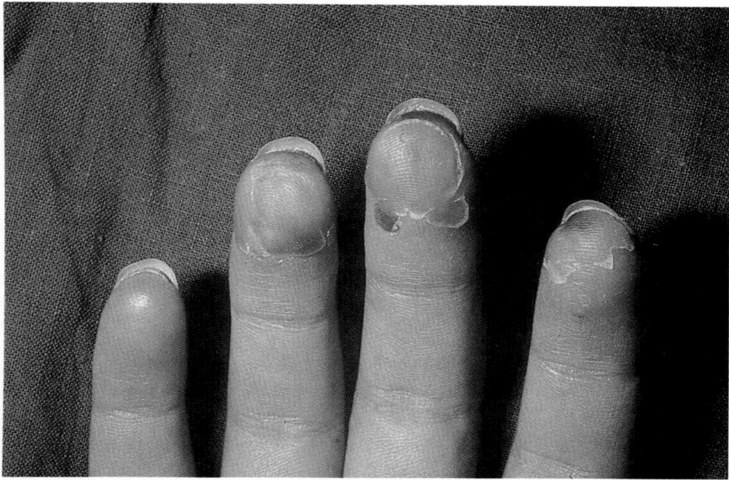

Fig. 8.8 Toxic shock syndrome with vasculitis and rash peeling from finger tips.

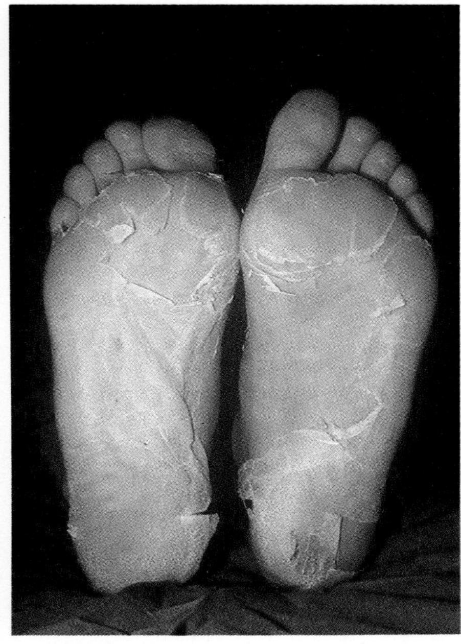

Fig. 8.9 Toxic shock syndrome with rash peeling from the soles.

from vagina, cervix, nose, blood, urine, or stool should be examined for phage type and toxin production.

Management

Treat as for septic shock. Remove tampons. Antibiotic therapy against *Staph. aureus* is indicated, but may not greatly help. Drain any absecess. Patients should change tampons frequently and avoid overnight usage.

Prognosis

The prognosis is generally good, especially if effective supportive care is available. Up to a third of patients may have another attack, but later episodes tend to be less severe.

INFECTIVE ENDOCARDITIS

Aetiology

Infective endocarditis involves heart valves and/or mural endocardium. Platelet-fibrin thrombi form on damaged endocardium, classically at the low-pressure side of an orifice. Organisms attach to the platelet-fibrin thrombi and become protected from the immune system by further thrombi deposited over them. Bacteria are responsible for 85 per cent of cases, streptococci in 70 per cent (usually *Strep. viridans*). The next most frequently encountered pathogen is *Staph. aureus*, particulary common in drug addicts. Non-bacterial endocarditis is a diverse group caused by fungae (for example, *Candida albicans*), rickettsiae (*Coxiella burnetii*), *Chlamydia* (for example, *C. psittaci*) and viruses (for example, coxsackie B), which are undetected by bacteriological culture.

Epidemiology

Although rare (about 0.1 per cent of hospital admissions), this disease causes considerable mortality and morbidity among those affected. Most patients have underlying, rheumatic, or congenital heart diesease or have had replacement cardiac surgery. Compared to the pre-antibiotic era, the mean age (50 years) has increased as fewer patients now have rheumatic heart disease, and intravenous drug abusers are as yet only a small group. The diagnosis is often missed in the elderly.

Clinical presentation

Onset is usually sudden with fever, chills, arthralgia, and myalgia. It can also be gradual with weakness, fatigue, pyrexia, night sweats, and weight loss. There may be a preceding precipitating event such as dental work or recent intravenous drug injection. Cardinal signs of bacterial endocarditis are pyrexia and heart murmur and the diagnosis is in doubt if one of these signs is absent. It should be considered whenever a patient develops both signs.

Most patients have a heart murmur involving the mitral or aortic valve. The increasing number of patients with tricupsid valve involvement reflects the increase in the number of drug addicts developing endocarditis. Endocarditis without a murmur can affect patients with acute disease of previously healthy valves, those with mainly endocardial involvement and some elderly patients.

As disease progresses, fragments of infected vegetations break off as emboli which may cause vascular occlusion of skin, bowel, kidney, heart, limb and cerebral arteries. Drug addicts with right-sided endocarditis often present with pneumonia from infected pulmonary emboli. Changing shape of vegetations as emboli fragments become detached can cause changing murmurs; valve obstruction or rupture can cause sudden heart failure.

Further progression produces additional signs, many of them immunologically mediated. Conjuctival and subungual haemorrahges, flame shaped retinal haemorrhages (Roth spots), haematuria, nephrotic syndrome, finger clubbing, Osler's nodes (painful cutaneous nodules), and Janeway's spots (non-tender subcutaneous, maculopapular lesions in the pulp of fingers) were common in the pre-antibiotic era but only splenomegaly is as common now as then. In non-bacterial endocarditis clinical findings are similar but less florid.

Investigation

Non-specific findings are elevated erythrocyte sedimentation rate, C-reactive protein, anaemia of severity related to the length of the disease, thrombocytopenia in patients with splenomegaly, reduced complement, and the presence of rheumatoid factor in over half of those with prolonged disease.

Repeated blood cultures are essential to confirm septicaemia which may be intermittent. In acute cases, four sets of blood culture can be taken in one hour at 15 min intervals before starting treatment. Where treatment is less urgent, six sets of blood cultures are taken over two days. If antibiotics have been given recently, septicaemia may be inhibited for many days and cultures must then be taken over a longer period. Echocardiography is useful in demonstrating valve vegetations, but negative results do not exclude endocarditis.

When blood cultures are negative one should suspect slow growing organisms (brucellae, *Haemophilus* spp.) or non-bacterial causes and arrange serological tests for coxiella, chlamydial, and viral antibodies. When acute viral disease is suspected faeces should also be examined for coxsackieviruses.

Management

The laboratory must be consulted to determine the minimal inhibitory concentration (MIC) and minimal bactericidal concentration (MBC) of appropriate antibiotics. Bacteria which are normally susceptible can be surprisingly resistant to treatment when surrounded by fibrin as in endocarditis, and serum levels ten times the MIC are often required. In acute cases therapy should be started immediately after taking blood cultures. In subacute cases, especially if the clinical presentation is unclear, one may often await blood culture results before starting treatment.

In acute cases where treatment is started before bacteriological results are available, intravenous penicillin and gentamicin are indicated. The choice of penicillin depends on the suspected organism: benzylpenicillin for streptococci, flucloxacillin for staphylococci, and ampicillin for faecal streptococci. Patients with a history of penicillin allergy should be given ciprofloxacin or cefotaxime. Other recommended treatments (and alternatives) for infective endocarditis are shown in Table 8.4. Antibiotics are usually given intravenously until clinical response is achieved and then orally for up to six weeks.

Surgical replacement of the affected valve may be required in patients with progressive heart failure, fungal endocarditis, repeated embolic episodes, or recurrent or resistant endocarditis. Efficacy of treatment and the possibility of superinfection need consideration in those with prolonged fever. Reactions to prolonged high-dose antibiotics are not uncommon: pyrexia, rashes, urticarial reactions, and interstitial nephritis. Small doses of corticosteroids and antihistamines may control these side-effects but otherwise the drug needs to be changed.

Prognosis

Appropriate antibiotic therapy gives rapid clinical improvement with very sensitive organisms, but this may take weeks with more resistant ones or where infection is well established. Untreated, recovery is rare; with treatment mortality remains around 30 per cent (50 per cent in those with prosthetic valves). Factors

Table 8.4. *Treatment of infective endocarditis*

Organism	Treatment	Alternative
Penicillin-sensitive streptococci	Benzylpenicillin + gentamicin	Cephalosporin
Penicillin-resistant streptococci	Ampicillin + gentamicin	Vancomycin
Staphylococci	Flucloxacillin + gentamicin, or + fusidic acid	Ciprofloxacin + rifampicin cephalosporin; vancomycin
Haemophilus spp.	Cefotaxime	Chloramphenicol
Fungi	Fluconazole or amphotericin B	–
Coxiella; *Chlamydia*	Tetracycline	Erythromycin

associated with poor prognosis are listed in Table 8.5. Patients with Coxiella infection have higher mortality whereas those with viral endocarditis have better prognosis. The outlook in tricuspid endocarditis in intravenous drug users is generally good.

Heart failure is the commonest cause of death. Renal failure, rupture of mycotic aneurysms, strokes, emboli, and abscess formation in the valves or myocardium also cause considerable morbidity and mortality.

Prevention

In patients with abnormal heart valves (including prosthetic valves), infective endocarditis may be prevented by prophylactic treatment. The recommended measures are: oral amoxycillin, erythromycin or clindamycin one hour before dental surgery or ear-piercing; intramuscular ampicillin and gentamicin before urogenital or abdominal surgery; intravenous gentamicin and fusidic acid before cardiac surgery.

Table 8.5. *Poor prognostic factors in infective endocarditis*

Prosthetic valves	Elderly
Fungal infection	Multiple-valve involvement
Gram-negative bacteria	Multiple bacterial infection
Congestive heart failure	Resistance to treatment
Severe valve destruction	Delay in starting therapy

Prosthetic endocarditis

Infection may be early (within two months of valve replacement) or late (after two months). The former is believed to be due to infection around the time of the operation and has higher mortality than the latter. Patients with infected prosthetic valves have higher mortality than those with normal valves. Unexpected or persistent pyrexia, new murmurs, cardiac failure, or embolic phenomena suggest endocarditis. The microbiology of prosthetic endocarditis mainly shows unusual organisms or skin commensals such as *Staph. epidermidis*, *Staph. aureus*, *Strep. viridans*, diphtheroids, and fungi. Significant valve dysfunction or congestive cardiac failure are indications for valve replacement. Some feel that all infected valves should be replaced. Patients should be given prophylactic antibiotics for any procedure likely to produce a bacteraemia.

9

Fever, tuberculosis, brucellosis, and Q fever

FEVER

Infections often, but not always, cause fever by releasing 'exogenous pyrogens' (toxins and other similar proteins that induce inflammatory reactions). These stimulate polymorphs, monocytes, and macrophages to produce 'endogenous pyrogens' which in turn induce changes in the hypothalamus to affect the thermoregulatory centre and reset the body's 'thermostat' unstably at a higher mean level. Many non-infective processes can also stimulate leucocytes to produce endogenous pyrogen and cause fever. Tuberculosis, brucellosis, and Q fever are described here because they can manifest themselves mainly as fever, often of long duration and sometimes puzzling to diagnose.

Measuring and interpreting changes in body temperature

Body temperature can be measured in various ways, most usually from axillary skin with the thermometer kept in place by the arm held over the chest, or from buccal mucosa with the thermometer under the tongue. The former is more hygienic and safe. The rectal temperature, normally 0.5 °C above axillary temperature, is convenient to measure in very young babies and essential in patients who are hypothermic.

To exclude or monitor fever, single or occasional readings are inadequate and an approximately four-hourly temperature chart, carefully kept, is helpful for managing serious infections. Body temperature in otherwise healthy persons varies throughout the day and can also be affected by exposure to extreme weather conditions, by clothing, activity, surgical and other trauma, and ovulation. Fever can sometimes be caused by antibiotic drugs or suppressed by corticosteroids. 'Factitious fever' is when patients manipulate their temperature recordings; 'fraudulent fever' is real but deliberately induced, for example, by injection of infected material.

Fever may therefore be physiological, pathological, factitious, or fraudulent, but levels above normal usually reflect disease (Table 9.1).

Fever patterns (Fig. 9.1)

Fever may be *continuous* when temperature remains high both day and night with little variation, or *intermittent* when wider swings are interspaced with periods of

Table 9.1. *Factors affecting body temperature*

Raised (fever)
Infection
Other disease, especially neoplasia or connective tissue disorder
Ovulation
Excessive clothing (especially in infants)
'Drug fever', including reactions to vaccines
Blood transfusion
Lowered (hypothermia)
Unprotected exposure to cold (e.g. in the elderly or mountaineers)
Hypothyroidism
Misleadingly 'normal'
Immunosuppression including overwhelming infection, uraemia, corticosteroid therapy, and 'old age'
Bed rest—may resolve fever in tuberculosis

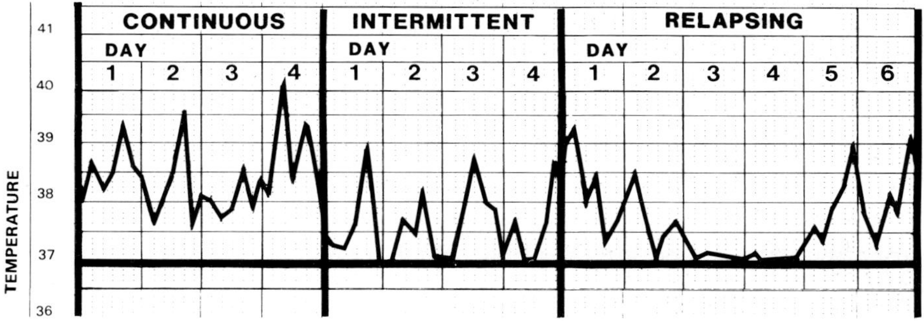

Fig. 9.1 Typical fever patterns.

normal temperature. Febrile episodes separated by long afebrile periods are said to be *relapsing*, as in leptospirosis where the afebrile period may last up to a week, or in chronic meningococcal septicaemia where it is much longer.

Clinical presentation

It is unusual to find no clinical or epidemiological pointers to the cause of fever. Initially, it is wise to presume that infection is responsible. Urgent antimicrobial therapy may be necessary or the risk of cross-infection may warrant isolation, hygienic precautions, or the tracing of an environmental or other source of infection.

Infection is suggested by acute onset of fever, an obvious focus of inflammation or sepsis, or a predisposing cause such as immunosuppression. Epidemiological features pointing to infection include recent overseas travel, recent exposure to ill family members or other human or animal contacts, or onset of fever during an outbreak of, for example, influenza, bronchiolitis, measles, or Legionnaires' disease. At an early stage, dangerous pathogens hazardous to attendants, laboratory staff, or other contacts should be considered—for example, the uncommon Lassa fever after a recent visit to West Africa.

Helpful associated features

In tonsillitis, high fever early in the illness suggests streptococcal disease; more insidious onset suggests infectious mononucleosis. In tuberculosis, fever may resolve with bed rest. In established malaria, rigors are usual. Brief, high spikes of fever are sometimes seen in patients with bedsores, presumably from intermittent pyrogen release. Fever without associated rise in pulse rate is sometimes seen in brucellosis, typhoid, or factitious fever.

Rigors. One should distinguish rigors (very unpleasant, uncontrollable shivering, lasting up to half-an-hour, usually followed by profuse sweating) from 'chills' (brief episodes of shivering). Rigors are associated with a very rapid rise in temperature and the vigorous muscular movements may cause pain and even be confused with an epileptic fit. Rigors especially raise suspicion of malaria, lobar pneumonia, cholecystitis, pyelonephritis, or septicaemia.

Headache. This is common with fever, and usual in some infections such as influenza and enteric fever. Persistent, generalized, or severe headache with fever may indicate meningitis or other intracranial infection. Babies cannot complain verbally of headache but may be irritable or have a pained expression with wrinkled forehead. Localized headache with fever may be due to focal infection such as sinusitis or mastoiditis.

Delirium. Confusion, hallucinations, and nightmares are seen, especially in children but also in adults, more so in certain febrile illnesses such as lobar pneumonia and septicaemia. Abrupt withdrawal of alcohol due to illness in those habituated may also contribute to confusion and delirium.

Vasodilatation. The characteristic flushed face of fever may extend to the whole body. Erythematous rashes, if present, become more prominent. Vasodilatation may also cause hypotension, 'lightheadedness', or fainting on standing up or exertion.

Meningism. This syndrome of headache, stiff neck, and sometimes photophobia, discussed in Chapter 6, can be due to intracranial infection, but in children and young adults is often observed in other febrile conditions such as streptococcal tonsillitis, pneumonia, and urinary tract infections.

Convulsions. Up to about the age of six years, 'febrile convulsions' can occur in measles, shigella dysentery, or other infections, especially viral and sometimes after vaccinations. They are very alarming for parents. In the first year of life they are often atypical, but in older children are usually of grand mal type. Treatment and prevention are based on slowing rapidly rising fever by tepid sponging, fanning, and antipyretics. Anticonvulsants are not usually necessary for one episode but can be used prophylactically if convulsions are recurrent. A child with no previous history of epilepsy or cerebral disease who has a febrile convulsion is prone to recurrence in subsequent fevers during childhood but is not more liable than average to convulsions in later life.

Abdominal pain. In children, abdominal pain is often associated with fever. Alterations of intestinal motility or tender intra-abdominal lymph nodes may be responsible. Other causes of acute abdominal pain such as appendicitis must be excluded.

Clinical examination

Special attention should be paid to the upper respiratory tract (especially middle ear in children) including paranasal sinuses, lungs and lymph nodes. Examination should include spleen for enlargement, skin and buccal mucosa for rashes, abdomen for tenderness, neck for involuntary stiffness, eyes for photophobia, and, in infants, the fontanelle for bulging.

Laboratory investigations (Table 9.2)

Laboratory identification of an infective cause of fever may be difficult and slow. There may be reluctance to seek confirmation of a clinical diagnosis if management is unlikely to be altered. However, delay in initiating tests can make it more difficult or impossible to establish a specific diagnosis later when the causal organism may become undetectable and the pattern of antibodies indistinguishable from the IgG legacy of some previous infection.

Also, surveillance of the natural history, current epidemiological and resistance patterns of organisms usually requires their isolation, and useful guidelines on control and management are only possible when this information is kept up-to-date (Chapter 17).

Management

If the cause of fever is not quickly demonstrated, allowing specific measures to be instituted, several other responses are possible:

(a) *Treat symptoms and observe closely*. This is often appropriate in the not very ill child, who may recover rapidly.

(b) *Keep the patient under constant review* while further results of investigations are awaited. This can be appropriate in the moderately but not dangerously ill patient in whom antimicrobial therapy may mask an important diagnosis.

Table 9.2. *Laboratory investigation of fever*

Investigations that provide information rapidly

Blood for total differential white-cell count, blood film for morphology and malarial parasites.

Urine microscopy for pus cells and organisms.

Chest radiography for pulmonary consolidation.

Throat swab and sputum for organisms, especially corynebacteria and acid-fast bacilli.

Cerebrospinal fluid microscopy for cells, organisms and biochemical analysis (if signs suggest meningitis, or in infants if fever remains unexplained).

Serum amylase (may be raised in mumps).

Immunofluorescent studies on respiratory secretions.

C-reactive protein (particularly raised in bacterial infections)

Investigations that provide information later

Blood, urine, stool, sputum, throat swab and cerebrospinal fluid culture.

Antibody titres to suspected agents including: *Brucella*, *Coxiella burnetii*, *Mycoplasma pneumoniae*, *Toxoplasma gondii*, *Chlamydia*, cytomegalovirus, EB virus.

Biopsy (e.g. bone marrow, lymph node, pleura, spleen, liver) for culture and histology.

(c) *Start empirical therapy* after initial specimens have been taken. This is often necessary in potentially life-threatening circumstances when the choice of treatment is based upon judgement of possible infective causes. Antimicrobial, including antituberculous, therapy, can be used and sometimes the response is taken as confirmatory evidence of a diagnosis. However, antimicrobials are rarely sufficiently specific to identify with certainty which organism(s) they have acted against.

PROLONGED FEBRILE ILLNESS

Pyrexia of unknown origin

This somewhat nebulous term describes febrile illness of unidentified cause that has failed to resolve, despite 'routine' investigations and often empirical antibiotics. It is customary to reserve 'pyrexia of unkown origin' (PUO) for febrile illnesses that have persisted for at least three weeks and are still unexplained after a week of hospital investigations. In practice, after a detailed history and examination it is rare to find no clues to the likely nature of the problem. Febrile illness that remains unexplained for three weeks warrants a systematic approach to further investigations which vary according to what has already been done. Table 9.3 lists some of the possibilities to be considered in addition to those already mentioned. In about one-third of pyrexia of unknown origin infection is found, in one-third neoplasia, and in the remainder collagen-vascular (immunogenic) disease, drug reactions, or granulomatous disease such as sarcoidosis. Two important causes, tuberculosis and brucellosis are described below, while typhoid fever is discussed in Chapter 15. Sometimes no cause is found.

Table 9.3. *Some diseases causing prolonged fever*

Imported infections (Chapter 10)	Any severe debilitating illness (metabolic, endocrine)
Brucellosis	Recurrent pulmonary emboli
Tuberculosis	Drug induced fever
Chronic pyelonephritis	Factitious fever
Bacterial endocarditis	Fraudulent fever
Q fever	Neoplasia, especially (a) lymphoma; (b) leukaemia; (c) deep-seated tumours (e.g. hypernephroma)
Osteomyelitis	Collagen-vascular (immunogenic) disease, e.g. polyarteritis nodosa, cranial arteritis, systemic lupus erythematosus, rheumatoid arthritis
Deep-seated abscess	
Rheumatic fever	
Sarcoidosis	
Inflammatory bowel disease	
Cirrhosis of liver	
Chronic hepatitis	

TUBERCULOSIS

Tuberculosis can involve most organs in the body and manifests itself in many different ways. World-wide it most often presents with fever and pulmonary features. In many countries, especially in Asia and Africa, it is very common, difficult and expensive to treat. The cost of drugs and long duration of treatment encourage partial cures, relapse, and development of drug resistance. In developed countries such as Britain, tuberculosis declined greatly in recent decades but there is a disturbing flow of new and reactivated cases in the indigenous population and among immigrants. The incidence of extrapulmonary disease, which is often more difficult to diagnose, has not decreased. The proportion of imported cases presenting with non-pulmonary features is greater when health screening procedures before emigration include chest radiography to detect pulmonary diseases requiring treatment before leaving the home country.

Aetiology

Many strains of mycobacteria cause human disease, but most commonly *Mycobacterium tuberculosis*. *M. bovis*, usually contracted from raw milk, is especially important where cattle are not routinely screened for this infection. 'Atypical' mycobacteria such as *M. avium-intracellulare* can cause disease which is usually localized, for example in cervical lymph nodes or skin.

Epidemiology

Currently, the incidence of tuberculosis is highest in lower socio-economic groups, in developing countries and in non-white ethnic groups. Quiescent tuberculosis can be activated by immune suppression by, for instance, corticosteroids or other immunosuppressive drugs, or by infection by such viruses as measles or HIV. Tuberculosis is now one of the commonest presentations of AIDS, especially in Africa and Asia, a serious threat since it is estimated that about one-third of the world population carries tuberculous infection. In immunocompromised persons atypical mycobacteria, including BCG vaccine (see below), can readily cause systemic infection. In Britain, respiratory tuberculosis and tuberculous meningitis are now found mainly in adults.

The usual source of human infection is the 'open' case from whom bacilli are passed on through respiratory droplets or sputum (Fig. 9.2a and b). Organisms ingested in food, especially milk (as with *M. bovis*) (Figs. 9.3 and 9.4), can cause abdominal or glandular disease. Mycobacteria inoculated directly into skin cause nodules or ulcers and this route of infection is used to vaccinate with the attenuated live Bacillus Calmette–Guérin (BCG) strain (Chapter 16).

'Open' pulmonary disease may be asymptomatic or insidious in onset and can infect many contacts before being detected. There can be much cross-infection when the patient works, for example, in a school, factory, or public place such as a swimming pool. Previously treated or naturally resolved disease may become silently reactivated and passed on, for example from grandparent to grandchild. Fulminant disease or reactivation can follow immune depression induced for instance by corticosteroid therapy, diabetes mellitus, alcoholism, measles in malnourished children, or AIDS.

Evidence of previous infection may be first detected when chest X-ray performed for other reasons shows apical lung fibrosis, cavitation, or calcification, or when a tuberculin sensitivity test is found to be positive (see below).

Clinical presentation and diagnosis

Primary infection. The consequence of the first exposure to the organism may be completely asymptomatic, with a small 'Ghon focus' of inflammation, often in the periphery of a lung, with enlargement of draining lymph nodes (Fig. 9.5). Malaise, fever, weight loss, and sometimes erythema nodosum (Figs 13.4 and 13.5) may be presenting complaints. Four to six weeks after infection the

(a)

(b)

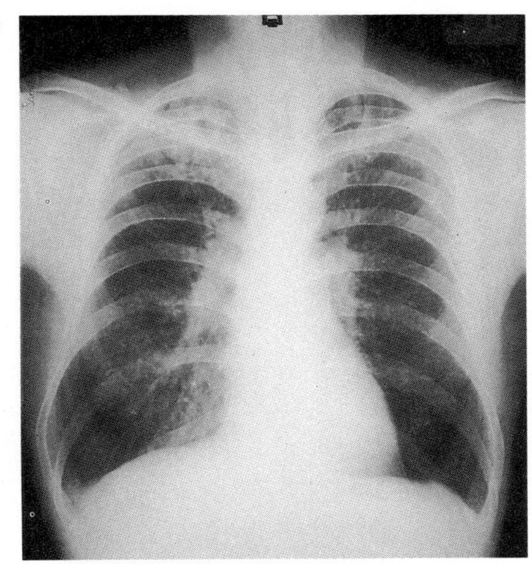

Fig. 9.2 **(a)** Asymptomatic mother holding child with miliary tuberculosis; **(b)** cavitation in mother's right lung apex.

tuberculin skin test becomes positive and, because material for culture is rarely available at this stage, diagnosis is usually based on this immunological evidence together with clinical, epidemiological, or radiological findings. Primary tuberculosis can resolve spontaneously, perhaps encouraged by rest. However, it may progress by local spread to cause pleurisy or bronchopneumonia, or through the bloodstream to cause miliary disease, meningitis, or both. Treatment is therefore essential. When primary infection involves the gut, meninges, regional lymph nodes, or skin, a high index of suspicion is necessary to avoid delay in diagnosis.

Fig. 9.3 Infected, unpasteurized milk is the usual source of *M. bovis* infection.

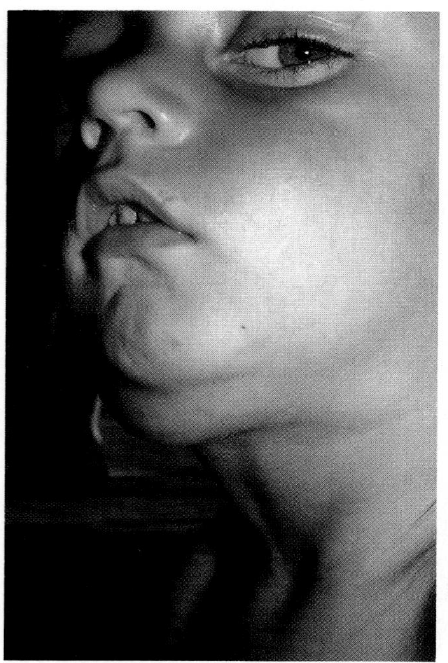

Fig. 9.4 Primary infection with enlarged submandibular lymph nodes that subsequently suppurated. *M. tuberculosis* isolated.

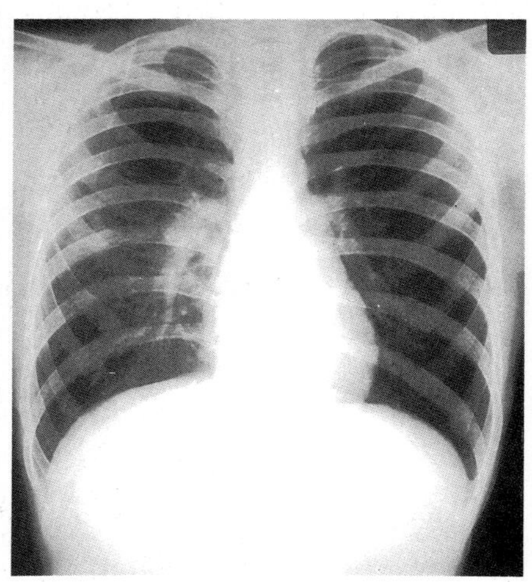

Fig. 9.5 Primary tuberculosis with Ghon focus in right mid-zone and hilar lymphadenopathy.

Miliary tuberculosis. Widespread blood-borne infection of many organs can follow primary infection or occasionally reactivation. Cell-mediated immunity as measured by the tuberculin test is often depressed or absent. The onset can be rapid, over a week or two, with fever, progressive weight loss, and debility followed by shortness of breath as pulmonary lesions and anaemia develop. Occasionally, choroidal tubercles are seen in the optic fundi but the diagnosis is more often suspected from finding numerous small radiographic opacities in both lung fields (Fig. 9.6). These widely and evenly distributed granulomas may be found in liver or marrow biopsy material, which can also be cultured. In contrast to tuberculous bronchopneumonia, sputum is usually absent. Organisms can sometimes be cultured from urine though they may not be seen by direct microscopy.

Tuberculous meningitis. This can be a manifestation of primary or miliary infection or result from direct or blood-borne spread from an old focus of tuberculosis. It is discussed in Chapter 6.

Tuberculous pleural effusion. This develops, usually in children and young adults, soon after primary infection but rarely before tuberculin tests become positive. It is distinct from the pleurisy associated with tuberculous broncho-pneumonia or cavitation. Often the only clinical sign is effusion, and there may be few symptoms except perhaps some weight loss, mild fever, and rarely dyspnoea

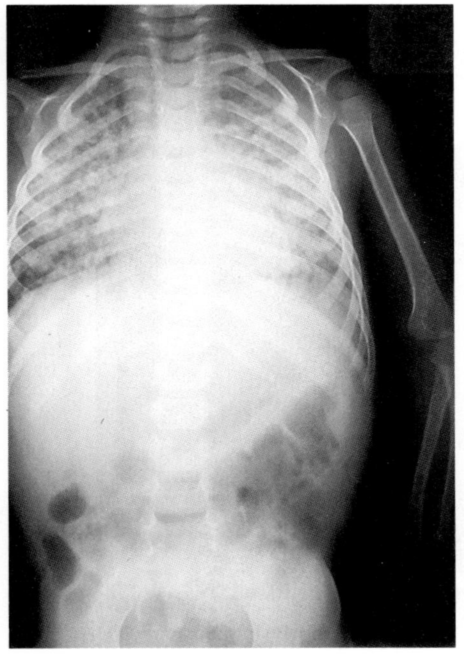

Fig. 9.6 A 'snowstorm' appearance of lungs in miliary tuberculosis.

from large effusions. The diagnosis is made by examining the straw-coloured pleural fluid which contains almost solely lymphocytes. Organisms are not often seen but may be cultured from the fluid or pleural biopsy. Histological examination of pleura usually reveals granulomas.

Tuberculous pericarditis. This usually follows spread of infection from adjacent mediastinal lymph nodes. As with tuberculous involvement of the pleura, lymphocytic effusion may be the presenting feature, suggested by a globular heart shadow on X-ray and confirmed by echocardiography. Earlier stabbing or 'pleuritic' chest pain may have been present. Sometimes the pericardial surfaces are adherent, with early calcification detectable by radiography. There may be cardiac tamponade from effusion or constrictive pericarditis. Constrictive pericarditis may be a late complication or sequel years later. Diagnosis is confirmed by surgical biopsy of the pericardium, histology, and culture. Corticosteroids given with antituberculous chemotherapy may reduce effusion and the risk of constrictive pericarditis developing. Surgical aspiration with or without pericardectomy may be necessary.

Abdominal tuberculosis. Infection may be confined to intra-abdominal viscera or lymph nodes (from which the peritoneium may become infected—Fig. 9.7). Classically, this form of disease results from *M. bovis* infection, but *M. tuberculosis* in swallowed sputum may be responsible. Inaccessability of abdominal lymph nodes for histological or mycobacterial examination make this disease difficult to confirm. Liver biopsy may show granulomas, but this is usually a feature of miliary disease rather than infection confined to the abdomen. Tuberculous ascites gives the abdomen a characteristically 'doughy' feel on

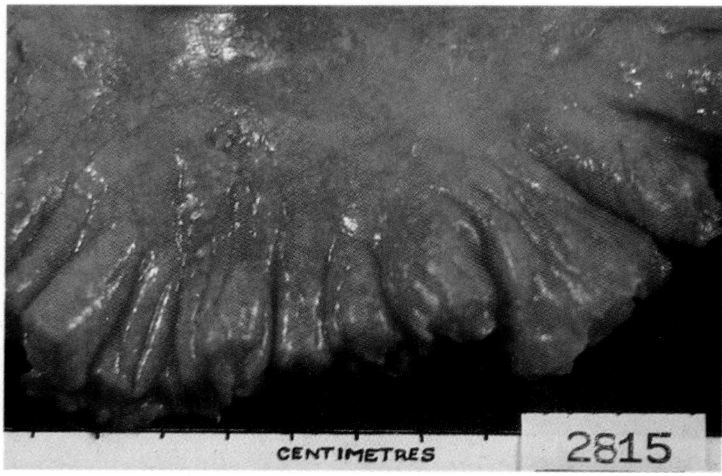

Fig. 9.7 Extensive tuberculous involvement of the peritoneum.

palpation. The fluid is usually high in protein and contains many white blood cells, mainly lymphocytes. Computed tomography can identify enlarged nodes and peritoneal biopsy can confirm diagnosis by histology and culture.

Tuberculous lymphadenopathy. This can be associated with disease elsewhere or be on its own, most obviously in cervical lymph nodes (Fig. 9.8). Any type of mycobacterium may cause cervical lymphadenopathy, particularly *M. bovis* where infected milk is drunk. In countries where *M. bovis* is rare, 'atypical mycobacteria' are often responsible. The patient complains of swelling and sometimes tenderness, usually insidious in onset, but few other symptoms. With cervical node involvement, pyogenic or dental abscess may be suspected initially. Biopsy usually shows granuloma or caseation and the organisms may be seen on microscopy or grown on culture. A sinus may open through skin either spontaneously or through a biopsy incision. Antituberculous therapy is indicated for *M. bovis* or *M. tuberculosis*; atypical mycobacteria respond unpredictably to drugs but spontaneous resolution is usual.

Tuberculous bronchopneumonia and cavitation. Primary infection may rapidly progress to bronchopneumonia, especially in the malnourished, but organisms may lie dormant asymptomatically for many years. Post-primary bronchopneumonia is the commonest clinical manifestation of tuberculosis. Bronchopneumonia (Fig. 9.9) usually results from direct spread of organisms to other parts of the lungs or air passages. Cavitation, caseation, and calcification usually affect the apices. The infection produces fever, copious infectious sputum, haemoptysis, and sometimes pneumothorax. Sputum may still be infectious in chronic, localized, and unsuspected pulmonary tuberculosis where coughing

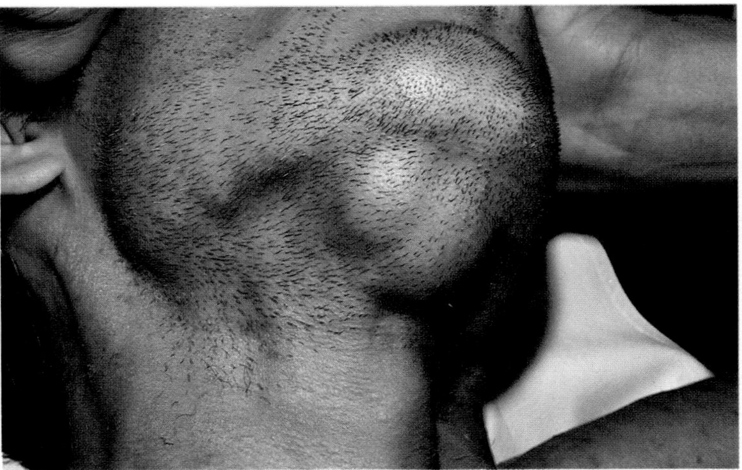

Fig. 9.8 Enlarged submandibular lymph nodes due to mycobacterium.

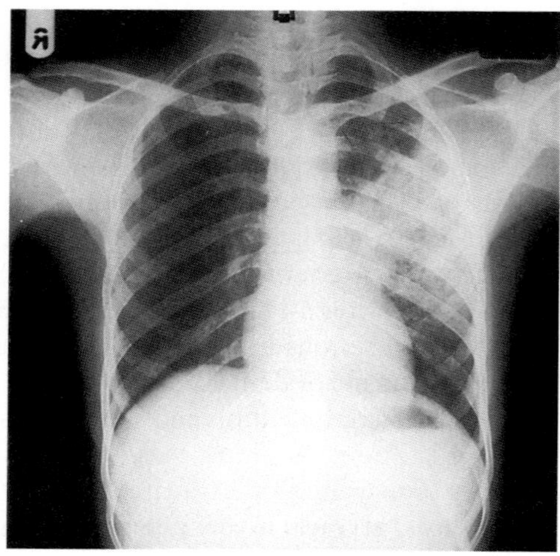

Fig. 9.9 Post-primary tuberculous bronchopneumonia (highly infectious).

may be disregarded or considered 'chronic bronchitis'. There may be secondary infection with Gram-negative and anaerobic organisms. Diagnosis is by direct microscopic examination and culture of sputum, laryngeal swabs, or gastric washings. Occasionally, biopsy to exclude other lung pathology such as neoplasia reveals tuberculous infection.

Urogenital tuberculosis. Tuberculosis of the kidney results from blood-borne infection. It may present in isolation or as part of more widespread disease. Symptoms include frequency of micturition and dysuria. Haematuria may be only microscopic and 'sterile' pyuria (pus with no growth of usual urinary tract pathogens) is strongly suggestive. A deformed kidney shown by intravenous pyelography can be due to tuberculous abscess or granulomas, and there may be calcification in long-standing disease. For specific diagnosis, early morning (concentrated) urine should be cultured.

Genital tuberculosis in the male, usually secondary to renal disease, affects the epididymis or testes. In the female infection of fallopian tubes or uterine endometrium may cause menorrhagia and infertility; endometrial curettings contain granulomas or organisms.

Osteomyelitis and arthritis. Onset of bone infection is often insidious and relatively symptomless, especially when a joint is not involved. The vertebrae are most often affected, less often the hips, knee, ankle, elbow, and other sites. Pain, localized swelling, limitation of movement in joints, or effusion may be evident. Sinus formation may be a presenting feature. The specific diagnosis is usually made histologically or by culture of discharge or biopsy material. The infection is blood-borne and tuberculosis may be present elsewhere in the body. In addition to antituberculous therapy, splinting may prevent deformities, particularly when vertebrae are involved. Tuberculous joints are unlikely to regain useful movement.

Cold abscesses. These usually cause discharging sinuses without the usual features of redness, heat and tenderness (Fig. 9.10). There may be no systemic symptoms. They often arise from lymph nodes (Fig. 9.11a), bone or thoracic cavity (Fig. 9.11b) and can be confirmed by microscopy and culture. When discharging from unusual sites such as breast abscess or middle ear, diagnosis is often delayed.

Investigation

Microscopy and culture. The best evidence of current infection is isolation of the organisms from specimens such as sputum, urine, gastric washings, or biopsy material. Visualizing acid-fast bacilli is strongly suggestive but, particularly from the urinary tract, non-tuberculous bacilli (for example, *M. smegmatis*) may look similar. Culture also allows drug-sensitivity studies to guide management. Haematological findings may include leucopenia, monocytosis, or occasionally a leukaemoid reaction. The erythrocyte sedimentation rate is usually raised. An underlying immunocompromising condition such as HIV infection must be considered.

Histological evidence. Caseation is suggestive and helps to differentiate tuberculosis from sarcoidosis where cellular histology (focal granuloma with epithelioid and giant cells) is similar. Acid-fast bacilli are not always easily seen.

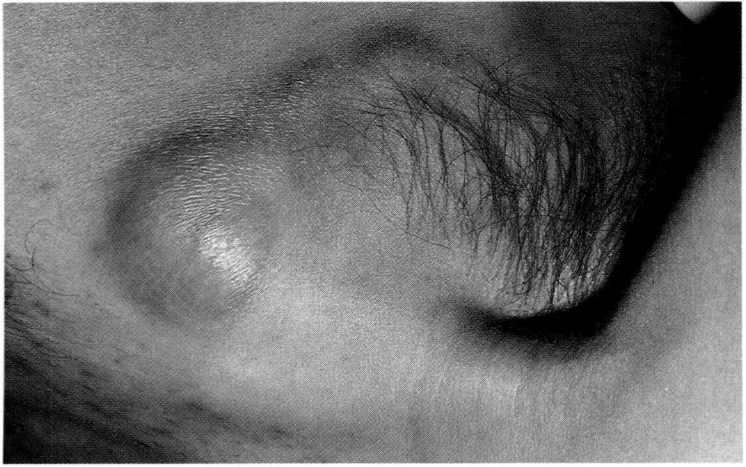

Fig. 9.10 'Cold' abscess of chest wall. Note enlarged axillary lymph nodes.

(a)

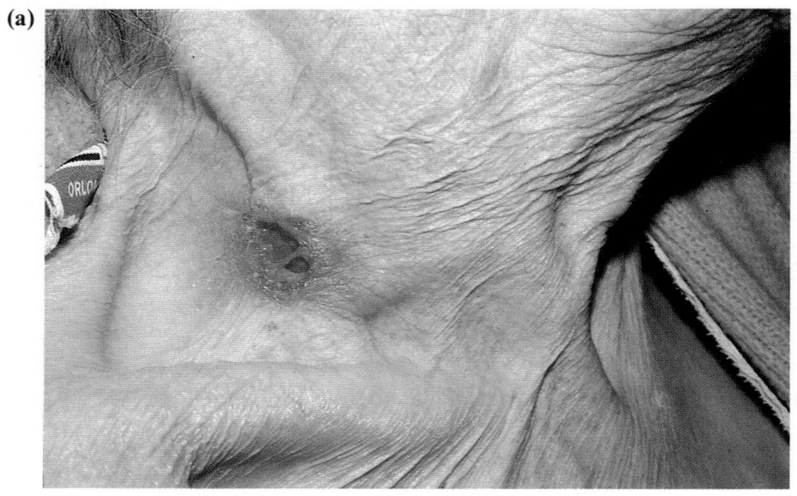

(b)

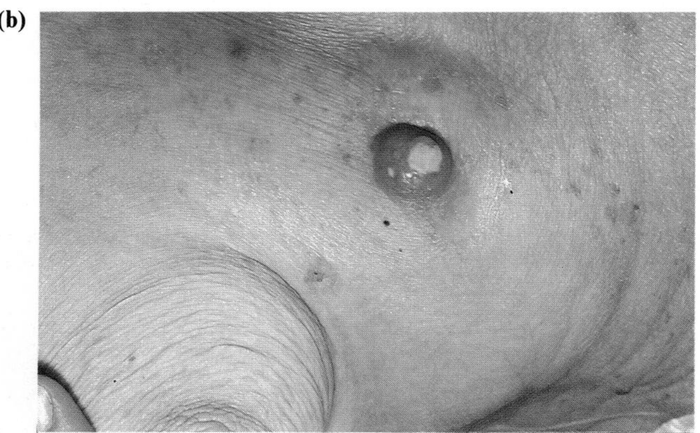

Fig. 9.11 Tuberculous sinuses over **(a)** cervical lymph node; **(b)** chest wall medial to right breast.

Tuberculin skin testing. A *strongly* positive reaction (Fig. 9.12) suggests active or recent infection, especially in those not constantly exposed. This, combined with suggestive clinical evidence of disease, may warrant a trial of therapy if culture material is unavailable or results are awaited. Weaker positive reactions are less helpful and may be due to previous infection, possibly with atypical mycobacteria, or BCG vaccination. A negative reaction usually excludes

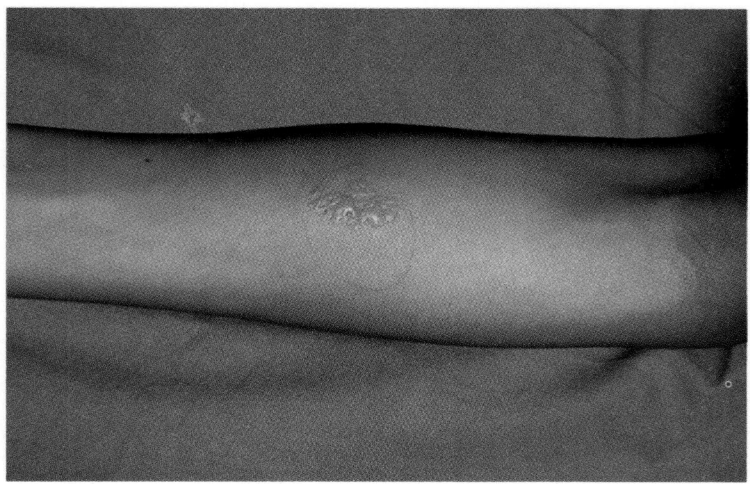

Fig. 9.12 A strong positive Mantoux test: vesiculation, surrounding inflammation and oedema suggest active infection.

infection, but positive reactions often revert to negative when cell-mediated immunity is depressed as during measles, immunosuppressive therapy, or disseminated tuberculous disease itself.

Treatment

Anti-tuberculous chemotherapy now provides the basis of treatment, supported occasionally by surgery. Atypical organisms however can be very resistant to drugs. Treatment can rarely await culture results and choice of drugs is usually empirical once specimens have been collected. Reappraisal of the management of any predisposing illness such as AIDS, diabetes mellitus or alcoholism may be required.

When a choice of drugs is available and financial resources are adequate, it is usual initially to use three or four drugs in combination (Table 9.4). After a satisfactory response, one or two of the drugs can be discontinued. Both side-effects (which are common) and effectiveness need consideration. Isoniazid (with pyridoxine to prevent the side-effect of peripheral neuropathy) is cheap but can cause a systemic lupus-like syndrome. Streptomycin must be given parenterally and causes VIIIth nerve damage if dosage is too high (a particular danger in children, the debilitated, and those with renal failure). It has been largely replaced in Britain by oral rifampicin, which quickly renders patients non-infectious but is expensive, hepatotoxic, especially when combined with isoniazid, and discolours urine red. Ethambutol occasionally reduces visual acuity from central scotoma; during treatment regular ophthalmic observations are wise. Pyrazinamide can precipitate gout and is also hepatotoxic. All can cause anorexia, nausea, and vomiting.

Table 9.4. *Drug treatment of tuberculosis in Britain (based on recommendations by Joint Tuberculosis Committee of British Thoracic Society*, 1990)*

First two months—three or four drugs:
Rifampicin + isoniazid (+ pyridoxine) + pyrazinamide (+ ethambutol if isoniazid resistance is likely)

Thereafter (adjusted if necessary according to results of sensitivity tests):
Rifampicin + isoniazid (+ pyridoxine)
for 4 months (total 6 months treatment)
9 months, for lymph node infection (total 11 months)
10 months, for meningitis (total 12 months)

*Schedules may be varied in other countries according to availability and cost of drugs and practicalities of administration—for instance, twice weekly dosage; longer courses with older drugs.

It can be difficult to persuade a patient who feels better that prolonged medication is necessary to prevent relapse. Compliance with treatment can therefore be poor especially where the price of drugs is high and availability limited. Various regimes with intermittent dosage, for example twice weekly, have been successful, can be supervised and reduce cost. Recent studies suggest that short courses such as 6 months for pulmonary disease are effective so long as rifampicin is used throughout the course of treatment.

Prevention

Isolating patients with infectious sputum and checking close contacts at home and at work can prevent secondary cases or detect them early. Chest radiography and tuberculin testing (especially in children) can be used. For tuberculin sensitivity testing the intradermal (Mantoux) test or the technically simpler multiple puncture (Heaf or Tine) tests are used (Fig. 9.13a and b). BCG vaccination (Chapter 16) is not given to contacts immediately since it spoils the tuberculin tests for diagnostic purposes; also its value is doubtful in those already incubating the disease. Tuberculin-positive young children must have had recent or current primary infection and, since miliary disease is especially feared in this age group, even in the absence of disease treatment is indicated. Tuberculin-negative child contacts of an infectious source (for example, parent or schoolteacher) can receive chemotherapy by isoniazid alone together with BCG vaccination using an isoniazid-resistant vaccine strain. Contact tracing may reveal the original source of the patient's infection.

Improvements in housing, nutrition, and hygienic conditions at school and work lower the incidence of tuberculosis.

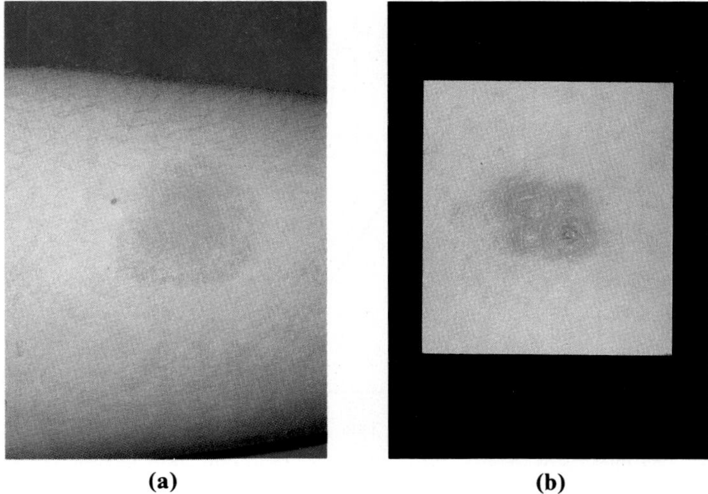

Fig. 9.13 Positive tuberculin tests: **(a)** single intradermal injection (Mantoux); **(b)** multi-puncture (Heaf).

BRUCELLOSIS

This infection can present with fever and few abnormal physical signs.

Aetiology and epidemiology

The principal cause is *Brucella abortus* (source: cattle), sometimes *Br. melitensis* (source: goats), and occasionally *Br. suis* (source: pigs). Human to human spread is extremely rare. Brucellosis is endemic throughout the world except where efforts have successfully eradicated it from animal sources. Where prevalent, farmers and veterinarians are at special risk, through close contact with animals at parturition when infected material can be inhaled, ingested, or inoculated through abraded skin. The fetus and placenta of aborted animals can remain infectious for long periods and require careful disposal. Unpasteurized milk and milk products can transmit infection. In recent years, *Br. abortus* has been largely eliminated from British cattle by regularly testing herds and slaughtering infected animals (Fig. 9.14). *Brucella melitensis* sometimes infects holiday visitors to Southern Europe who consume raw goat milk or fresh soft cheese, but is not indigenous in Britain.

Clinical presentation, complications, and differential diagnosis

Some infections are asymptomatic; others, after an incubation period of a few weeks or months, cause acute or subacute illness with fever that may be intermittent and prolonged, sometimes for many months. Associated features

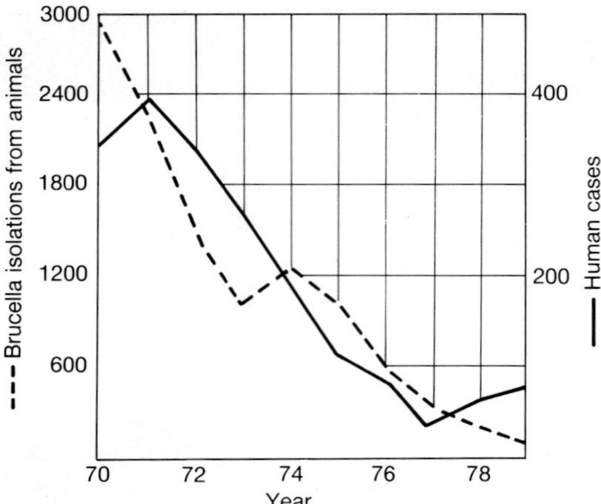

Fig. 9.14 The decline of brucellosis in Scotland following skin-testing and slaughter of infected cattle.

can include marked sweating, arthralgia, back and muscle pains, and exhaustion. If brucellosis becomes 'chronic', exhaustion, recurrent fevers, headaches, weight loss, and depression can persist for years. There may be no abnormal physical signs except sometimes splenomegaly, hepatomegaly, or lymphadenopathy. Occasionally, acute septicaemia presents as fulminant, toxic illness. Complications and persistent features can include arthritis (including spondylitis), endocarditis, orchitis, hepatitis, and peripheral or central nervous system signs.

Depending on the presentation, other illnesses may have to be considered, including tuberculosis, bacterial endocarditis, enteric fever, Q fever, and non-infective diseases such as lymphomas or collagen disorders.

Investigations

The total white-cell count in peripheral blood is usually normal or reduced. Blood or bone marrow cultures are helpful but special culture media and methods necessitate prior consultation with the laboratory. Serological tests are also useful: raised antibody titres are usual in acute illness, and common but less helpful in diagnosing chronic disease. Liver or lymph node biopsy can show granulomas without caseation, suggesting brucellosis when tuberculosis and sarcoidosis are unlikely.

Management and prognosis

Response to antimicrobials may be slow and unpredictable. Tetracyclines or co-trimoxazole with streptomycin or rifampicin given together or alternately for at least six weeks are usually helpful in acute illnesses but less so in chronic disease. Residual complaints can be due to disturbed function of affected organs and relapse is not infrequent. In chronic illnesses, the patient and family may need psychological support.

Prevention

This depends upon controlling the infection in animals and avoiding exposure to them and to animal products. Travellers to countries where brucellosis is endemic should not consume milk, unless pasteurized or boiled, or cheese and butter unless known to have been similarly treated during manufacture.

Q FEVER

This is another zoonosis, world-wide in distribution, infecting many animals including sheep and cattle and also some ticks, though these are not important vectors. The causal rickettsia, *Coxiella burneti*, is very hardy, stable when dried in dust, and spread mainly by the airborne route from excreta and birth products of infected animals or from objects contaminated from these. It can be found in, and occasionally be transmitted by, milk from infected animals in which infection is often silent and chronic. Spread from person to person rarely, if ever, occurs.

Infection is often silent, especially in young children. In older persons fever and headache are common. Atypical pneumonia affects half the patients. Acute and chronic hepatitis with hepatosplenomegaly is not uncommon. There may be pericarditis or endocarditis, especially in older subjects or those with previously diseased heart valves in which chronic, progressive disease with vegetations may develop, ultimately fatal unless recognized and treated.

Specific diagnosis is made serologically. Antibodies to antigenic phase I organisms suggest chronic infection, whereas antibodies to phase II antigen are present in both acute and chronic infection.

Specific treatment is by high dosage of chloramphenicol or a tetracycline which should be continued until there is clinical response and no fever for at least 24 hours. Valvectomy and prosthetic replacement may be required in Q fever endocarditis.

10

Infections of the genital tract

INTRODUCTION

Many infections and infestations of skin and soft tissues discussed in Chapter 3 can affect the genitalia. Mumps, brucellosis, tuberculosis, and leprosy can involve the gonads. In endemic areas, epididymitis and perineal or scrotal lymphangitis are often due to filarial infection. This chapter is concerned, however, with infections principally affecting the genital tract which are commonly acquired by sexual activity, usually because they are poorly infectious except by intimate contact.

Infections transmitted sexually can sometimes spread by other means, as when syphilis is contracted congenitally from the mother or when other household members become infected from highly infectious oral syphilitic chancres. Gonococcal and chlamydial infections can be transmitted to the newborn during parturition. Sexually transmitted disease is commoner among, but not confined to, sexually promiscuous persons including homosexuals, in whom syphilis, hepatitis B, and the acquired immunodeficiency syndrome (AIDS, see Chapter 12) are especially prevalent. Alimentary infections can also spread through anal intercourse and, in addition to herpes simplex and classical sexually transmitted diseases, cryptosporidiasis, giardiasis, and salmonellosis can contribute to the diarrhoea which can be especially troublesome in progressive HIV infection. Sexually transmitted infections are found throughout the world with individual diseases posing different problems in different regions. For example, chlamydial urethritis is now the commonest cause of non-gonococcal urethritis in Britain, penicillin-resistant gonococci are well-established in the Far East, and syphilis is more serious where socio-economic circumstances cause delayed or inadequate treatment. A small but important problem is transmission of drug-resistant infections across international boundaries.

Aetiology and presentation

Table 10.1 summarizes the organisms causing the usual presenting symptoms of urethritis, urethral and vaginal discharge, and genital ulceration. Diseases caused by the commonest of these are discussed below. Uterine and intra-pelvic sepsis generally require specialized gynaecological investigations and management not detailed in this book.

Genital infections can also be manifest elsewhere in the body, sometimes with

Table 10.1. *Infective causes of genital-tract disease*

Vaginal discharge	
Candida albicans	Common. Pruritis marked discharge may be slight
Trichomonas vaginalis	Offensive discharge. May be pain and local oedema
Gardnerella vaginalis	Offensive greyish discharge without inflammation
Chlamydia trachomatis *Neisseria gonorrhoeae*	More often cause urethritis or pelvic inflammatory disease
Staphylococcus aureus	Usually infects high-absorbency tampon, associated with toxic shock (Chapter 8)
Herpes simplex *Treponema pallidum* Viral warts	Rarely, if involving cervix
Urethritis and urethral discharge	
Non-specific agents *Chalamydia trachomatis* *Neisseria gonorrhoeae* *Trichomonas vaginalis* *Candida albicans* *Ureaplasma urealyticum*	No clearly distinguishing clinical features
Genital ulceration	
Herpes simplex	Multiple vesicles or ulcers, often recur
Treponema pallidum	Primary chancre
Varicella zoster	Unilateral, painful
Scabies	Small papules may be secondarily infected
Chancroid	Painful multiple ulcers
Tuberculosis Granuloma inguinale Lymphogranuloma venereum	Lymph-node involvement may predominate

no obvious pointers to the entry site of infection, for example AIDS, generalized syphilitic rash involving the hands and feet, and painful arthritis in gonorrhea. Non-genital parts of the body surface such as the mouth and rectum occasionally are primary sites of these infections.

Investigations

The most commonly performed investigations for genital infection are shown in Table 10.2. Since acquisition of any sexually transmitted infection indicates the

Table 10.2. *Useful investigations for genital infection*

Microscopy and culture of urethral discharge
Microscopy and culture of vaginal discharge (not efficient for detecting gonococcal infection)
Culture of cervical and/or rectal swabs (to detect gonococcal and chlamydial infection)
Culture of swabs from ulcers
Dark ground microscopy of exudate from lesions for *Treponema pallidum*
Serology for *Treponema pallidum*
Biopsy of unexplained lesions, enlarged lymph nodes or uterine currettings for histology and culture
Chlamydial antigen tests in early morning first-catch urine

possibility of having contracted others, perhaps at the same time, investigations should not stop once one infection has been diagnosed. It is particularly important to exclude the serious but treatable infections by syphilis and gonorrhoea which may be clinically inapparent at the time of examination.

Contact tracing and surveillance

Tracing sexual contacts is important for control by detecting other infected persons, with or without symptoms, who may not have sought treatment. It is difficult for reasons of confidentiality and because patients are treated not only at special clinics with specially trained staff, but also by general and private practitioners and sometimes with drugs obtained surreptitiously. For the same reasons surveillance is difficult. Increasing attendance at special clinics may partly reflect greater willingness to seek advice openly rather than increased incidence of disease. For syphilis, diagnostic tests have been available for many years and antenatal screening is routine, so laboratory reports reflect incidence changes more reliably than, for example, for genital papillomaviral or chlamydial infections where increased reporting largely reflects new diagnostic techniques.

SPECIFIC GENITAL TRACT INFECTIONS

Candidiasis

Aetiology and epidemiology

Candida albicans is a fungus, often commensal in humans, which under certain circumstances causes disease. It can infect many parts of the body, especially the vagina but also the gastrointestinal tract (Chapter 4), skin, urinary tract, and occasionally the bloodstream with dissemination to internal organs. Mere isolation of this usually commensal organism does not necessarily indicate that it is causing disease.

In genital infection *C. albicans* is mainly important as a cause of vulvovaginitis and balanitis, usually sexually transmitted. Commensal *Candida* may proliferate in association with diabetes mellitus, broad-spectrum antibiotic therapy, immunosuppression (including HIV infection), pregnancy, and debility. A mother with or without symptoms may infect her infant during parturition.

Presentation and differential diagnosis

Only genital candidiasis is considered here. In women a usual complaint is itching or a burning sensation in the vulva and vagina. There may be dyspareunia. Discharge is variable, often scanty and cheesy. In severe infections pain can make examination difficult and there may be marked discharge with haemorrhage from infected surfaces. In men, irritation with erosions and sometimes pustules may affect the prepuce and glans penis (Fig 10.1). Both males and females occasionally develop urethritis.

Investigations

Gram-positive yeast cells visualized by microscopic examination of smears from infected areas are identified as *C. albicans* by culture. Urine should be tested for glucose, and other predisposing causes (see above) considered. Gonorrhoea, trichomoniasis, and primary syphilis may need to be excluded.

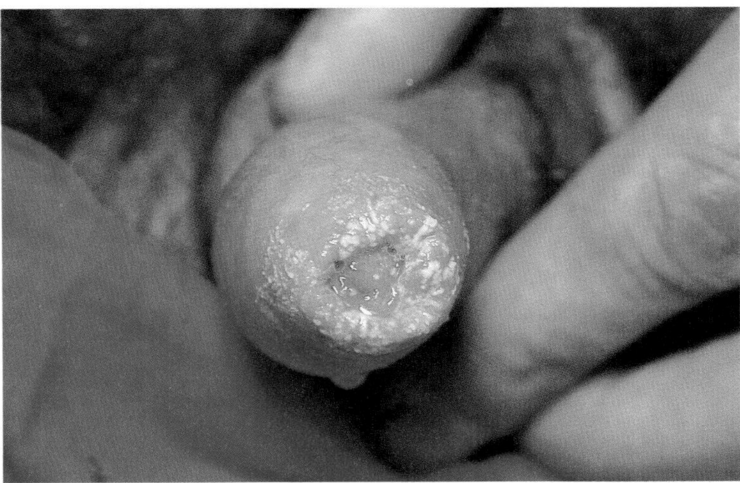

Fig. 10.1 Severe *Candida* balanoprosthitis presenting in a patient with juvenile-onset diabetes.

Management

Pessaries and creams containing an imidazole such as clotrimazole are effective in treatment. Single-dose oral fluconazole is convenient and often more rapidly effective. Treating the alimentary tract and/or the spouse, even if asymptomatic, can help to prevent relapse.

Trichomoniasis

Aetiology and epidemiology

The flagellate protozoon *Trichomonas vaginalis* is usually transmitted sexually. It particularly causes symptoms in women of child-bearing age but is often asymptomatic in males. The natural reservoir is man and the distribution is world-wide.

Clinical presentation and differential diagnosis

Women may have copious, frothy, offensive vaginal discharge, sometimes with excoriation of the perineum, pain, dyspareunia, dysuria, or cystitis. Symptoms often begin or are exacerbated at menstruation or after intercourse. Mild or asymptomatic cases are common, with pruritus the only complaint. The vaginal walls appear oedematous and inflamed, sometimes with petechial haemorrhages. The cervix may show a 'strawberry' appearance. Men may have urethral discharge, dysuria, sometimes difficulty in micturition and painful prostatitis.

Investigations

Motile parasites can be identified by immediate microscopic examination of a wet preparation of vaginal or urethral discharge. Swabs for culture (if required) should be taken into Stuart's transport medium. After treatment swabs should be taken from the uterine cervix to exclude concurrent *Neisseria gonorrhoeae* infection which may have been masked.

Management

Oral metronidazole is given concurrently to the patient and sexual partner.

Gonorrhoea

Aetiology and epidemiology

This disease is caused by *N. gonorrhoeae*, a Gram-negative intracellular diplococcus. Confined to humans, it is usually contracted by direct contact causing genital, anal, or pharyngeal infection. Infection at birth can cause neonatal conjunctivitis. Asymptomatic carriage, especially in females, is an important source of infection.

Clinical presentation

The incubation period is usually less than one but may be up to 6 weeks. In males, purulent urethral discharge (Fig. 10.2) usually follows a brief period of dysuria. Infection can spread to para-urethral structures: Cowper's gland at the base of the penis, seminal vesicles and epididymis perhaps leading to sterility, bladder, and occasionally inguinal lymph nodes.

About one-third of females have initial symptoms—dysuria, increased frequency of micturition or, rarely, vaginal discharge. The remainder are asymptomatic. Cervicitis only occasionally causes vaginal discharge which can be offensive, sometimes with back pain. Infection may progress to the uterus, fallopian tubes, and peritoneum, causing pelvic sepsis or perihepatitis. This is more likely during menstruation, the puerperium, following cervical dilatation or the introduction of an intrauterine contraceptive device.

Proctitis in both sexes is usually asymptomatic. In men it usually follows homosexual practices; in women it usually follows spread of infection from vaginal discharge and may cause pain on defecation, bleeding, and occasionally anal discharge. Pharyngitis may result from orogenital contact and is usually asymptomatic.

Investigations

The organisms can usually be grown from pus, collected by swabs impregnated with charcoal to aid adherence of organisms. Samples are collected from the

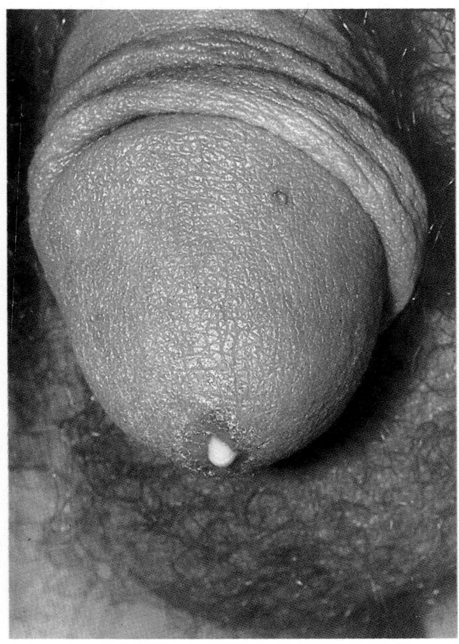

Fig. 10.2 Purulent urethral discharge in gonorrhoea.

urethra, cervix (requiring direct visualization), rectum, or pharynx and ideally inoculated directly on to culture plates (modified Thayer–Martin selective medium or modified New York City medium) and immediately incubated. A second set of cultures should be taken if the first are negative. If direct plating is not available, swabs should be placed immediately in Stuart's or Amies' transport medium, to maintain viability of organisms up to 24 hours without refrigeration (Fig. 10.3). Alternative special media (for example, modified Thayer–Martin transport medium) allow even longer transit periods. The gonococcal complement fixation test is not sufficiently reliable to be recommended.

Complications

Neisseria gonorrhoeae is one cause of Bartholin's abscess. Acute salpingitis causes colicky lower abdominal pain, fever, nausea, and vomiting. Bimanual examination causes pain, and a tender adnexal mass may be palpable. Salpingitis must be differentiated from acute appendicitis, ectopic pregnancy, and other causes of 'acute abdomen'. Occasionally, gonococcal salpingitis and endometritis present insidiously with menorrhagia and dysmenorrhoea; adhesions may cause sterility in females.

Blood-borne dissemination of gonococci may cause fever and skin pustules from which gonococci may be isolated. Painful septic arthritis, with organisms in synovial fluid, may cause joint destruction. Occasionally the meninges and endocardium are involved. Intermittent fever and symptoms can mimic chronic meningococcaemia (Chapter 7). Ophthalamia neonatorum results from gonococcal infection during childbirth (Chapter 11).

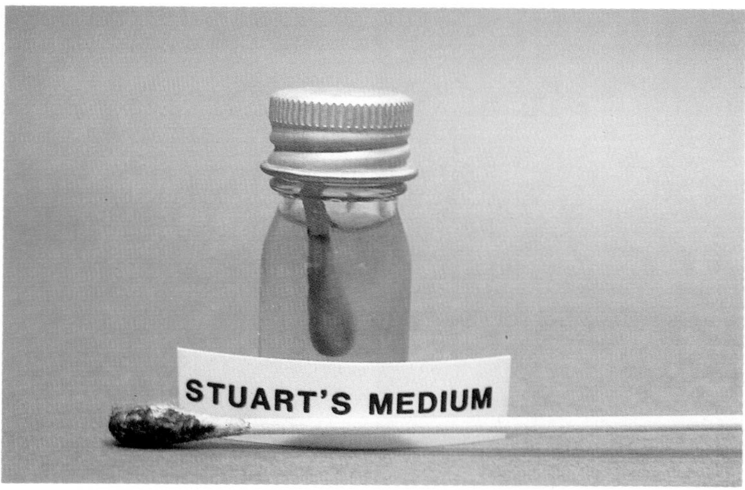

Fig. 10.3 Swabs put into transport medium and end broken off. Unbroken swab also shown.

Management

Treatment can be by one intramuscular injection of procaine penicillin together with oral probenicid, or alternatively a large dose of amoxycillin or doxycycline hydrochloride together with probenicid. It is important to use adequate single dosage to avoid failures from poor compliance. Resistance of gonococci, especially to penicillin, is becoming commoner in many parts of the world—hence the importance of culture whenever possible. The infection can be treated with spectinomycin or a modern cephalosporin, effective against β-lactamase producers. Occasionally a Bartholin's or Cowper's abscess requires drainage and deep-seated infections need longer courses of treatment. Reinfection from asymptomatic infected sexual partners is likely unless contacts are also effectively treated.

Chlamydial urethritis (and lymphogranuloma venereum)

Aetiology and epidemiology

Chlamydia trachomatis causes about half the cases of non-gonococcal urethritis in males and is an important cause of pelvic inflammatory disease in women. Serotypes A–C cause trachoma; serotypes D–K cause mainly urogenital infections but also conjunctivitis, mainly in babies. Serotypes L1–L3 cause *lymphogranuloma venereum* (LGV), a common venereal disease in tropical areas and seaports, occasionally imported into Europe (see below, p. 258).

These organisms are transmitted sexually, by fingers to the eyes of adults or from the mother's genital tract to the eyes of her newborn infant (Chapter 11).

Clinical presentation, differential diagnosis, and complications

Chlamydial urethritis has an incubation period of about one to four weeks. It primarily affects men, with burning on micturition and mucopurulent discharge, less purulent than in gonorrhoea. In women cervicitis and salpingitis may be asymptomatic. *Chlamydia trachomatis* should be suspected when tests for gonococcus, trichomonas and candida infections are negative. Complications include epididymitis, cystitis, and rarely urethral strictures. Lymphogranuloma vernereum is described below.

Reiter's syndrome comprises arthritis with ocular lesions, mainly conjunctivitis, often with lesions of skin, mouth, and connective tissues. It is associated with non-gonococcal urethritis, infective diarrhoeal illness, or sometimes other infections. 'Reactive arthritis' (Chapter 4) is the term used when conjunctivitis is absent. Commoner in males, it can entail considerable systemic upset, but spontaneous recovery is usual after 2–3 months. Most sufferers have tissue type HLA-B27 and many cases are probably unrecognized.

Investigations

Chlamydias may be demonstrated by specific immunofluorescence staining using monoclonal antibodies, by ELISA tests or by isolation from urethral or endocervical samples. Special transport medium is required and specimens should be sent immediately to the laboratory. Serological tests are rarely used but can show high antibody titres.

Management

Tetracyclines, alternatively erythromycin, are effective but treatment may need to be prolonged.

Genital herpes

Aetiology and epidemiology

This condition is usually caused by type 2 strains of herpes simplex virus but increasingly by type 1 strains which more often affect the face and mouth. Direct or indirect orogenital contact can transfer either virus type between the mouth and genitalia. Genital herpes is usually sexually transmitted. About 20 per cent of adults have serological evidence of herpesvirus type 2 infection with a higher proportion in the sexually promiscuous. Type 2 genital infections more often recur than type 1 infections.

Clinical presentation and complications

After an incubation period of up to 12 days, a tender, erythematous patch appears on the skin or mucous membrane, followed quickly by development of vesicles that progress to pustules and crusts. In females primary infection often involves the cervix as well as the vulva. In males the glans penis is the usual primary site (Fig. 10.4); herpetic proctitis may affect a passive male homosexual

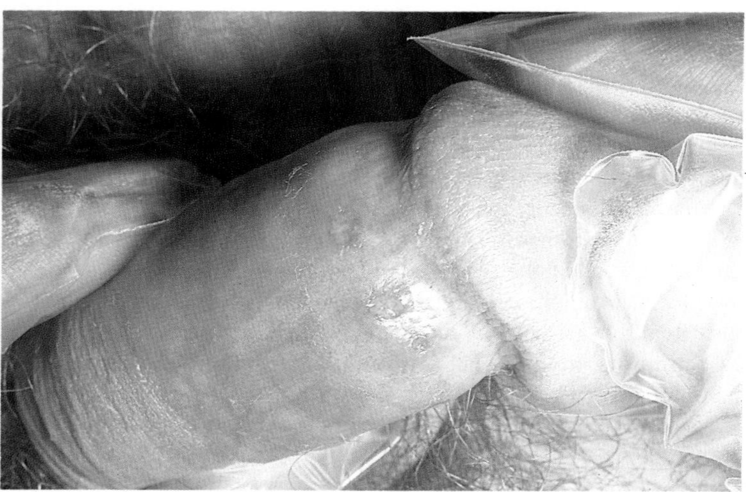

Fig. 10.4 Ulcers due to herpes simplex genitalis.

partner. In both sexes adjacent areas may be involved. Sometimes infection is disseminated with skin lesions elsewhere on the body and, rarely, complications such as meningoencephalitis. Local lymph nodes may be tender and enlarged. Healing of primary infections and loss of infectivity take about 7–21 days, or longer when secondarily infected. The extent of lesions varies considerably from one or two that may not even be noticed to numerous lesions with inflammation and swelling, sometimes causing severe dysuria and urinary retention.

Recurrent disease generally involves the vulva, cervix, or glans penis and repeated attacks may be fairly frequent, especially in the first year after primary infection, less often as time goes on. Sometimes they are associated with menstruation. Herpesvirus type 2 infection has been postulated as a cause of cervical cancer but this has yet to be substantiated; both are associated with such factors as early sexual intercourse and multiple sexual partners.

Diagnosis

This is often clinical but virus isolation and typing can be performed using tissue cultures. It is essential to exclude other sexually transmitted diseases such as syphilis, gonorrhoea, and chlamydial infection.

Management and prevention

Oral and topical acyclovir are used with parenteral therapy for severe infections. Prophylactic acyclovir can be used in those with frequent recurrences; inosine pranobex, an immunological stimulant, may be of benefit. Unfortunately, these drugs do not eradicate latent virus or prolong the intervals between recurrences. Drug resistance is not yet recognized clinically but may become a problem. Sexual intercourse should be avoided when lesions are present. Active genital herpes at delivery may require caesarian section (Chapter 11). Vaccination is not yet available.

Syphilis

Aetiology and epidemiology

Syphilis is widespread throughout the world and is caused by *Treponema pallidum.* It spreads by direct contact with infective lesions on skin or mucous membranes, usually on genitalia, in the oral cavity, or at the angles of the mouth. Spread can also be transplacental or via infected blood.

Syphilis is characterized by its systemic nature and prolonged course. Initial symptoms and infectivity during early and early latent stages (see below) may last intermittently for 2–4 years. Early use of penicillin abolishes infectivity and reduces further spread. Later signs and symptoms may not be easily linked to the original infection which may have been asymptomatic. Degrees of immunity can develop but do not necessarily eradicate the disease. Transplacental spread leads to primary congenital infection which may be lethal (Chapter 11), but primary acquired disease is rarely so.

Clinical presentation and complications

Syphilis may present in early or late forms. After an incubation period of about three weeks the primary chancre, usually solitary, develops at the site of inoculation: a macule becomes a papule and then a painless, moist indurated, circular ulcer about 1 cm in diameter. This usually appears on the external genitalia (Fig. 10.5), skin, vaginal or urethral mucosa, uterine cervix, anal canal, or lips. Chancres are highly infectious, persist for 4–6 weeks if untreated, and regional lymph nodes are usually enlarged, discrete, rubbery and painless. Non-genital lesions require an astute clinician to suspect their real nature.

About two months after initial infection (which may have been asymptomatic) a rash develops with variable shape, colour, and distribution. It is characteristically generalized and symmetrical and often affects the palms and soles (Fig. 10.6a and b). Papular lesions may resemble warts, called condylomas ('condylomata lata'). Sloughing condylomas and ulcerated or bleeding lesions are likely to be highly infectious (Fig. 10.7), especially on mucous membranes where they coalesce as 'snail track ulcers'. Generalized lymphadenopathy is common. Early syphilis should always be considered when a skin rash has not otherwise been satisfactorily explained. Untreated secondary features usually disappear spontaneously within a few weeks but may reappear intermittently for up to four years.

In late disease, after varying periods of latency, some patients develop further lesions which are characteristically extragenital. After about five years, chronic granulomas with central necrosis (gummas) may develop in the skin and mucosae, especially of mouth, nose, and larynx and in many internal organs. After about 10 years, arteritis can affect cerebral, spinal, and arterial vessels

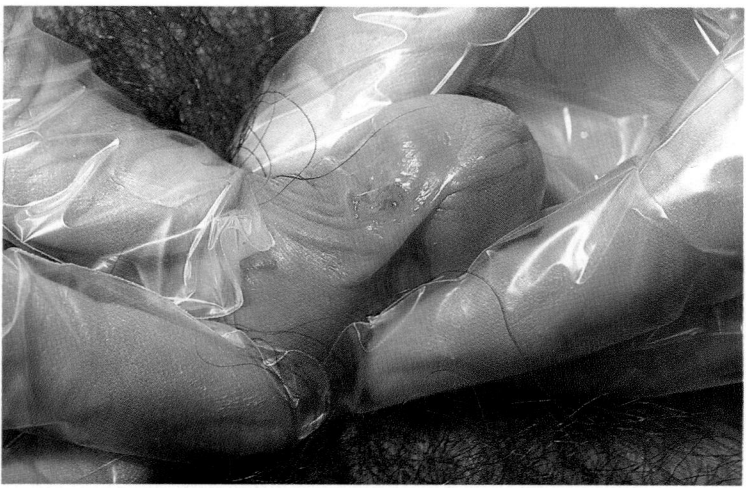

Fig. 10.5 The painless primary chancre of syphilis.

(a)

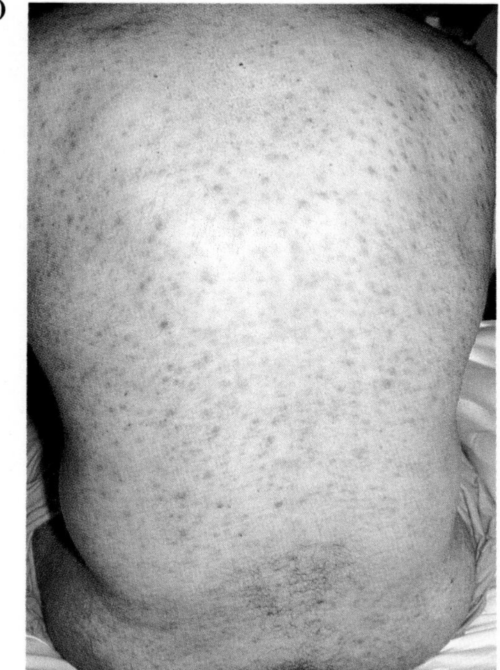

(b)

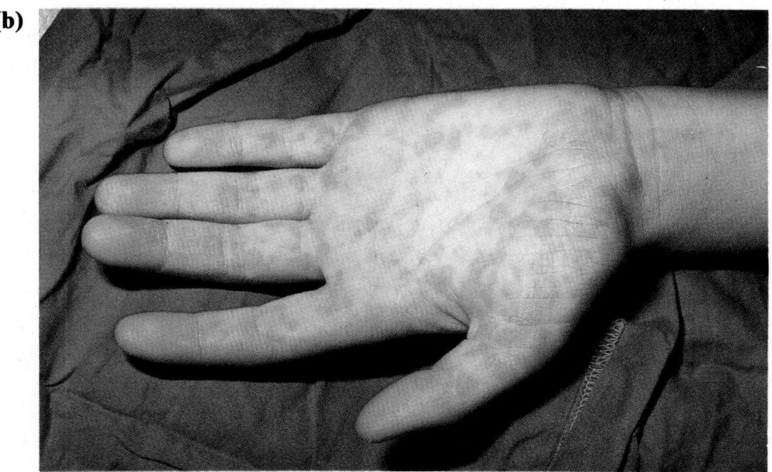

Fig. 10.6 (a) and **(b)** The rash of early syphilis with typical involvement of hands.

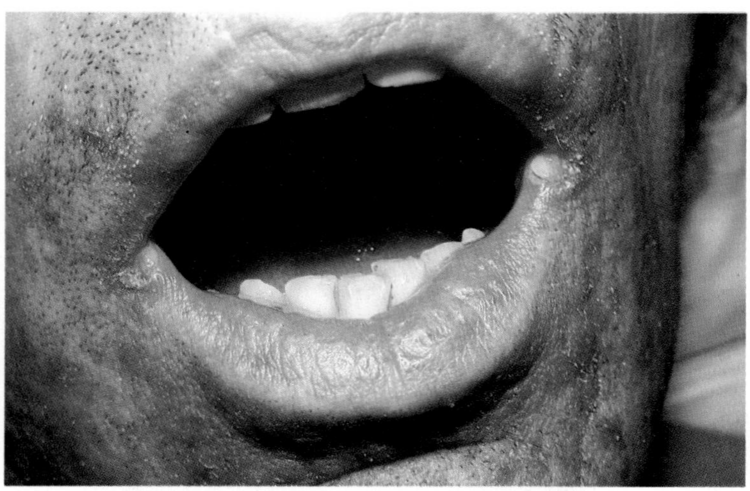

Fig. 10.7 Angular stomatitis due to *T. pallidum*—highly infectious.

(meningovascular syphilis). Aneurysm formation classically affects the ascending throacic aorta with coronary artery insufficiency and aortic incompetence.

Neurosyphilis usually presents after more than 15 years and may be due to vascular damage or direct invasion of nerve tissues. Abnormalities of the pupil include an irregular margin and the small 'Argyll Robertson' pupil, which reacts to accommodation but not to light. Degeneration of the posterior columns (tabes dorsalis) causes sensory loss, ataxia, loss of tendon reflexes, and pain, characteristically intense and stabbing, usually in the legs. Autonomic involvement can cause vomiting, rectal, bladder, and laryngeal symptoms. Paresis and/or dementia ('General Paralysis of the Insane') may not be easily distinguishable from many other causes of neurological and psychological disease.

Investigations

Spirochaetes may be seen in exudates or lymph node aspirates by dark field or phase-contrast microscopy. Negative serological tests in early disease are not unusual; repeated tests for up to a year after exposure may be necessary to detect diagnostic changes. Serological tests based upon treponemal antigens are preferred and diagnosis must not depend solely upon a non-specific test (Table 10.3). Cerebrospinal-fluid changes (Table 10.4) are generally reversible by adequate treatment.

Treatment

Penicillin is the treatment of choice, preferably by daily injections of procaine penicillin for two weeks for early or three weeks for late disease. Oral treatment is

Table 10.3. *Serological tests for syphilis*

Non-specific antibodies
  Veneral Disease Research Laboratory test (VDL)—varying titres can reflect activity
  Rapid plasma reagin (RPR)—a quick screening test
Specific antitreponemal tests
  Treponema pallidum haemagglutination (TPHA)—simple but negative in early disease
  Fluorescent treponemal antibody test (FTA)—useful in early disease, 'problem cases' and confirmation
  Treponema pallidum immobilization (TPI)—a complex test mainly for reference and research purposes

Table 10.4. *The cerebrospinal fluid in neurosyphilis*

Raised lymphocyte count		
Raised total protein and sometimes IgG		
Pressure raised in meningovascular disease		
VDRL test	—	many false negatives
TPHA and FTA tests	—	usually positive (more sensitive tests) but may fail to return to normal after effective treatment

not recommended, because of uncertain absorption and compliance. An alternative is a single large dose of long-acting benzathine penicillin. Erythromycin or tetracycline can be given when the patient is allergic to penicillin. On starting treatment fever, rigors, and general aches and pains often occur after the first dose in early syphilis, an example of the Jarisch-Herxheimer reaction, an allergic response to dead or dying organisms. These reactions are rarely dangerous.

Prognosis and prevention

With adequate treatment the outlook is good if long-term follow-up confirms reversion of specific serological reactions to negativity or very low levels without titre rises to indicate reinfection.

However, in late disease with structural damage to functioning organs recovery is less certain, and any improvement is usually evident within two or three months. Pain, gastric crises, urinary symptoms, and trophic changes in skin and joints may require symptomatic treatment or occasionally surgery. As with other sexually transmitted diseases, prevention depends upon education about the risks of promiscuity; early diagnosis; adequate treatment; follow-up, and effective contact tracing. Antenatal screening can help to prevent congenital infection. No immunization is available.

Genital warts

Human papillomaviruses cause genital warts (condylomata acuminata). The types of these viruses that transmit and cause genital infections apparently differ from those that cause skin warts in other sites. Warts occur anywhere on the penis, the urethral meatus, the vulva, vagina, and cervix (Fig. 10.8), also in the rectum and perianal region especially but not exclusively in homosexuals (Fig. 10.9). The warts are generally larger than common skin warts and must be differentiated from syphilitic chancres. They are treated by local application of podophyllin, by cryosurgery (freezing), electrocautery or surgical (scissor) excision. Papillomaviruses have been associated with pre-malignant and malignant disease of both the male and female genital tract. Colposcopy may be required to detect some cervical lesions.

OTHER CONDITIONS

Molluscum contagiosum, scabies, and lice (see Chapter 3) can also affect the genitalia and be spread sexually. Pediculosis pubis due to the crab louse, *Pthirius pubis* is usually confined to pubic hair and transmitted sexually. The main symptom is itch, and the lice are usually large enough to see with the naked eye. As with other lice, γ-benzene hexachloride or malathion provide effective treatment.

The following conditions are mainly encountered in tropical areas:

Chancroid

This is the commonest cause of genital ulceration in Africa and south-east Asia. After an incubation period of up to three weeks painful papules and pustules usually appear. Local lymph nodes enlarge, become matted, painful, and may suppurate with sinus formation. Relapse is common. The responsible organism, *Haemophilus ducreyi*, may be visualized by Gram staining material or pus from ulcers or affected lymph nodes. Special culture methods are required to confirm diagnosis. Other causes of genital ulcers such as syphilis must be excluded. Treatment with cotrimoxazole or erythromycin is usually effective.

Lymphogranuloma venereum

This chlamydial infection is less common than chancroid and occurs throughout the tropics. The incubation period is from one to four weeks. The initial lesion is

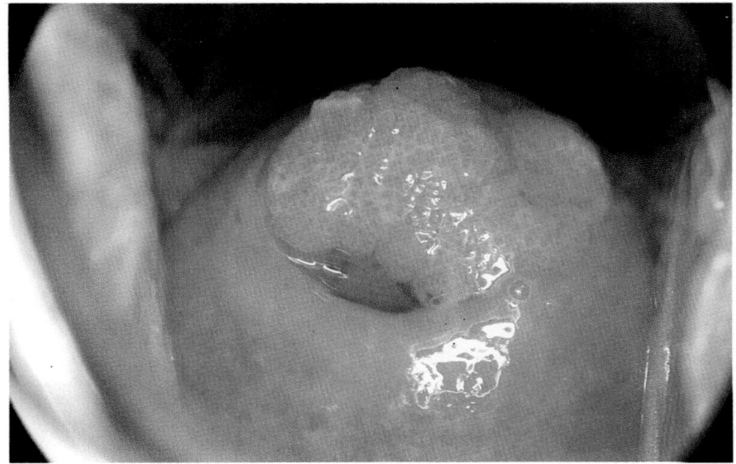

Fig. 10.8 A large cervical wart due to papillomavirus.

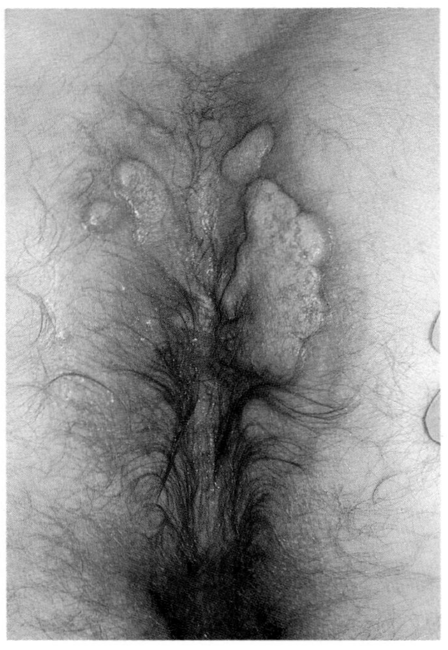

Fig. 10.9 Perianal warts are also spread sexually.

usually a solitary, painless papule or vesicle which may be confused with herpes. This lesion may not be noticed and disease may then present with discrete, firm tender lymphadenopathy progressing to matting, suppuration, and sinus formation (Fig.10.10). There may be systemic spread. Other causes of lymphadenopathy must be excluded. Diagnosis is confirmed by isolating

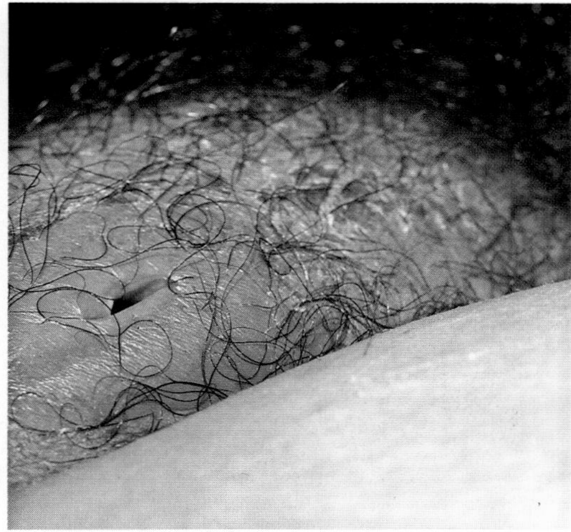

Fig. 10.10 Lymphogranuloma venereum: discoloured, indurated inguinal mass above thigh with discharging sinus.

Chlamydia trachomatis (LGV strains) from aspirates, or by immunofluorescent or complement fixation tests. Treatment is with tetracycline or erythromycin.

Granuloma inguinale

Caused by *Calymmatobacterium granulomatis*, this is most often seen in south-east Asia. After an incubation period from a few days to three months, the primary lesion grows from a small papule to a beefy granuloma often with 'rolled edges', usually painless and bleeding easily. It can affect extragenital sites such as inguinal skin, thighs and anus. Lesions spread only slowly without lymphadenopathy but there may be secondary infection. Diagnosis is by demonstrating characteristic 'Donovan bodies' (bipolar capsulated bacilli within macrophages) in Giemsa-stained affected tissue. Treatment is usually with cotrimoxazole, tetracyclines, or chloramphenicol.

11

Infection in pregnancy and neonates

INTRODUCTION

The uniquely intimate relationship between the developing fetus or neonate and the mother makes transmission of certain organisms between them likely. During pregnancy transmission is usually blood-borne via the placenta. Once labour starts, organisms within the dilating cervix and vagina inevitably infect the previously sterile fetus. After delivery, intimacy continues during normal handling and breast-feeding and the newborn infant becomes infected (or colonized), though not necessarily with symptoms, at many sites, particularly the skin, the gastrointestinal and upper respiratory tracts. Normally these organisms come mainly from the mother. Protective factors such as IgG antibodies against many of them will already have been transferred from the mother's blood via the placenta during pregnancy. Other protective factors such as secretory IgA continue to be transferred within breast milk.

Hospital delivery brings newborn children together to be handled by many 'strangers' carrying their own mixture of commensals and potential pathogens, in an environment likely to induce drug resistance in organisms. This increases the chance of meeting organisms against which only immature immunological defences are available. Likewise, the delivered mother has a raw uterine surface accessible to her own commensals and to organisms from the hospital environment, so that postpartum infection is common.

In addition to the closeness of mother–baby relationship and the type of infecting organisms, the normal defences of both mother and newborn are low—the former perhaps relating to her need to carry an immunogenically distinct being, the latter from immunological immaturity. In the mother during pregnancy both cell-mediated and antibody-mediated immunity are mildly depressed. The fetus itself can produce IgG immunoglobulin from week 12 and IgM from around week 30. It also receives passive protection by maternally-derived IgG but not from maternal IgM, which cannot cross the placenta. Inadequacy of several defence mechanisms like neutrophil locomotion or antibody production may be exaggerated in premature or ill neonates.

Thus both subnormal defences and intimate contact predispose to certain infections (Table 11.1) and may increase their severity. However, infection is probably responsible for less than 10 per cent of all congenital defects, and most infections during pregnancy have no long-term consequences.

Table 11.1. *Infections causing special problems in pregnancy and the postpartum period*

Infection	Main problems
Rubella	Fetal—cataract, cardiac anomalies, deafness
Cytomegalovirus	Fetal—mental retardation, generalized infection
Toxoplasmosis	Fetal—posterior uveitis, generalized infection
Hepatitis B	Infantile: chronic carriage
Herpes simplex	Neonatal—generalized infection
Varicella	Fetal—scarring Neonatal—generalized infection Maternal—pneumonia
Enteroviruses, especially coxsackieviruses	Neonatal—generalized infection, carditis
Parvovirus	Fetal—hydrops foetalis, abortion
Human immunodeficiency virus (HIV)	Diagnosis, chronic HIV carriage, early AIDS
Syphilis	Fetal—stillbirth Neonatal—generalized infection
Tuberculosis	Neonatal infection
Malaria	Fetal—abortion, stillbirth, prematurity Maternal—anaemia, death in non-immunes
Listeriosis	Fetal—abortion, stillbirth Neonatal—septicaemia, meningitis, respiratory distress Maternal—septicaemia, meningitis
Enterobacteriaceae *Strep. agalactiae*	Neonatal—septicaemia, respiratory distress, meningitis
Tetanus	Neonatal—rigidity, spasms, mortality
Chlamydia	Neonatal—conjunctivitis, pneumonia
Gonococcus	Neonatal—conjunctivitis, arthritis, meningitis
Septic abortion	Abortion, maternal septicaemia
Postpartum intrauterine sepsis	Maternal septicaemia
Breast abscess	Neonatal infection
Urinary tract infection	Maternal—pyelonephritis Fetal—prematurity, low birthweight

SPECIFIC INFECTIONS

Congenital rubella syndrome

Prior to rubella immunization, congenital rubella was estimated to be responsible in Britain for 15 per cent of childhood nerve deafness and 20 per cent of congenital heart defects in *non*-epidemic years—an estimated 300 cases of severe impairment per year. More recently in Britain after the introduction of MMR vaccine in the second year of life, reported cases have fallen to single figures, up to a half with multiple defects. Also, because MMR vaccine reduced transmission of wild rubella virus, the number of therapeutic abortions associated with rubella has fallen below 100 annually.

Rubella viraemia is normally significant only during primary infection, i.e. before antibodies are present. Once the virus has reached the maternal bloodstream it can cross the placenta at any time of pregnancy, particularly in early and late stages. The presence of 15 i.u. (International Units) of antibodies derived from natural infection or vaccination is deemed protective against viraemia. Reinfection is usually detected only by a rise in antibody levels not by clinical symptoms, and infrequently causes damaging congenital infection.

Rash in the mother indicates that viraemia has occurred. The effect on the fetus of maternal rubella with rash depends first on the fetus becoming infected (Table 11.2). The effect on the infected fetus varies mainly with gestational age. The virus slows both replication of cells and their differentiation into different cell types so that major defects, usually multiple, follow infection during the first trimester when organogenesis takes place. Infection early in the second trimester may cause deafness but those infected after week 16 suffer no major abnormalities (Table 11.2). Intrauterine confirmation of fetal infection from maternal rubella may be obtained by fetoscopy. Congenital infection is considered to have

Table 11.2. *Risks of damage to fetus by maternal rubella during pregnancy*

Stage of gestation (weeks) when mother infected (symptomatic plus asymptomatic)	Percentage of fetuses infected	Percentage of infected fetuses damaged	Overall risk of damage to fetus (per cent)
<11	90	100	90
11–16	55	37	20
17–26	33	0	0
27–>36	53	0	0

E. Miller, J. E. Cradock-Watson, and T. M. Pollock, *Lancet*, 1982, **ii**, 781–4.

occurred if the infant has IgM antibodies shortly after birth or if IgG antibodies persist for more than six months (when maternally derived antibodies would have waned).

Clinical presentation of congenital rubella

At birth, affected infants may be light-for-dates and have multiple defects (Fig. 11.1) such as the classical triad of patent ductus arteriosus, cataracts, and sensorineural deafness. Slight hearing loss may be the only defect and auditory evoked responses should be tested at 3 months of age; long-term follow-up is advisable since special education may be required. Children with congenital rubella can continue to excrete the virus, particularly in oral secretions and urine, and even at one year virus can be isolated from almost 10 per cent. Since they may be a source of infection, contact with non-immune, possibly pregnant women should be avoided.

In congenital rubella the child's IgM response is higher and more prolonged than in postnatal rubella. The IgG response falls more rapidly, may disappear in childhood, and such patients should not be considered immune but offered rubella immunization by the routine schedule.

Clinical problems in the pregnant mother

Two problems arise: (i) the pregnant mother who develops a rubelliform rash (Fig. 11.2), and (ii) the pregnant mother who comes into contact with someone with a rubelliform rash (Fig. 11.3). It is wise to assume initially that the rash of the contact was due to rubella—this can be evaluated by serological tests of the contact. Transmission is much more likely when contact was close or with a

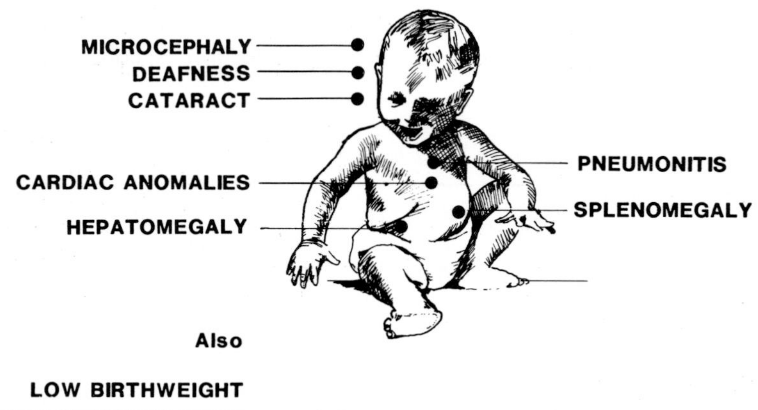

Fig. 11.1 Major clinical signs of congenital rubella. Many combinations of these signs may be present.

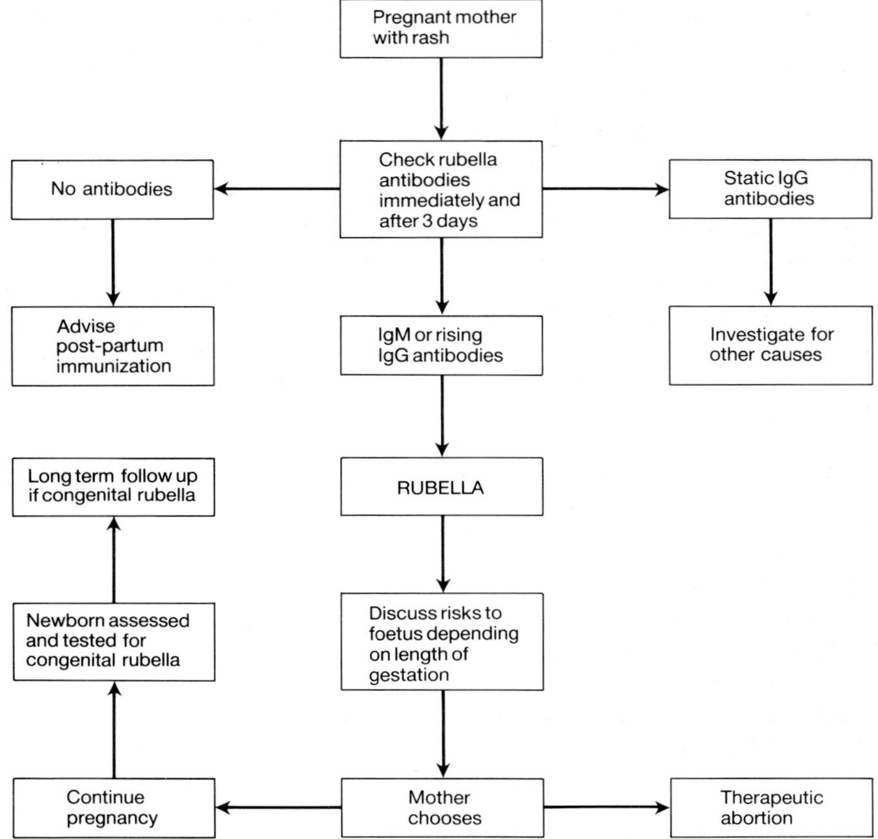

Fig. 11.2 Management of a pregnant mother with rubelliform rash.

family member. Management (Figs 11.3 and 11.4) may involve administration of hyperimmune rubella globulin: since this protects only about 40 per cent of cases it is only considered for those who do not wish therapeutic abortion should rubella develop. It is recommended only within five days of contact and is not normally recommended after week 20 of pregnancy, by when risk of serious fetal damage is negligible. Hyperimmune globulin may prolong the incubation period of rubella for up to three months.

Prophylactic immunization

The immunization programmes aim to prevent primary rubella in pregnant women and thus eliminate congenital rubella. Two effective and safe live attenuated vaccine strains are RA27/3 and Cendehill. RA27/3 is more immunogenic, results in higher antibody levels, longer protection, and lower rates of reinfection, but causes slightly more reactions (malaise, fever, mild rash,

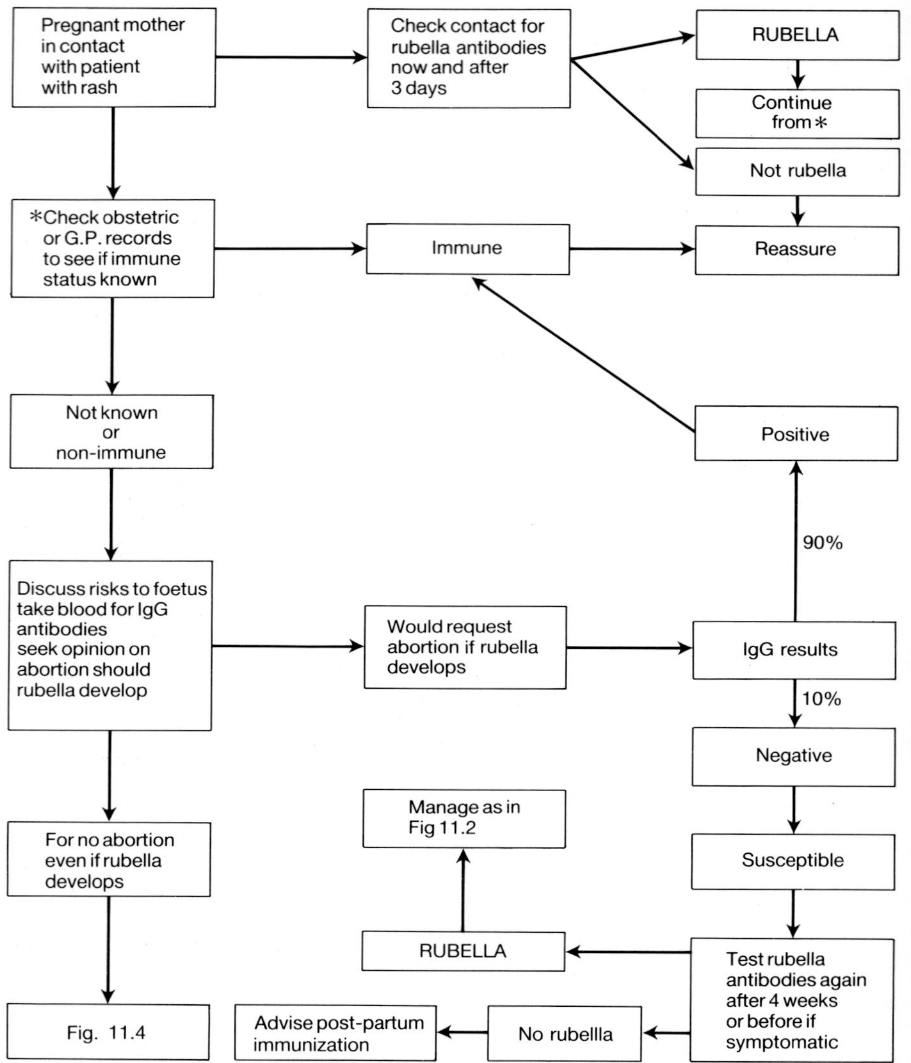

Fig. 11.3 Management of a pregnant mother in contact with rubelliform rash.

and arthritis) and is used in Britain: seroconversion follows in over 95 per cent. Immunity is long-lasting and only a few lose protective antibodies after 16 years. Although fetal infection has followed inadvertent administration of vaccine during pregnancy, no fetal damage has been recorded and therapeutic termination is not recommended in those circumstances. However, as a precaution, pregnancy is considered an absolute contraindication to rubella immunization and recipients should be advised not to become pregnant over the

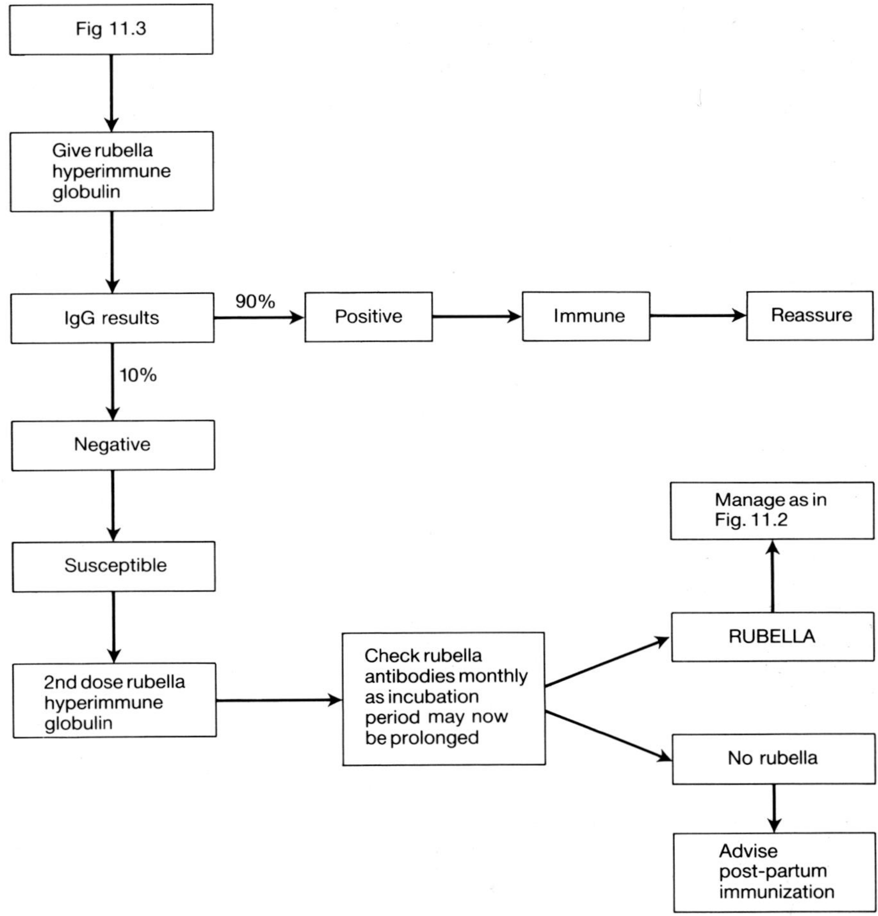

Fig. 11.4 Further management of pregnant mother with rubelliform rash who would not seek therapeutic abortion.

next three months. The vaccine strains are not transmitted from person-to-person.

From 1988 live attenuated measles, mumps and rubella (MMR) vaccine has been offered to all children in Britain in the second year of life and current uptake is about 90 per cent. This is similar to US policy where reported rubella and congenital rubella syndrome have been reduced to low levels (Fig. 11.5). UK figures now show a similar trend. Meanwhile the previous UK policy of immunizing 11- to 14-year-old girls will be maintained at least until those receiving MMR reach this age group.

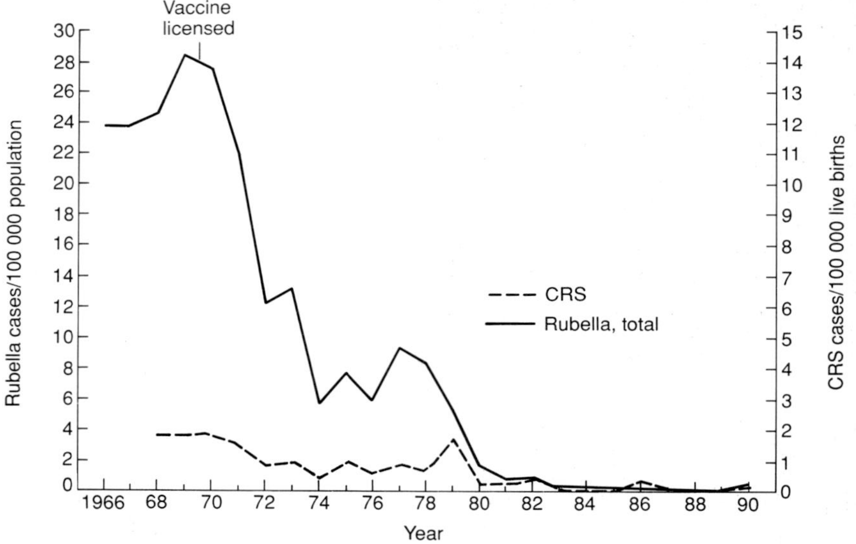

Fig. 11.5 Incidence rate of reported rubella and congenital rubella syndrome (CRS) cases—United States, 1966–90. *Morbidity and Mortality Weekly Report* (US Department of Health), 1991, **40**, 93.

Cytomegalovirus infection

Cytomegalovirus (CMV) is the commonest recognized cause of congenital infection (in 1 per cent of pregnancies). It is transmitted transplacentally. After primary infection in pregnancy up to 50 per cent of newborns show evidence of infection (IgM specific antibodies or isolation of CMV from throat or urine in the first two weeks of life) as opposed to 6 per cent following reinfection or reactivation (like other herpes viruses, CMV remains latent after primary infection). After reactivation or reinfection in the mother, the infant is less likely to show clinical effects, whereas about 10 per cent of infants infected during primary maternal infection are ill. Typical features of severe illness are low birthweight, microcephaly, hepatosplenomegaly, jaundice, and petechial rash (Fig. 11.6). Long-term sequelae are varying degrees of intellectual impairment and hearing-loss developing in the pre-school years—most prominent in those symptomatic at birth but also affecting 10 per cent of initially symptomless children. Infants infected by primary maternal infection also excrete much more virus in their urine and are potential sources of infection for up to two years.

Routine screening for CMV infection is not performed during pregnancy because of cost, and because interpretation is difficult, since the presence of antibodies does not prevent reactivation. Because most primary maternal infections are asymptomatic in the mother, infection in neonates is only

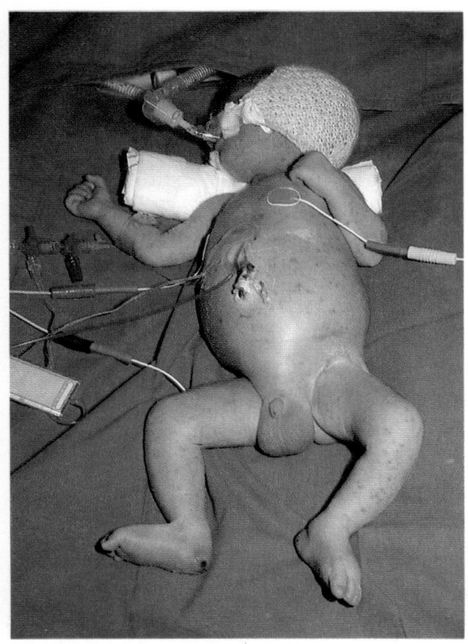

Fig. 11.6 Cytomegalovirus, congenital infection, day 5: hepatosplenomegaly, haemorrhagic rash, and jaundice present from birth.

recognized in the few that show symptoms. For these, long-term follow-up is indicated to assess normal development and provide appropriate care.

Direct infection of infants during labour and the postpartum period is also common, as the virus is readily excreted from the cervix and in breast milk; but these infections are usually asymptomatic. They may also be infected by blood transfusions.

Toxoplasmosis

Aetiology and epidemiology

Toxoplasma gondii is distributed world-wide with many vertebrate hosts. Its sexual cycle occurs only in the intestinal mucosa of cats and, although tissues of other vertebrates may contain cysts, their faeces are not infectious. Human infection is mainly acquired by eating poorly cooked meats, especially pork and mutton, and by faecal–oral spread from cats. Only *primary* infection in the pregnant mother may result in transplacental transmission of *T. gondii*, estimated at 2 per 1000 pregnancies in Britain. Fetal infection occurs in about half of these primary maternal infections and is more likely in late pregnancy, though the consequences of early infection are more severe.

Clinical presentation and investigation of congenital toxoplasmosis

Fetal infection in early pregnancy is likely to cause spontaneous abortion. About one-fifth of the fetal infections show clinical features at birth, usually minor such as retinal scars. Intracerebral calcifications may be found radiologically. Major signs such as hepatosplenomegaly, hydrocephaly, or microcephaly and chorioretinitis are infrequent. As most maternal infections are asymptomatic, most congenital infections are not detected. Where the newborn infant shows signs the diagnosis is confirmed serologically. Specific IgM antibodies are detectable in 75 per cent of infected infants at birth. Undetected congenital infection usually becomes symptomatic in childhood or even adult life as, for instance, chorioretinitis.

Acquired postnatal toxoplasmosis

Common features of the few symptomatic acquired infections are lymphadenopathy, malaise, and fever, sometimes with maculopapular rash, lasting several weeks or months. More severe or fatal disease with encephalitis may affect immunocompromised patients, and latent toxoplasmosis may be reactivated by immunosuppression. Disseminated infection may follow organ transplants.

Management

Pyrimethamine and sulphonamide treatment is not often required or helpful in post-natal toxoplasmosis in the immunocompetent, but given to the infected mother may reduce the risk of congenital infection. Pyrimethamine should not be used in the first trimester because of potential toxicity but spiramycin can be used alone or with sulphonamide. Congenitally infected neonates should be treated.

Prevention

Meat should be well-cooked before consumption. Contact with cat faeces should be avoided and hands washed after handling meat or gardening, *especially* during pregnancy. In some countries, but not the UK, routine antibody screening is performed in female school-leavers or in early pregnancy to check for immunity. Serological screening is repeated in the second and third trimesters, any primary infections treated, subsequent newborns assessed for infection, and long-term follow-up arranged. In Austria this reduced the incidence of congenital toxoplasmosis over 10 years by 50–70 per cent.

Hepatitis B

Transplacental infection with hepatitis B virus (HBV) is uncommon. Most transmission occurs during parturition, probably from materno-fetal bleeding. Those at greatest risk of transmitting infection are mothers who are HBe antigen positive from recent acute infection or from chronic carriage: about 90 per cent of their infants become infected. Less than 10 per cent of infants of chronic HBsAg carriers with anti-HBe become infected.

Almost all neonatal infections are asymptomatic but lead to a much higher chronic carriage rate than infection acquired later in life. Nearly all those infected by HBeAg positive mothers become chronic carriers; few of those infected by anti-HBe-positive mothers do so. Early HBV infection and subsequent chronic HBsAg carriage is associated with development of glomerulonephritis, cirrhosis, and primary hepatocellular carcinoma in adult life. Prevention is thus desirable but, to be cost-effective in areas of low endemicity, is only possible if HBsAg carrier mothers are first identified. In a British population with low carriage rates, routine screening to detect these carriers would be very expensive though now recommended in the USA. However, those with acute or chronic liver disease can be tested, as also mothers from groups with high carriage rates; for example, Chinese or drug abusers. In areas of intermediate or high prevalence (2–20 per cent HBsAg carriage rates) all neonates should be immunized.

When a carrier mother has been identified, labour-room staff should use gloves, masks, and goggles, with appropriate cleaning or disposal of contaminated instruments and linen. The newborn infant should be washed thoroughly and given hyperimmune globulin as soon as possible. Hepatitis B vaccine (Chapter 16) should be given at a different site at the same time or within 48 hours, with boosters at one and six months. This regime reduces the risk of infection in infants of HBe positive mothers to below 20 per cent, although many of those infected still become chronic carriers. Infants to whom this regime has been given should be routinely tested for HBsAg and HBsAb at 15 months to check for chronic carriage. HBsAg can be found in breast milk but it is not known whether this is an important source of infection.

Herpes simplex infection

Transplacental transmission of herpes simplex virus (HSV) (see also Chapters 3, 4, and 10) is uncommon and infection is usually acquired from the birth canal during labour. Since two-thirds of mothers whose infants develop neonatal HSV infections have neither signs nor symptoms of disease, those infants who might benefit from prophylaxis cannot easily be identified. Transmission from mothers with overt genital herpes can be reduced by caesarean section within four hours of membrane rupture. About half of those infants delivered through a birth canal actively infected with herpes develop infection, and prophylaxis with acyclovir is advisable though its efficacy is yet to be proved. Mothers with reactivated herpes are less likely to transmit infection, possibly because of transfer of neutralizing antibody to the fetus and the lower infecting dose.

Clinical presentation and management

Neonatal herpes begins in the first and second weeks of life and is rarely asymptomatic. It may initially be localized to skin, eye, or mucous membranes but in about 80 per cent spreads to the meninges or more widely. Untreated mortality for localized intracranial infection is 50 per cent and for disseminated disease 85 per cent. The most readily recognizable sign is the herpetic skin vesicle (Fig. 11.7). Vesicles are present at onset in most but do not always appear or do so

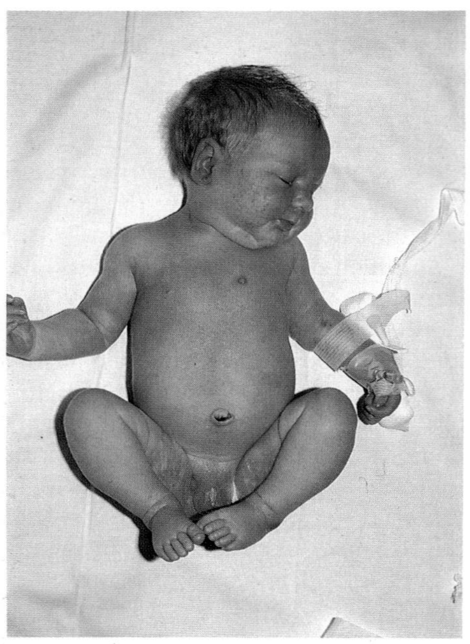

Fig. 11.7 Herpes simplex. Small cluster of vesicles on chest wall and ear lobe. Lesions also present on scalp and axilla.

only later after dissemination. The diagnosis should be considered in any very ill infant, especially with no other obvious or bacterial cause, and confirmed by viral cultures of throat and conjunctival swabs, CSF, urine, and any vesicle fluid. Maternal vaginal swabs should also be cultured for HSV.

Parenteral acyclovir is effective and should be used immediately the disease is suspected. Its prophylactic use should be considered in neonates delivered vaginally during acute infection.

Varicella

Maternal varicella in pregnancy is often more severe than in other adults, reflecting depression of cell-mediated immunity. The rash may be denser and evolve more slowly. Development of pneumonitis is life-threatening and more likely in smokers and in the second and third trimesters. Shortness of breath initially develops on exertion alone but also at rest as the process evolves. Auscultation is often normal despite radiological evidence of widespread interstitial shadowing (Fig. 11.8). Blood gases reveal hypocapnia and hypoxia. Further evolution can lead to diffuse interstitial fibrosis and death. Acyclovir given early speeds resolution and appears to prevent fetal damage. Scattered calcification may develop later in areas of pneumonitis.

Varicella-zoster virus can cross the placenta at any stage of pregnancy. Very

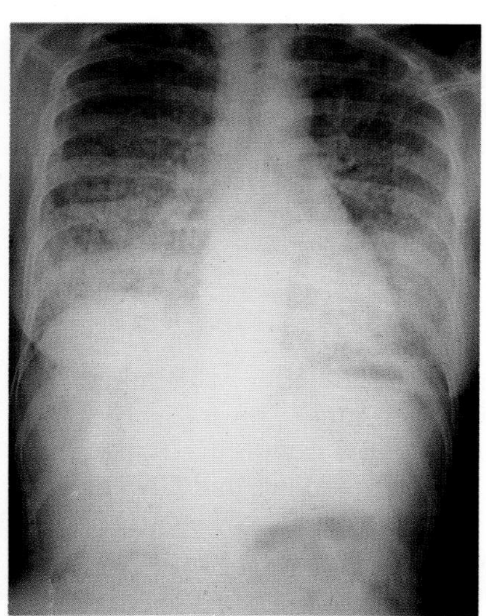

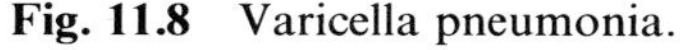

Fig. 11.8 Varicella pneumonia.

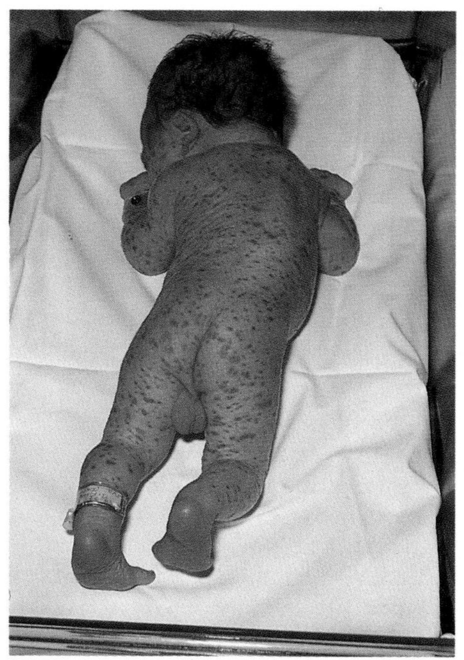

Fig. 11.9 Congenital varicella. Moderately heavy rash but normal appearance of lesions.

rarely, varicella in early pregnancy causes fetal limb hypoplasia, skin ulceration, and scarring. More often neonatal problems (Fig. 11.9) result from maternal varicella infection shortly before or after term. Disease is more severe in the neonate the closer to term maternal infection arises—especially in the last four days when neonatal skin lesions are bigger and slower to resolve (like varicella in the immunocompromised) and mortality rates have been over 30 per cent. Apparently in such cases maternal viraemia leads to placental transmission of virus but not of significant amounts of protective maternal IgG antibodies which have yet to be produced. Infants born to mothers who developed varicella in the previous four days should receive zoster immunoglobulin (ZIG). This should also be given to infants of mothers who develop varicella in the first few days postpartum, since viraemia may have been present before delivery. Where supplies of ZIG are adequate protection can be extended to neonates whose mothers developed varicella from one week before birth to four weeks after. Use of ZIG has virtually eliminated mortality. Acyclovir may also be used in neonatal varicella. It can be life-saving when varicella pneumonitis occurs in pregnancy. When varicella affects the fetus or neonate, zoster may follow at an early age.

Enteroviral infection

Epidemiology

Infection may be transmitted congenitally, at birth, or postpartum with neonatal illness following approximately within five days, 5–10 days, or later after birth, respectively. Most severe disease is transplacentally transmitted and the mother usually has a history of mild respiratory or flu-like illness a few days before birth. Coxsackievirus B infections have the highest mortality, followed by echoviruses and coxsackie A viruses. Neonatal infection with polioviruses is uncommon. Live oral poliomyelitis vaccine given during pregnancy has not been shown to be harmful, though this should be avoided.

Clinical presentation and mangement

Severe coxsackie B illness in the neonate may be biphasic with initial mild respiratory catarrh, vomiting or diarrhoea, followed later by severe, febrile illness as most body organs are involved, particularly myocardium, CNS, liver, and lungs. Severe echoviral disease affects liver, adrenals, kidney, and CNS and can be fatal. Milder, mainly asymptomatic illness is more usual, particularly in older infants. Disseminated bacterial or herpes simplex infections must be excluded. Diagnosis is supported by isolation of virus from faeces or throat swab from neonate or mother. In early onset disease the mother has high titres of IgM coxsackie B antibodies. Treatment is supportive. Normal human immunoglobulin may protect other neonates in the nursery, but strict attention to handwashing by attendants and other measures to minimize cross-infection are more important and can control threatened outbreaks.

Parvoviral infection

Human parvovirus appears to cross the placenta in up to a quarter of pregnant women with acute parvoviral infection. Most of these pregnancies continue normally to term with delivery of healthy babies and without apparent sequelae. A small excess risk of spontaneous abortion is apparent up to 18 weeks and a few infected fetuses later develop hydrops foetalis. This has been improved by intrauterine transfusions.

HIV infection (AIDS)

Pregnancy itself does not appear to speed disease progression in women with HIV antibodies or AIDS. However, viral transmission to the fetus, mainly intrauterine rather than perinatal, is common. Termination of pregnancy may be considered in HIV carriers. All the newborns have passively acquired maternal antibodies but only between a quarter and a third are infected. Infection may not be confirmed until 18 months by which time passive antibody is undetectable. Breast-feeding may transmit HIV infection, but is not a major route of transmission and, although in the developed world general advice would be against breast-feeding, in the developing world the advantages of breast-feeding greatly outweigh the small risk of transmission.

Syphilis

Fetal infection from the mother is most likely during early and early latent stages of syphilis. Late abortions and stillbirths are increased; of survivors about half show clinical features, the rest only serological evidence of infection. Signs that may be present at birth or develop shortly after are rashes (especially bullous), hepatosplenomegaly, and snuffles. The infant may be of low weight, fail to thrive, develop anaemia, and show osteochondritis and periostitis on X-ray. Signs may only appear later and then resemble acquired late stage (Chapter 10) syphilis, but interstitial keratitis, VIIIth nerve deafness, and Clutton's joints (hydrarthroses) are peculiar to congenital infection. Typical characteristics of the older patient with congenital syphilis result from inflammatory and healing processes: rhagades (fine scars radiating from mouth, nostrils, or anus, originally moist fissures), saddle nose (from cartilage destruction at the time of snuffles), altered dentition from damage to growing permanent teeth (Hutchinson's peg teeth, incisors with a prominent notch; or Moon's molars, rounded with ill-defined cusps), choroiditis, and sabre tibia—anterior bowing as a consequence of periostitis.

Management

Congenital syphilis is preventable by penicillin given to the infected mother. Because of the long course of the infection and variability of symptoms and signs in females, VDRL and TPHA tests are routinely performed in Britain on sera from mothers at the first antenatal booking visit. A positive result requires further clinical and laboratory assessment (Chapter 10). If current disease is diagnosed or if there is doubt, procaine penicillin should be given for two weeks; contact tracing should include previous offspring.

The newborn child should be examined for signs of infection and investigated serologically. Treponemes may be visualized by dark ground microscopy in nasal discharge. Most serological tests for syphilis are positive in infants of syphilitic mothers because of transfer of maternal IgG, but titres should fall over the next three months if the infant is uninfected. Rising titres or specific IgM strongly suggest congenital infection.

Early antenatal screening does not detect the few mothers who become infected in the second or third trimesters; they and their offspring go undetected unless clinical signs in either suggest the diagnosis. If diagnosed, *both* should be treated with procaine penicillin.

Tuberculosis

Asymptomatic maternal tuberculous disease, unless miliary, does not usually affect the outcome of pregnancy beyond the risks of any associated malnutrition. There is no particular risk from giving correct doses of standard antituberculous drugs in pregnancy, apart from the small risk of fetal ototoxicity with streptomycin which should not be given during pregnancy.

Mycobacterium tuberculosis can infect the placenta, but the main risk of transmission to the child is after birth from infectious maternal sputum. If the mother's sputum reveals acid-fast bacilli on microscopy, prophylactic isoniazid given for 3 months to the newborn allows mother and child to stay together. In Britain BCG is offered to newborns of infectious and non-infectious mothers. Normal BCG remains immunogenic even when isoniazid is being taken.

Malaria

Pregnancy may induce relapses of vivax malaria and cause falciparum infection in immune women to become symptomatic (Chapter 14). Falciparum malaria especially may mimic obstetrical conditions such as pyelonephritis or eclampsia. Acute illness may precipitate abortion or premature labour; haemolysis and subsequent anaemia can retard intrauterine growth. Congenital malaria is rare, although placental infection is common in endemic areas. Chemoprophylaxis (Chapter 16) should be taken by pregnant as by other travellers to endemic areas. The newborn infant in an endemic area should receive chemoprophylaxis from birth and continue breast-feeding.

Listeriosis

Listeria monocytogenes is widespread in animals, plants, soil, and water. Human faecal carriage rates can be up to 5 per cent of the general population. Reported cases increased over the last two decades but this was reversed in 1989 for reasons which are unclear. Asymptomatic infection can develop at any age. Susceptibility is increased in the second and third trimesters, and the organism has a predilection for the fetus. Maternal illness is often mild with little more than fever and sweating, sometimes meningitis and/or septicaemia. It usually responds readily to ampicillin, when pregnancy continues with eventual delivery of an uninfected child. Occasionally, even mild maternal illness causes abortion or stillbirth with a severely infected fetus. Pregnant women should avoid unwashed root vegetables and recognize a risk of infection in soft cheeses.

Listeria can be a vaginal commensal. Disease may appear early in the neonate infected during labour or later in the early postpartum period. *Early disease* is associated with low birthweight and with factors increasing the risk of congenital infection such as prolonged rupture of membranes and genital carriage. It appears within the first 24–48 hours with respiratory distress and septicaemia.

Radiological evidence of pneumonia is often present and pink skin papules are typical. Mortality is high. *Late disease* presents as meningitis after about two weeks, usually in normal-birthweight infants. Diagnosis is confirmed by isolation of the organism from blood, cerebrospinal fluid, or respiratory secretions. Otherwise, serology may be helpful. Ampicillin should be included in initial empirical antibiotic cover for neonates with suspected infection.

Conjunctivitis

Epidemiology

Ophthalmia neonatorum is defined as conjunctivitis developing within the first 21 days of life. It is common and in Britain is now mainly caused by *Chlamydia trachomatis* and *Staph. aureus*, less often by herpes simplex and adenovirus. *Neisseria gonorrhoeae* is less common but remains highly prevalent in many countries.

Clinical presentation and management

Gonococcal infection presents within five days with pronounced periorbital swelling, chemosis, and purulent discharge. It may progress rapidly to cause keratitis, corneal perforation, and blindness. Chlamydial conjunctivitis presents between 5–10 days of life, ranging from mucopurulent discharge, reddened conjunctivae (Fig. 11.10a), and swollen eyelids, to 'sticky eye'. Chlamydia infection less often and less rapidly involves the cornea. Diagnosis is by identification of the agent in the discharge (Fig. 11.10b). Local treatment is with chloramphenicol for *Staph. aureus* and penicillin for *N. gonorrhoeae*. Oral erythromycin is effective in chlamydial conjunctivitis and also clears nasopharyngeal carriage. Carriage of chlamydia strains occasionally leads to chlamydial pneumonitis at 1–3 months of age, with dry paroxysmal cough, no fever, eosinophilia, raised immunoglobulins and diffuse pulmonary infiltrates on X-ray (Fig. 11.11). Tetracycline or erythromycin therapy is appropriate for the postpartum mother with chlamydial infection (see Chapter 10), and erythromycin should be given for 3 weeks to the neonate.

Prevention

Where the prevalence of gonococcal ophthalmia neonatorum is high, routine prophylaxis can be given by silver nitrate eye-drops. Silver nitrate causes mild, self-limiting chemical conjunctivitis and may not prevent chlamydial conjunctivitis.

SEPTIC ABORTION

Since the 1967 Abortion Act, illegal abortions have become unusual in Britain and the incidence of septic abortion has declined. After legal termination about

(a)

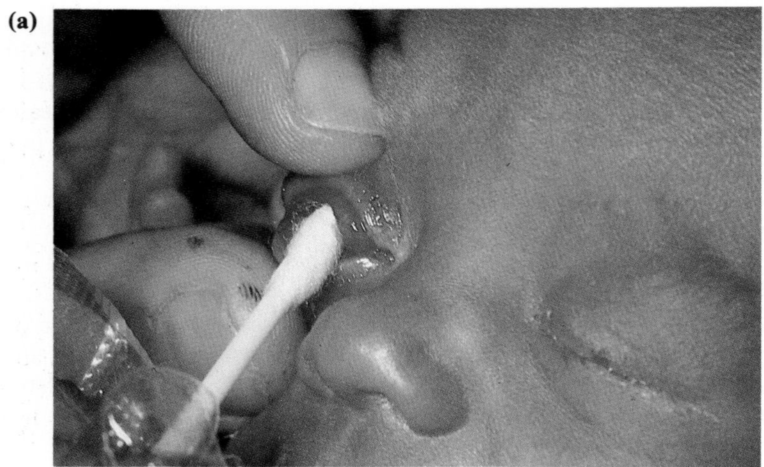

(b)

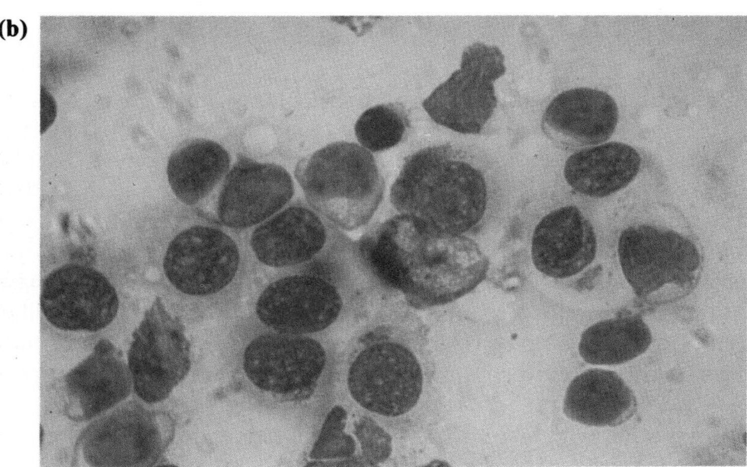

Fig. 11.10 Chlamydial conjunctivitis: **(a)** swabbing the injected conjunctival mucosa; **(b)** Giemsa stain of smear from swab with perinuclear inclusion bodies.

1 per cent have clinically significant uterine sepsis. Mixed infections are commonly responsible, usually commensals in the genital tract such as anaerobes, especially *Bacteroides* spp., *Strep. agalactiae* (group B streptococcus), *E. coli*, and *Staph. aureus*; *Pseudomonas aeruginosa* is significant in the tropics. The clinical features are fever, lower abdominal pain, and uterine tenderness and discharge. Septicaemia can cause deterioration. Management involves resuscitation of shocked patients, broad-antibiotic cover such as amoxicillin, gentamicin, and metronidazole combined, and prompt evacuation of residual uterine contents.

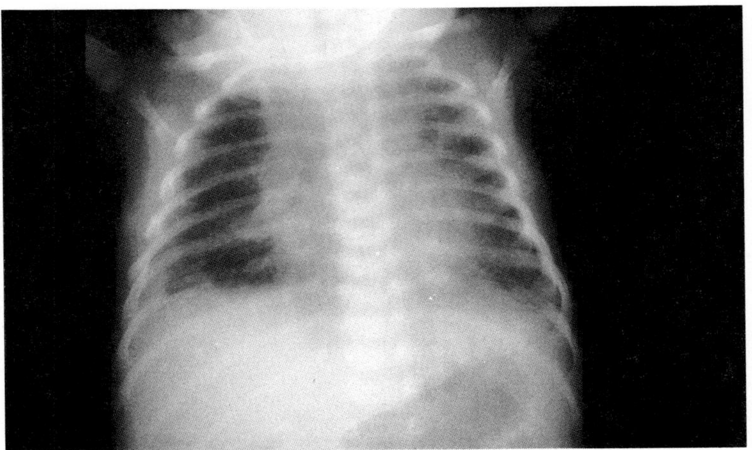

Fig. 11.11 Chlamydial pneumonitis.

NEONATAL BACTERIAL SEPSIS

Aetiology and epidemiology

Septicaemia and meningitis are commonest in the neonatal period and cause significant morbidity and mortality. Approximately 1 in 500 live-born children develop significant neonatal bacterial infection. The incidence and severity vary inversely to birthweight and maturity. Increase in intensive management of very premature infants adds its own burden of infection as already immature defences are breached by intravenous lines, umbilical catheters, endotracheal tubes, or chest drains, for example. Other factors that predispose to infection are prolonged labour or rupture of membranes, multiple vaginal examinations, and intrauterine monitoring.

Infections presenting within the first week are more severe and generally caused by maternal organisms. Those presenting later are usually acquired from other sources such as adult handlers or other neonates. *Streptococcus agalactiae* (group B *Streptococcus*) and *L. monocytogenes* (see above) can cause early or late onset disease. *Early onset* disease, often within 6–12 hours of birth, implies disseminated septicaemic infection with high mortality. *Late onset* disease is usually a meningeal infection, commonly by *E. coli* or *Strep. agalactiae*. Many other bacteria are involved less frequently, as shown by isolates from neonatal meningitis (Table 11.3).

Clinical presentation and management

Classical signs of infection such as fever may be absent, especially in premature infants. The baby may simply develop feeding and respiratory difficulties, sometimes with more obvious signs of local sepsis, pulmonary, or meningeal

Table 11.3 *Laboratory isolates from neonatal bacterial meningitis*

Organism	Percentage
E. coli	33
Strep. agalactiae	30
Other Gram-negative rods including *Klebsiella, Citrobacter, Enterobacter, Serratia, Salmonella, Proteus*	12
Other Gram-positive cocci including *Strep. pneumoniae* and *Staph. aureus*	10
L. monocytogenes	7
Pseudomonas	3
N. meningitidis	2
H. influenzae	2

Communicable Disease Surveillance Centre, 1985. *Br. Med. J.* **290**, 778–9. Total isolates = 920.

involvement. Since infection may progress rapidly one must treat early on suspicion alone. Bacteriological swabs from all sites—throat, nose, umbilical stump, and rectum—and specimens of blood, CSF, urine, and tracheal aspirate (if intubated) should be taken before antimicrobial therapy. For many years benzylpenicillin and gentamicin have been used, but newer antibiotics, particularly ureido-penicillins like azlocillin or mezlocillin, or third-generation cephalosporins like ceftazidime or cefotaxime, have improved the effectiveness of therapy, especially against Gram-negative organisms. Ampicillin remains most effective against *Listeria* and faecal streptococci. For proven infections 5–7 days' treatment is usually sufficient, except that meningitis requires a two-week course. If not responding and with no evidence of bacterial infection, enteroviral aetiology should be considered.

Prevention

Although potential pathogens can be identified by isolation from vaginal flora during pregnancy this does not necessarily enable neonatal disease to be prevented. For instance, *Strep. agalactiae* can be detected in 20 per cent of women at delivery but antenatal screening may not accurately predict those who will be positive at delivery. Only half of those neonates delivered to carrier mothers become colonized at birth and of these only 1 in 100 will develop clinical illness due to *Strep. agalactiae*. Intramuscular penicillin given to the mother during labour may not eliminate the organisms and, to prevent occasional neonatal disease, would be required for large numbers. At present no prophylactic method based on screening can absolutely prevent severe bacterial sepsis in the newborn. Reduction of cross-infection, skilled obstetrical and neonatal care, and prompt and appropriate treatment on early signs of illness provide the basis for lowering its incidence and mortality.

POSTPARTUM INTRAUTERINE INFECTION

Aetiology and epidemiology

Amniotic fluid is normally sterile except during labour. Even then the membranes provide a good barrier whilst intact. Once membranes have ruptured, the risk of amniotic fluid infection increases progressively and can be 25 per cent after 24 hours. This infection does not correlate directly with subsequent development of puerperal sepsis, which is most likely after caesarean section, prolonged labour, intrauterine monitoring, instrumentation, and/or repeated vaginal examination.

The organisms responsible are vaginal and rectal commensals, especially *Strep. agalactiae*, *E. coli*, *Gardnerella vaginalis*, Gram-negative anaerobes like *Bacteroides* spp., and Gram-positive anaerobes like *Peptococcus* and *Peptostreptococcus* spp. *Mycoplasma hominis*, *Ureaplasma urealyticum*, and *Chlamydia trachomatis* are often involved; multiple infections are common.

Clinical presentation and management

Typically there are high fever, uterine tenderness and foul-smelling lochia. *Strep. agalactiae* infections commonly present with high fever soon after delivery, with few other signs. Mycoplasmal infection usually causes high fever and mild uterine tenderness but no severe illness. *Chlamydia trachomatis* causes mild uterine tenderness and fever usually 2–5 weeks after delivery. Low-grade fever in the first 24 hours after delivery is common and in the absence of other signs does not warrant treatment.

Differential diagnosis of fever in the postpartum period includes urinary-tract infections, pelvic thrombophlebitis, and abdominal or episiotomy wound infections. Uterine sepsis is difficult to confirm because of the problem of obtaining uncontaminated endometrial swabs; but blood cultures may be positive.

Antibiotics must cover the likely mix of organisms (for example, combined metronidazole, gentamicin, and ampicillin). Single agents, even cephalosporins like ceftriazone, may leave small gaps in coverage such as *Strep. faecalis*. Severe sepsis requires resuscitative measures (Chapter 7).

URINARY TRACT INFECTION

Five per cent of pregnant women have asymptomatic bacteriuria (see also Chapter 5). This should be treated because it predisposes to pyelonephritis and possibly premature labour and low-birthweight infants. Pyelonephritis is commoner in pregnancy from increased urinary stasis in the dilated ureters and renal collecting system (probably from mechanical obstruction at the pelvic brim) and increased vesico-ureteric reflux. The organisms responsible and management are as for the non-pregnant, except that folate antagonists like trimethoprim alone or in combination are best avoided; if used, folinic acid supplements should be given as well. Ampicillin and nitrofurantoin are suitable alternatives for sensitive organisms.

12

Infections in the immunocompromised subject

The normal immune defences are constantly exposed to challenge. This may be intense in unhygienic conditions and often in hospital environments, especially if the patient is also subject to invasive procedures such as intravenous or bladder catheterization. Matters may be made worse if therapy also produces immunocompromise. Nevertheless, such circumstances do not normally cause serious problems unless there is severe humoral and/or cellular dysfunction, either innate or acquired (Table 12.1). Problems of infection in pregnancy and the neonate are discussed in Chapter 11. In this chapter, congenital and acquired immunocompromise are considered together, but because of its importance human immunodeficiency virus (HIV) infection is considered in more detail in a separate section.

Table 12.1. *Types of immunocompromise*

Physiological:	Neonate Elderly Pregnant	Acquired:	Infection Malignancies, especially heamatological Drugs and radiation Hyposplenism Others, e.g. burns, diabetes, malnutrition
Congenital:	Granulocyte disorders T-lymphocyte disorders B-lymphocyte disorders Complement disorders Infections, e.g. HIV Others, e.g. cystic fibrosis		

AETIOLOGY

Congenital immunocompromise

Genetically-determined immunodeficiencies present in childhood with recurrent infections, the severity and type depending on the nature of the defect. Normal immunological function is discussed in Chapter 1.

Granulocyte disorders. The principal genetic abnormalities are neutropenia and neutrophil dysfunction. Neutrophil counts below 1.0×10^9/litre usually entail increased susceptibility to infection. Infection is common in severe neutropenia with counts below 0.5×10^9/litre. Patients with neutrophil dysfunction often have normal counts but their cells are abnormal in function; for example, G6PD deficiency causing defective phagocytosis or chronic granulomatous disease in which there is phagocytosis but not the intracellular killing of bacteria.

T-lymphocyte disorders. Normally T-lymphocytes comprise 40–80 per cent of circulating lymphocytes. The normal lower limit for total lymphocyte count is 1.5×10^9/litre, varying with age and much higher in the first years of life. Low numbers of T-lymphocytes are found in thymic aplasia (Di George syndrome) and in severe combined immunodeficiency (in which there are low numbers of B cells also). In Wiskott–Aldrich syndrome, an X-linked recessive disorder, abnormalities in lymphocyte and platelet plasma membranes cause malfunction of both B and T cells.

B-lymphocyte disorders. In X-linked hypogammaglobulinaemia there are no circulating B cells but T-cells are normal in number and function. Infections increase after the age of six months when passively transferred maternal antibody has disappeared. In late-onset hypogammaglobulinaemia there are few or no B-cells which, if present, can produce only IgM. The commonest selective immunoglobulin deficiency is of IgA (1 in 700 of the population) but most of these patients do not require treatment.

Complement disorders. Complement is important in the inflammatory response and for phagocytic removal of foreign organisms. Severe infections are particularly evident in patients with C3 and C2 deficiencies, but repeated *Neisseria meningitidis* and *N. gonorrhoeae* bacteraemia have been associated with C6, C7, or C8 deficiencies.

Other types. Several other inherited diseases may create conditions that predispose to infection; for example, the viscid bronchial secretions of cystic fibrosis lead to recurrent chest infections. The outlook for these patients is usually better than in other congenital immune disorders.

In Down's syndrome, multiple congenital abnormalities and respiratory infections are particularly common in early life and death is usually due to infection in later years, the life expectancy being about 40 years.

Acquired immunodeficiency

Aetiology

Infection. Many infections reduce the ability of the patient's immune system to deal with other infections; for example, cell-mediated immunity is reduced in measles infection and may lead to re-activation of tuberculosis, especially in the under-nourished. Most such abnormalities are temporary and without serious effects: although viral infections are the commonest cause of acute neutropenia, this is transient, usually not serious and without secondary bacterial infection. Occasionally the defect is permanent; for example, depression of IgA and IgG production in congenital rubella. Epstein–Barr virus infection in the X-linked lymphoproliferative syndrome can cause fatal infectious mononucleosis, hypogammaglobulinaemia, or lymphoma. Infection with the retrovirus HIV causes the acquired immunodeficiency syndrome (AIDS) discussed below.

Haematological malignancies. Leukaemias and lymphomas are characterized by abnormal proliferation of leucocytes which do not function normally. Despite very high white-cell counts patients may have very few normal cells. About half of them die from overwhelming infection.

Drugs. Many drugs can cause neutropenia as a rare side-effect in susceptible individuals. Some produce neutropenia more regularly; for example, phenothiazines, chlorpropamide, and phenylbutazone. Others such as penicillamine and phenytoin may cause hypogammaglobulinaemia. More importantly, whole families of drugs designed to treat malignancies can damage the immune system by acting on dividing cells (normal or abnormal). Thus, use of cytotoxic drugs such as cyclophosphamide, azathioprine, and cytosine arabinoside is associated with episodes of infection.

Corticosteroids, which are extensively used in illness attributed to an overactive immune system, specifically depress immune function. They reduce inflammatory response, depress cell-mediated immunity, predispose to reactivation of tuberculosis and impair wound healing.

Irradiation. Exposure to excessive ionizing radiations and radioactive chemicals can damage the immune system severely, with infection as one of the main life-threatening consequences. Radiation therapy for cancers damages both cancer and normal cells, and one should monitor marrow function in these patients by frequent white-cell and platelet evaluations.

Hyposplenism. Hyposplenism, either from splenectomy or splenic atrophy, predisposes to severe infection such as overwhelming septicaemia, especially pneumococcal. The precipitating cause of splenectomy influences the likelihood of severe infections: patients with splenectomy for traumatic damage have fewer problems than those with underlying disease such as chronic leukaemia or malignant lymphoma. Splenectomized children require both prophylactic penicillin and pneumococcal vaccine, if possible given before spelenectomy.

Other factors. World-wide, protein–calorie malnutrition is probably the commonest cause of immunodeficiency. Cell-mediated immunity is particularly affected and may partly explain the high mortality rate of measles in malnourished children. Resistance to infection is also reduced by disruption of physical barriers that protect the body from the environment as in burns and exfoliative dermatitis. Chronic illnesses such as diabetes, chronic renal failure, and alcoholism, are also associated with abnormal immune function. Immunity declines in later life and the elderly are vulnerable to severe or fatal infections, for example, legionellosis, and salmonellosis.

CLINICAL PRESENTATION AND INVESTIGATION

In immunocomprised patients the usual signs of infection may be attenuated. As they are less able to mount an inflammatory response, pyrexia, pus, and pain are often absent or considerably reduced.

Infections may be more frequent, more severe, and more often caused by organisms reactivated from latency than in normal persons. Thus tuberculosis, herpes simplex, varicella-zoster, and cytomegalovirus infections provide serious problems.

Much information can be gained by careful clinical examination of ill immunocompromised patients with special attention to: (a) skin, especially at sites of intravenous or urinary catheters; (b) oropharynx, especially teeth and gums; (c) chest; (d) nasal sinuses; (e) ears; (f) ano-genital area. The site of infection can usually be established and its aetiology determined after systematic investigation such as described in Chapter 9 for undiagnosed fevers. Often, however, these patients must initially be treated empirically.

Most infections of immunocompromised patients are caused by common organisms which should be considered before those more exotic. Bacterial infections are commonest, most amenable to treatment, and should always be considered early. The important pathogens in immunocompromised patients are listed in Table 12.2. Although certain host defence abnormalities are associated with particular infections there may be multiple abnormalities and many possible pathogens.

As the immunocompromised patient may deteriorate rapidly from disseminated infection, investigations must be prompt. Many hospitals have established protocols for dealing with such patients. This has the advantage of being comprehensive, but generates a large work-load for the laboratory and belittles clinical acumen.

In most immunocompromised patients the factors predisposing to infection are known; for example, cytotoxic drugs, radiation, or splenectomy. When immunodeficiency is suspected the immune status should be determined by a scheme such as that shown in Fig. 12.1. Neutropenia is the commonest and most easily tested abnormality. Qualitative defects of neutrophils and lymphoctyes are more difficult to assess, and are best tested for when the patient is free from infection. Screening tests for the complement pathway are widely available.

Table 12.2. *Host defence abnormality and associated infective agents*

Host defence	Infection type	Species
Humoral	Bacterial	*Str. pyogenes* *Ps. aeruginosa* *Strep. pneumoniae* *H. influenzae* *N. meningitidis* *L. pneumophila*
Cell-mediated	Fungal	*Candida* spp. *Aspergillis* spp. *Cryptococcus* spp. *Pn. carinii*
	Viral	*H. simplex* Varicella-zoster Cytomegalovirus Measles
	Intracellular bacterial	Mycobacteria spp. *L. monocytogenes*
	Protozoal	*T. gondii*
Phagocytic	Bacterial	*Ps. aeruginosa* *E. coli* *Staph. aureus* *Strep. pneumoniae* *L. pneumophila*
	Fungal	*Candida* spp. *Aspergillus* spp. *Cryptococcus* spp. *Pn. carinii*
	Protozoal	*T. gondii*

Patients who show no evidence of abnormal immune function on first testing should be re-tested after an interval of a few weeks.

The following investigations are usually necessary:

(a) White cell total and differential count.

(b) Chest X-ray.

(c) Urine microscopy and culture.

(d) Blood culture.

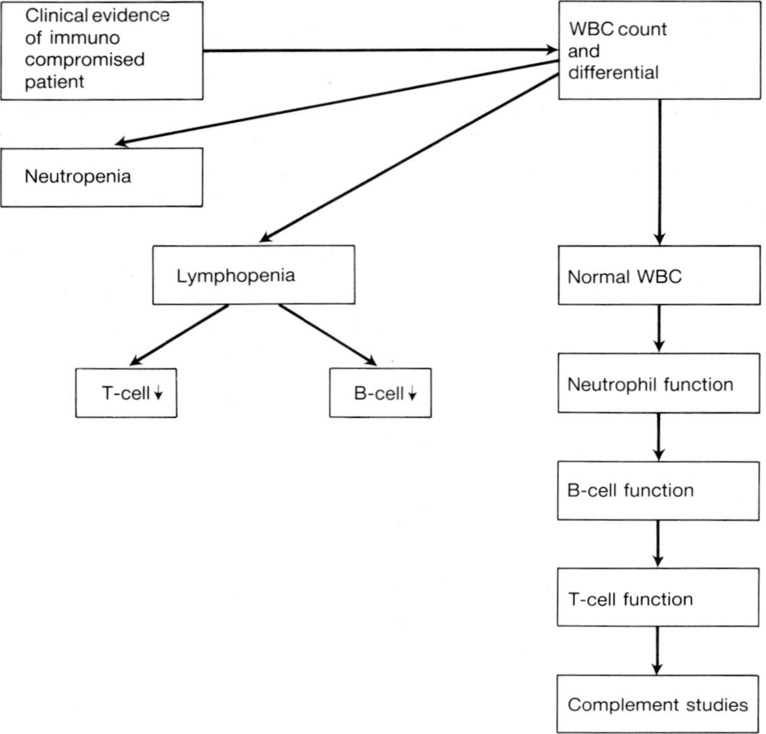

Fig. 12.1 Laboratory investigation of immune status.

(e) Microbiological (including virological) examination of throat swab (nasopharyngeal aspiration in children), sputum or induced sputum, urine, and material from any suspicious lesions (Chapter 19).

(f) Serological examination of the blood. Demonstration of an antibody response may not be possible in immunocompromised patients. In some cases such as *Candida* sp. the infectious agent may be demonstrated in the blood (Chapter 19).

When these investigations fail to define the site of infection or causative agent, further and invasive examinations may be required, such as barium swallow or endoscopy for fungal oesophagitis, percutaneous transtracheal aspiration, bronchoscopy with lavage, or transbronchial lung biopsy for *Pneumocystis carinii* infection. Radioisotope scans occasionally demonstrate a focus of infection; for example, intra-abdominal abscess.

A common problem associated with high mortality in the immunocompromised is combination of pulmonary infiltrates and pyrexia. Many agents alone or in combination can produce this clinical picture, commonly bacterial, viral (Fig. 12.2), fungal, and protozoal organisms (Table 12.3). Many non-infective causes of pulmonary infiltrates are listed in Table 12.4. The commonest differential diagnosis is between *Pn. carinii* pneumonia (Fig. 12.3), CMV pneumonitis, and idiopathic interstitial pneumonitis (steroid-responsive). If the patient's condition is stable or only slowly deteriorating, there may be time to identify causative organisms from initial investigations. However, rapid deterioration may necessitate open-lung biopsy before the patient's condition becomes irreversible.

In the immunocompromised, incorrect diagnosis is easy because of the many possible causes of the symptoms. Mistakes often arise from diagnosis based solely on sputum culture. Multiple agents often colonize the oropharynx but do not necessarily cause symptoms. Non-infective causes of pyrexia (Table 12.4),

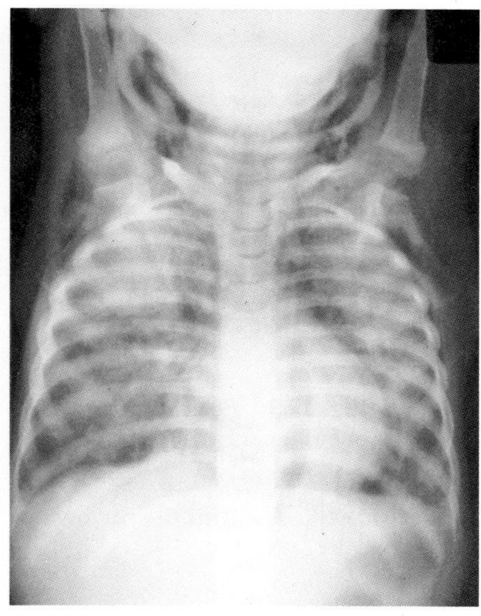

Fig. 12.2 Viral pneumonia due to measles in a child with leukaemia.

Table 12.3. *Causes of pneumonia in immunocompromised persons*

Bacterial	Viral	Fungal	Protozoal
Gram-positive, especially *Staph. aureus*, *Strep. pneumoniae* Gram-negative, especially *E. coli*, *H. influenzae*, *Pseudomonas* spp. *Mycobacteria* spp. *L. pneumophila* *Mycoplasma pneumoniae*	Cytomegalovirus Measles Varicella-zoster	*Candida* spp. *Aspergillus* spp. *Cryptococcus* spp. *Pn. carinii*	*T. gondii*

Table 12.4. *Non-infective causes of pulmonary infiltrates (±pyrexia) in the immunocompromised*

Extension of underlying disease to lungs (esp. lymphoma)
Idiopathic interstitial pneumonitis
Drug fever with pneumonitis (e.g. busulphan)
Radiation pneumonitis
Pulmonary emboli
Pulmonary oedema (e.g. intravenous cytosine arabinoside)
Leucocyte transfusions and systemic amphotericin B

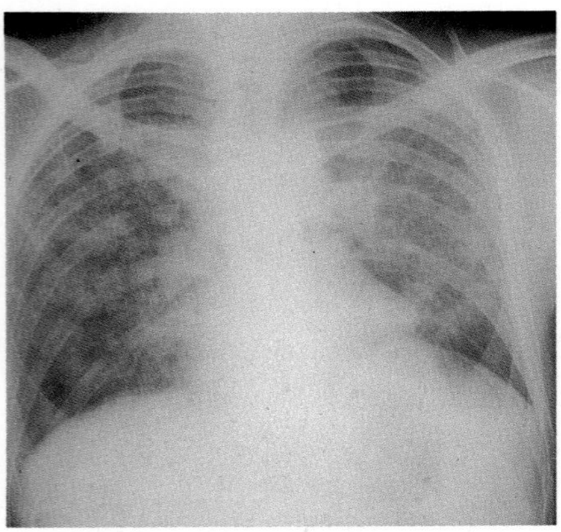

Fig. 12.3 *Pneumocystis carinii* pneumonia.

especially underlying disease, are often overlooked. Continual review of diagnosis and treatment is essential.

MANAGEMENT

Infection in immunocompromised patients is less responsive to antimicrobials and more rapidly progressive than in those with normal immune function. As the patient is often seriously ill, treatment should be aggressive using several different therapeutic approaches such as:

Antimicrobial agents

Antibiotics. Delay in starting antibiotic treatment increases mortality. Consider treatment in any neutropenic patient with pyrexia >38.5°C for 2 hours, unless following transfusion with blood, platelets, or plasma. Choice of the combination of antibiotics depends upon factors such as known sensitivities, personal allergies, the probable infecting organisms, and any local antibiotic policies. A synergistic combination of antiobiotics is better than a single agent or non-synergistic combination. Initially a combination of aminoglycoside and beta-lactam antibiotics may be effective. Combinations in use include aminoglycoside (gentamicin/amikacin/netilmicin) plus penicillin (ticarcillin/azlocillin/piperacillin) or cephalosporin (cefotaxime/lactamoxef). Treat neutropenic patients with antibiotics directed against the most dangerous pathogens—Gram-negative organisms, especially *Pseudomonas aeruginosa* (Fig. 12.4) and *Staph. aureus*. Some clinicians have advocated newer antibiotics (ceftazidine, imipenem, ciprofloxacillin) as single agents, but most favour a combination of antibiotics.

If the patient responds, continue treatment for two weeks or until the granulocyte count returns to normal, with modifications in the light of results from cultures. In a minority fever does not respond to initial treatment; a third antimicrobial agent may then be required.

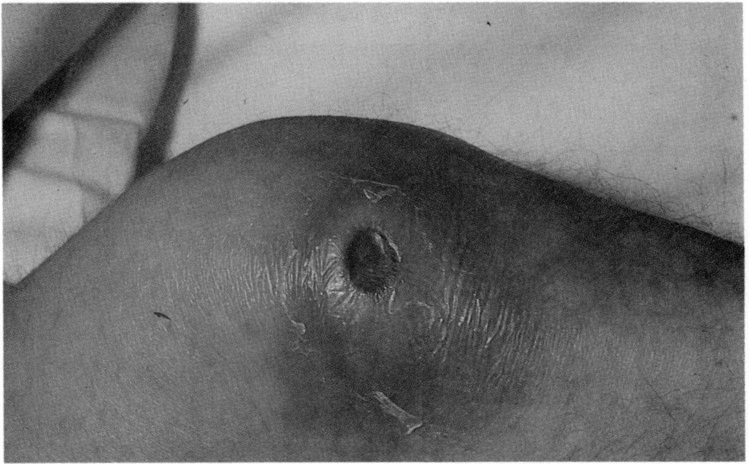

Fig. 12.4 *Pseudomonas* skin infection.

Antifungal agents. Defects of neutrophil or T-cell function predispose to fungal infections, usually with pyrexia unresponsive to empirical antibiotics, or relapse 2–3 weeks after initial response. Fluconazole is the drug of choice, possibly supplemented by amphotericin B or 5-flucytosine in candidiasis (Figs 12.5 and 12.6) or cryptococcosis. Infection with *Pn. carinii* is particularly common in the immunocompromised and requires treatment by high-dose cotrimoxazole or pentamidine.

Antiviral agents. Herpes simplex (Fig. 12.7) and varicella-zoster (Fig. 12.8) can easily become disseminated in the immunocompromised. Both viruses are

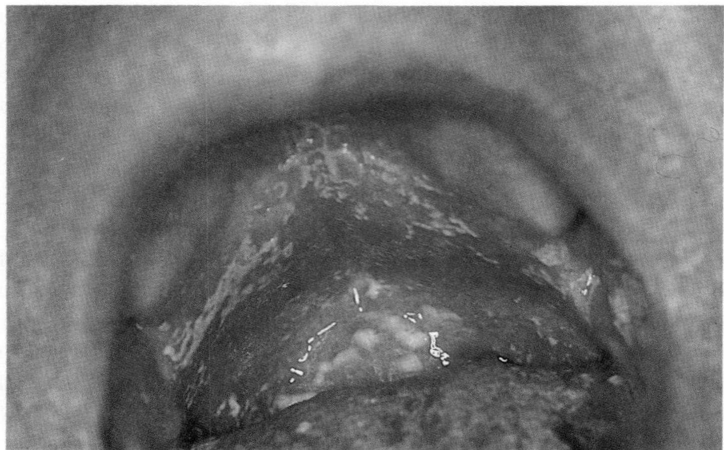

Fig. 12.5 Oral candidiasis in a patient with lymphoma.

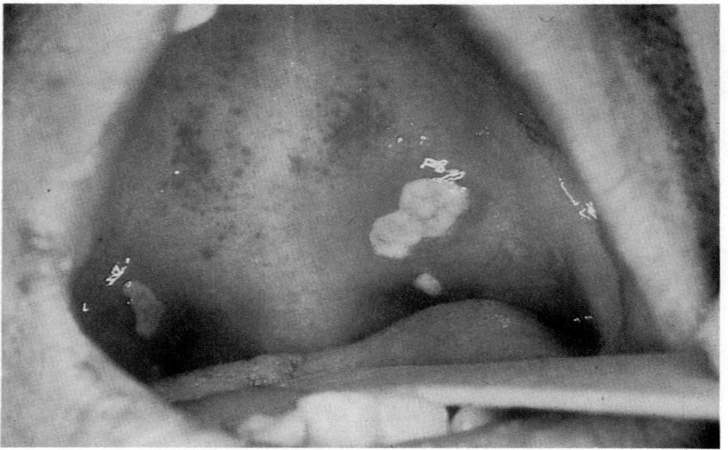

Fig. 12.6 Oral candidiasis in a patient with acute leukaemia.

sensitive to acyclovir but with doubled dose in varicella-zoster infections. All immunocompromised patients with localized herpes simplex or varicella-zoster infection should be treated. In this way the incidence of disseminated infection has been greatly reduced. Gancyclovir can be used for severe cytomegalovirus infections and ribavirin considered for measles pneumonia.

Other specific therapies. Toxoplasmosis can be treated with pyrimethamine, sulphonamide and folinic acid. Clindamycin or co-trimoxazole can be substituted for sulphonamide. Newer drugs such as azithromycin are being evaluated.

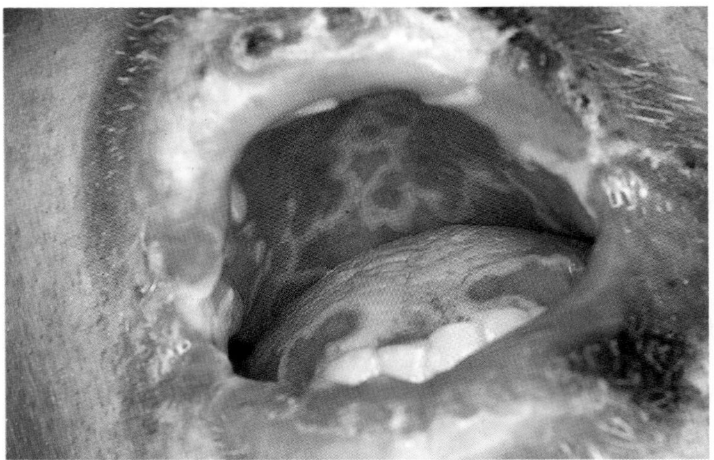

Fig. 12.7 Severe herpes simplex reactivation with intra-oral lesions in a patient treated with steroids.

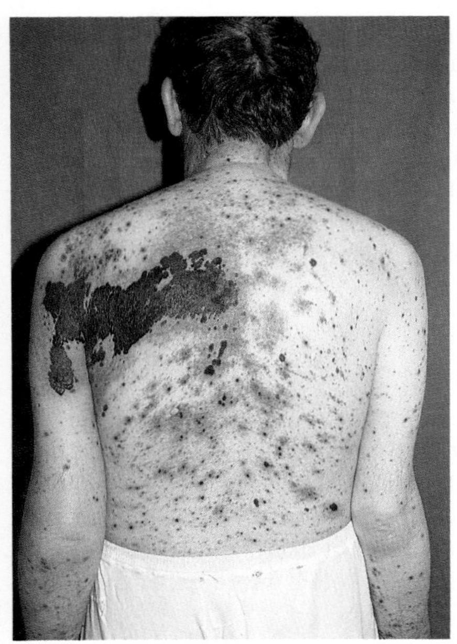

Fig. 12.8 Disseminated varicella-zoster infection in a patient with acute leukaemia.

Granulocyte and other transfusions

Granulocyte transfusions are used selectively, being expensive and potentially hazardous to the recipient. The best results are obtained in severely neutropenic patients with bacterial infection which has resisted appropriate antibiotic therapy. Patients with immunoglobulin deficiencies should receive immunoglobulin or fresh frozen plasma at regular intervals.

PROPHYLAXIS

Since the consequences of infection are so serious in the immunocompromised, prevention is important. Before starting prophylactic treatment, one should be clear how long it will be required, what is the defect in the patient's immune system, whether there are multiple defects and whether the advantages of prophylaxis outweigh its side-effects.

Mild immunodeficiency

Most serious infections result from a combination of several factors. Mild immunodeficiency with only small disturbances in tests of immune function does not require prophylaxis.

Single defects such as hypogammaglobulinaemia can often be corrected easily by regular immunglobulin treatment; prophylactic therapeutic measures are not usually required but episodes of infection are treated.

Immunocompromised patients can influence the number and severity of their infections by behaviour. They should avoid potentially infectious situations such as contact with friends and relatives with infections and report promptly the first signs of infection to their doctor. Those likely to be compromised for some time should also be screened for endogenous pathogens by skin, throat, and nasal swabs. The findings are often useful in subsequent pyrexial episodes as most infections originate from the patient's own flora.

Severe immunodeficiency

The approach is different in seriously immunocompromised patients with profound or multiple disturbances in immune function. This group, especially transplant patients and those being treated for malignancies such as leukaemia, is rapidly increasing because severe immune suppression is important in treating these conditions and infection is the major cause of death. For these patients, prophylaxis is as important as treatment but is often more expensive, time-consuming, and inconvenient (Table 12.5).

Protective isolation. In the attempt to prevent nosocomial infection, various protective procedures range from reverse barrier nursing to the use of expensive laminar air-flow units. The former is of minor benefit whilst the latter is particularly useful in reducing fungal infection. The best results are obtained when protective isolation is combined with other procedures to reduce the patient's contact with potential pathogens. Thus the patient is put in a clean room

Table 12.5. *Prophylactic measures in the immunocompromised*

Protective isolation	Gut decontamination
Sterile diet	Systemic antimicrobial agents
Superficial decontamination	Immunization

in which all objects have been disinfected; all staff and visitors must wash their hands on entry and wear gloves, gowns, caps, and overshoes; all instruments used on the patients, such as thermometers and stethoscopes, are first sterilized and then left within the room. Other measures are listed in Table 12.5.

Diet. Non-sterile foods such as salads, fresh fruit, and dairy products should be avoided. Food should be thoroughly cooked, preferably in a microwave oven on the ward, and sterile water used for drinks or ice.

Superficial decontamination. Since skin, hair, and body orifices are major sources of potential pathogens, antiseptics such as chlorhexidine or povidone–iodine should be used for daily baths and 4-hourly mouth gargles, and antiseptic hair shampoo used on alternate days. The axillae and perineum require special attention—if initial swabs show nasal carriage of bacteria or fungi, treat both locally and systemically to eradicate carriage. Skin preparation before any invasive procedure demands special care.

Gut decontamination. Since the gut is a major source of infections in the immunocompromised, non-absorbable antibiotics have been used with limited success to attempt 'total' gut decontamination. However, immunocompromised children who have received oral co-trimoxazole prophylactically to prevent *Pn. carinii* infection, have fewer enterobacteria in their stool, and this 'selective' decontamination leaves the anaerobic gut flora intact and inhibits colonization by fungi and aerobic bacteria. Also co-trimoxazole is more palatable than the regimens for total decontamination.

Systemic antimicrobial agents. The case for prophylaxis is strongest for co-trimoxazole or inhaled pentamidine in prevention of *Pn. carinii* infection, mostly from reactivation. Where initial surveillance samples show fungal colonization, antifungal agents should be given during neutropenic periods. Prophylactic antibiotics are not indicated except for short periods in special circumstances (for example, in dental procedures and endoscopy). Granulocyte transfusions have no prophylactic benefit.

Immunization. Live vaccines are contraindicated in these patients. Pseudomonal vaccines are useful in burned patients, pneumococcal vaccines in splenectomized patients. Severely immunocompromised patients received little benefit from vaccines as they are often incapable of an antibody response; they may be helped by hyperimmune globulin against viruses such as cytomegalovirus, varicella zoster, and measles.

HUMAN IMMUNODEFICIENCY VIRUS (HIV)

Acquired immunodeficiency syndrome

This syndrome (AIDS) is the most severe manifestation of infection by human immunodeficiency virus. It is a syndrome of opportunistic infection, HIV-induced disease (such as encephalophathy or wasting syndrome) and/or associated cancers, but in the absence of any known cause of immunodeficiency other than HIV infection (Table 12.6). There should be positive laboratory tests for this infection and patients are excluded as AIDS cases if they have *persistently* negative antibody or cultural tests for HIV infection or have a normal number of T-helper lymphocytes. In countries with poor diagnostic facilities such as most of Africa, Central and South America, the diagnosis of AIDS is based on the World Health Organization's clinical definition of AIDS (Table 12.7). Similar criteria apply for the clinical definition of paediatric AIDS: two major signs (weight loss/abnormally slow growth; chronic diarrhoea >1 month; prolonged fever >1 month) plus two minor signs (generalized lymphadenopathy; oropharyngeal

Table 12.6. *Diseases which indicate AIDS in the absence of another cause of immunodeficiency even without laboratory evidence of HIV infection*

Parasites	Cerebral toxoplasmosis Cryptosporidiosis (diarrhoea >1 month)
Viruses	Cytomegalovirus (other than liver, spleen, lymph node) Herpes simplex virus (ulceration >1 month, oesophageal, pulmonary)
Bacteria	*M. avium/M. kansasii* (disseminated)
Fungii	*P. carinii* pneumonia Candidiasis (oesophageal) Cryptococcosis (extrapulmonary)
Neoplasia	Kaposi's sarcoma (<60 years) Lymphoma of brain (<60 years) Lymphoid interstitial penumonia (<13 years)
Others	Progressive multifocal leucoencephalopathy

Table 12.7. *World Health Organization clinical definition of AIDS in an adult*

Major signs	
	1. Weight loss >10 per cent of body weight
	2. Chronic diarrhoea >1 month
	3. Prolonged fever >1 month (intermittent/constant)
Minor signs	
	1. Persistent cough >1 month
	2. Generalized pruritic dermatitis
	3. Recurrent herpes zoster
	4. Oropharyngeal candidiasis
	5. Disseminated herpes simplex infection
	6. Generalized lymphadenopathy
Diagnosis:	Two major plus at least one minor sign in the absence of other causes of immunosuppression (e.g. severe malnutrition/malignancy). Kaposi's sarcoma or cryptococcal meningitis sufficient for diagnosis

candidiasis; persistent cough; repeated common infections; generalized dermatitis; confirmed maternal HIV infection), in the absence of known causes of immunosupression.

Aetiology

HIV specifically infects cells that express the CD4 molecule, especially T-helper cells. Other cells that carry CD4 are monoctye/macrophages, Langerhans' and dendritic cells. After infection, HIV can persist latently within cells. It can be reactivated in dividing CD4 lymphocytes which become functionally damaged and destroyed. The result of this is severe immunological impairment.

There are at least two types of HIV: HIV-1 and HIV-2. The latter is common in West Africa and is closely related to the simian immunodeficiency virus. The physical characteristics of the two types are very similar but the incubation period for HIV-2 appears to be much longer. HIV-2 is spreading outside West Africa.

The virus is readily inactivated by heat, hypochlorite, glutaraldehyde, ethanol, and by inactivation procedures used in preparing hepatitis B vaccine from human plasma.

Epidemiology

HIV has been isolated from blood, semen, genital secretions, cerebrospinal fluid, tears, saliva, urine, breast milk, and several body tissues. The major routes of transmission are by sexual contact, contaminated blood or blood products, sharing contaminated needles and/or syringes, transplacental transmissions, and from breast milk. The virus is not spread by the aerosol route, tears, saliva, or urine. Household members of patients with HIV are not at risk except by sexual contact. Transmission of HIV from mother to fetus or newborn is considered in Chapter 11. Rarely, infection has resulted from transplantation of infected tissues or from artificial insemination. Among health care workers, risks of occupational infection are low. In those with needle-stick injuries the risk of infection is about 0.3%. Presence of HIV antibody indicates infection with the virus but not necessarily AIDS.

Since the first descriptions of AIDS cases in 1981 and definition of the syndrome in 1982 there has been a dramatic growth in understanding of this world-wide infection. Groups at high risk of infection are homosexual or bisexual men, needle-sharing drug misusers and those with casual heterosexual contact. In the past, blood and blood products were a major source of infection and many haemophiliacs became infected. However, in Britain all blood donations and organ donors are now screened for infection, which should prevent this transmission. In the Western world most patients with AIDS have been homosexual or bisexual men, especially those who have had multiple partners; those who indulge in receptive anal intercourse are most at risk. In Africa, infection is principally transmitted by heterosexual, vaginal intercourse. Serological studies showed 60–80 per cent of female prostitutes in Zaïre, Zambia, Rwanda, and Kenya to be infected with HIV. World-wide, heterosexual transmission is probably the commonest route of infection. The position in the Western world is further complicated, as prostitutes in these areas are often drug misusers. As with many other infections, waves of infection which reflect at-risk groups are likely. Thus, the homo/bisexual group was first affected then intravenous drug misusers, and it is expected that the next wave will be the heterosexual group, and lastly those with congenital infection (Fig. 12.9).

The incubation period for seroconversion after HIV infection is 3–12 weeks, but individuals should be followed for a year after the last exposure.

Clinical presentation

After HIV infection most patients have asymptomatic infection. In a few, seroconversion to HIV is accompanied by an acute illness. This may resemble infectious mononucleosis or be an acute infection of the central nervous system such as encephalitis. In some cases a rash is present (Fig. 12.10). Whether or not patients have an acute illness, they proceed to become carriers of the virus, probably for life. The carrier state is frequently accompanied by persistent generalized lymphadenopathy (PGL) which is characterized by lymphadenopathy at two or more extra-inguinal sites for over three months. It may be asymptomatic or present symptoms such as malaise, fatigue, night sweats, fever, diarrhoea, and weight loss over 10 per cent or 7 kg. Often lymphadenopathy is caused by agents unrelated to AIDS which should be excluded. Many other

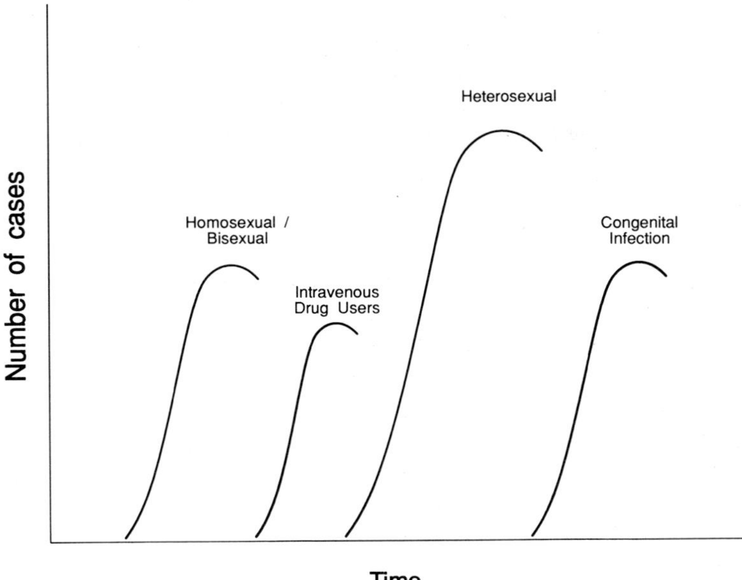

Fig. 12.9 Waves of HIV infection (see text) with declines related to health education.

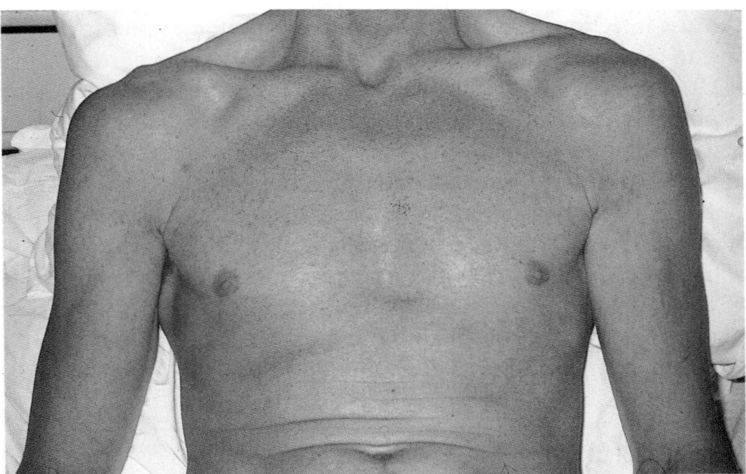

Fig. 12.10 Generalized rash in a patient who subsequently seroconverted to HIV.

complaints (constitutional, dermatological, gastrointestinal, neurological, haematological, respiratory, and psychiatric) either singly or in combination may accompany the carrier state (Table 12.8). Some of these conditions may be diagnostic of AIDS (Table 12.6). Thus a range of clinical features from

Table 12.8. *Clinical features (singly or multiple) of HIV carrier state*

System	Features
Constitutional	Asymptomatic, fever, malaise
Dermatological	Itching, dry skin, folliculitis, seborrhoeic dermatitis, herpes simplex, genital warts, *Molluscum contagiosum*, fungal infections
Gastrointestinal	Sore mouth, *Candida* spp., herpes simplex, gingivitis, hairy leukoplakia, oesophageal candidiasis, anorexia, weight loss, malabsorption, diarrhoea
Neurological	Paraesthesias, neuropathy weakness, unsteady gait
Haematological	Thrombocytopenic purpura, splenomegaly
Respiratory	Bacterial, fungal and viral pneumonia
Psychiatric	Anxiety, depression, loss of memory, hallucinations

asymptomatic to full-blown AIDS result from the increasingly severe immunological abnormalities.

Classically, AIDS patients present with serious opportunistic infections, Kaposi's sarcoma, or both, with no previous history of immunocompromise. Opportunistic infections provide the commonest presentation of AIDS. Organisms that can take advantage of defective T-cell function are particularly evident (Table 12.9). *Pneumocystis carinii* pneumonia is the commonest serious infection. Cytomegalovirus reactivation with pneumonitis or gastrointestinal involvement is frequent, sometimes with retinitis (Fig. 12.11). Perianal herpes simplex can be severe and locally invasive. Oral candidiasis is common and can readily spread to the oesophagus. Mouth ulcers can be severe and it may not be possible to identify the cause (Fig. 12.12). Latent *M. tuberculosis* may be reactivated and atypical mycobacteria such as *M. avium-intracellulare* (present in soil and water) can produce disseminated infections. The two commonest CNS infections are toxoplasmosis (especially encephalitis) and cryptococcosis (especially meningitis). Gastrointestinal infections are common (especially oral candidiasis), the

Table 12.9. *Common infections in AIDS patients*

Pn. carinii	*T. gondii*
Cytomegalovirus	*Cryptococcus*
Herpes simplex	*Salmonella* spp.
Candida spp.	Cryptosporidiosis
Mycobacteria	

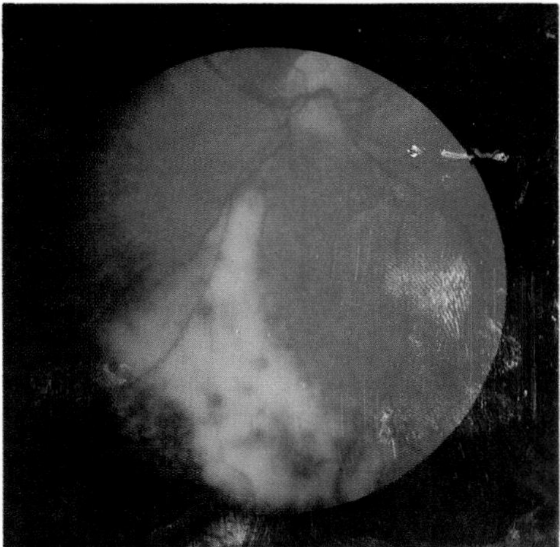

Fig. 12.11 Cytomegaloviral retinitis in an AIDS patient.

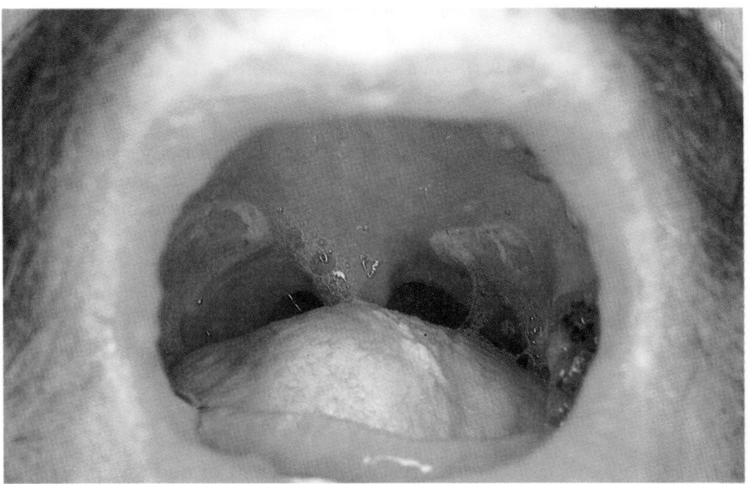

Fig. 12.12 Mouth ulcers of unknown cause in an AIDS patient.

most serious being cryptosporidiosis which can cause severe and prolonged diarrhoea, malabsorption, and weight loss. Malaria and cryptococcosis are frequent in African AIDS patients in whom *Pn. carinii* pneumonia is not common. In some countries in Central Africa where more than half the population may be infected by *M. tuberculosis*, HIV infection has provoked a dramatic increase in tuberculosis morbidity. A wasting syndrome (Slim disease) is common in Africa, with weight loss over 10 per cent of body weight plus either unexplained chronic diarrhoea (>1 month) or chronic weakness and unexplained prolonged fever (>1 month).

HIV encephalopathy is the presence of disabling cognitive and/or motor dysfunction interfering with activities of daily living, progressing over weeks to months, in the absence of a concurrent illness or condition other than HIV-infection to explain the findings.

AIDS patients with Kaposi's sarcoma may have widespread skin, visceral, and lymph-node disease. Skin lesions may be small (like petechiae), large (like bruises), nodular, or pigmented. Any viscus may be involved, most commonly gastrointestinal organs (especially stomach and colon). Typically, lesions are extensive with frequent appearances of new tumours. Other lymphoid tumours may also affect AIDS patients.

There have been several attempts to devise a staging system for HIV infection. That of the World Health Organization is shown in Tables 12.10 and 12.11. The

Table 12.10. *WHO clinical stages 1, 2, and 3*

Clinical stage 1	
	1. Asymptomatic
	2. Pesistent generalized lymphadenopathy (PGL)
	Performance scale 1: asymptomatic, normal activity
Clinical stage 2	
	3. Weight loss, < 10 per cent of body weight
	4. Minor mucocutaneous manifestations (seborrheic dermatitis, prurigo, fungal nail infections, recurrent oral ulcerations, angular cheilitis)
	5. Herpes zoster, within the last 5 years
	6. Recurrent upper respiratory tract infections (i.e. bacterial sinusitis)
	And/or Performance scale 2: symptomatic normal activity
Clinical stage 3	
	7. Weight loss, > 10 per cent of body weight
	8. Unexplained chronic diarrhoea, > 1 month
	9. Unexplained prolonged fever (intermittent or constant), > 1 month
	10. Oral candidiasis (thrush)
	11. Oral hairy leukoplakia
	12. Pulmonary tuberculosis, within the past year
	13. Severe bacterial infections (i.e. pneumonia, pyomyositis.
	And/or Performance scale 3: bed-ridden, < 50 per cent of the day during the last month.

Table 12.11. *WHO clinical stage 4*

Clinical stage 4	
	14. HIV wasting syndrome (see text)
	15. *Pneumocystis carinii* pneumonia
	16. Toxoplasmosis of the brain
	17. Cryptosporidiosis with diarrhoea, >1 month
	18. Cryptococcosis, extrapulmonary
	19. Cytomegalovirus (CMV) disease of an organ other than liver, spleen or lymph nodes
	20. Herpes simplex virus (HSV) infection, mucocutaneous >1 month, or visceral any duration
	21. Progressive multifocal leukoencephalopathy (PML)
	22. Any disseminated endemic mycosis (e.g. histoplasmosis, coccidioidomycosis)
	23. Candidiasis of the oesophagus, trachea, bronchi or lungs
	24. Atypical mycobacteriosis, disseminated
	25. Non-typhoid salmonella septicaemia
	26. Extrapulmonary tuberculosis
	27. Lymphoma
	28. Kaposi's sarcoma (KS)
	29. HIV encephalopathy (see text)
	And/or Performance scale 4: bed-ridden, >50% of the day during the last month

system may be refined by introducing laboratory results such as CD4 lymphoctye or total lymphocyte counts. Thus with CD4 lymphocytes ($\times 10^6$ l) each clinical stage may be subdivided into three categories: A (>500), B (200–500), and C (<200).

Investigations

The earliest sign of HIV infection is HIV antigen which then becomes undetectable in serum, only to reappear late in the illness (Fig. 12.13). Commercial kits are available for detecting HIV antigen (notably p 24) but are expensive. A few specialized laboratories have facilities for growing HIV. The diagnosis is usually made by demonstrating HIV-specific envelope antibody. As false–positive results may occur, it is essential to confirm all positive results by another type of test, such as two different types of ELISA test or Western blotting (Chapter 19). With Western blotting at least two different envelope bands must be present. In those with indeterminate bands, radio-immunoprecipitation assay (RIPA) may resolve the position; alternatively they may be followed up.

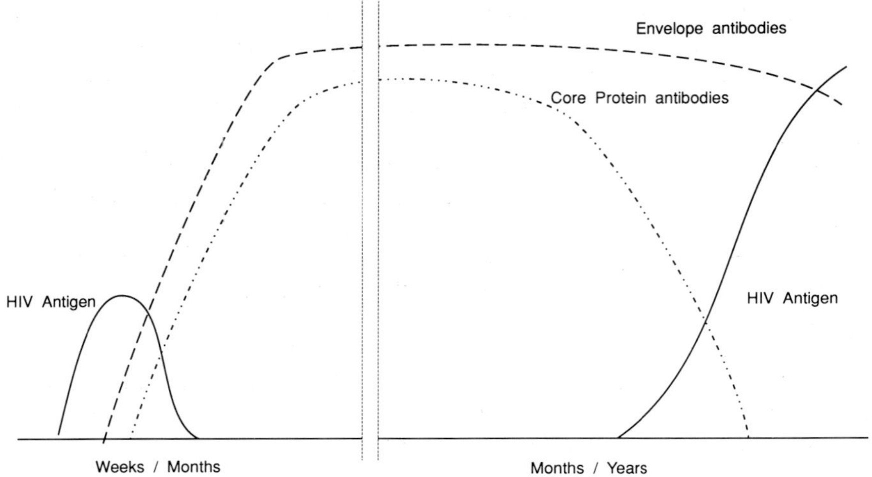

Fig. 12.13 Presence of HIV and antibodies during illness.

Although there is cross-reaction between HIV1 and HIV2, patients should be specifically tested for both HIV1 and HIV2.

Demonstration of HIV antigen or antibody does not prove that the patient has AIDS. Many patients in high-risk groups (homosexuals, drug addicts and haemophiliacs) have HIV antibody, indicating past exposure and probably persisting infection.

It is important to consider the implications of an HIV antibody result to the patient. It may have severe emotional and ethical implications, as well as financial consequences such as difficulty or inability to obtain life insurance, hire-purchase, or mortgage agreements. Patients should give informed consent to the test, which should probably not be done unless there are counselling facilities for patients with positive results. Definite indications for the test are donors of blood, semen, or organs, and pregnancy in a high-risk group.

In PGL the differential diagnosis includes other causes of lymphadenopathy such as Epstein–Barr virus, cytomegalovirus, and *T. gondii* infection. Because of immune dysfunction there is often failure to develop rising antibody titres to infection with agents such as cytomegalovirus and *T. gondii*. Taking and transportation of specimens to the laboratory are discussed in Chapter 19.

Treatment

Patients are preferably treated by clinicians with experience of HIV-infected persons. Many drugs are being developed, but the drug of choice is zidovudine (azidothymidine, AZT, Retrovir) which inhibits HIV reverse transcriptase. Treatment with this lowers mortality and the likelihood of opportunistic infections, reduces constitutional symptoms (fever, malaise weight loss), and progress of the disease. Unfortunately the drug is very toxic, especially to those with ill-health, anaemia, neutropoenia, or a low CD4 lymphocyte count. Side-effects include nausea, myalgia, insomnia, severe headache, and bone marrow toxicity. Concomitant dideoxyinosine or dideoxycytidine reduces resistance and side-effects.

Drugs are available for treating many of the opportunistic infections, but response may be poor. Attempts to rectify immunological abnormalities, for instance by bone marrow transplantation, have been unsuccessful.

Prophylactic measures aim to interrupt transmission by sexual and parenteral routes. Drug addicts should not share needles. Barrier methods of contraception (condoms, sheaths) are necessary. Among homosexuals, reduction in the number of sexual partners (particularly in casual sex) and certain sexual practices such as anal intercourse reduce exposure to infection. Health care personnel who have had a needle-stick injury from an HIV-infected individual or other severe exposure may require a course of zidovudine.

Prognosis

HIV infection is chronic, persisting for life (Fig. 12.14). The median time for progression to AIDS is 8 years, with 5–10 per cent in the first three years, and about half at 6–8 years after infection. The prognosis for AIDS patients not receiving antiviral therapy is poor: median survival about one year, with few surviving beyond five years.

In patients with Kaposi's sarcoma alone, AIDS is less severe, immune dysfunction less, and absence of opportunistic infection correlates with better survival (80 per cent survive one year). Yet, the median survival for these patients is only 18 months. Patients presenting with opportunistic infections are those with more severe immune dysfunction and a poorer prognosis. Those with both opportunistic infection and tumour have the worst prognosis.

Homosexuals, bisexuals, and drug misusers are discouraged from donating blood; all blood donations are checked for HIV antibody. Similarly, all organ

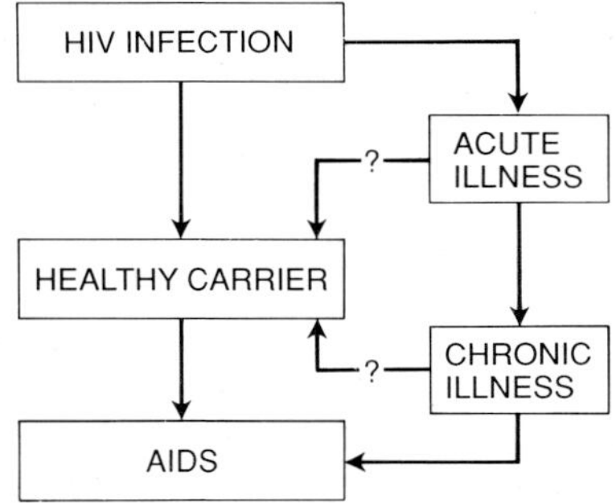

Fig. 12.14 Course of HIV infection.

(for example, kidney, marrow, and eye) and semen donors are tested for antibody. Blood products (Factors VIII and IX) are heat-treated to destroy HIV. Persons with HIV antibody should not normally be vaccinated since this may increase the viral activity and the likelihood of developing AIDS, but normal childhood vaccinations are advisable.

Immunological deterioration in diagnosed AIDS patients has not yet been reversed, but rarely those with antibody to HIV may lose detectable antibody and not progress to AIDS. Perhaps in future it may be possible to reverse or prevent the progressive immunological breakdown in patients diagnosed in the prodromal stages of AIDS. Demonstration that heat-killed whole virus vaccines can protect chimpanzees from infection is hopeful. However, because of the long incubation period of the disease, vaccine trials are likely to be protracted, but in future vaccines and more anti-HIV agents will probably become available.

13

Infection-related illnesses

INTRODUCTION

The infecting organism is usually directly responsible for the characteristic features of acute illness by its activities and early interactions with the host. Recovery from acute infection is usual, but in a few cases another illness appears 2–3 weeks after the initial infection (Table 13.1). These later illnesses are not usually infectious, but often simulate infection on presentation.

Infectious agents associated with such illnesses are shown in Table 13.1. The mechanisms by which these illnesses are produced are incompletely understood but an abnormal immune response is often involved. Two postulated immunological mechanisms for producing illness are *autoantibody production* and damage to tissues by *immune complexes*. In the former, the organism and a particular body tissue (for example, the heart in rheumatic fever) may share a common epitope (unit antigenic determinant). Antibodies produced against the organism become attached to the common epitope on the organ and cause damage. Immune complexes are soluble combinations of antigen with antibody and can become deposited in certain tissues such as the kidney in post-streptococcal glomerulonephritis. In most illnesses following acute infection, several immunological processes may be involved simultaneously or in sequence. Some individuals have a genetic predisposition, for instance tissue type HLA-B27 in Reiter's syndrome. Most cases are sporadic but a few can be epidemic (for example, myalgic encephalomyelitis).

Clinical features are varied and all systems can be affected. Pyrexia, malaise, and headache are frequent. The systemic nature of these diseases is also often shown by involvement of skin as well as various other organs. The range of clinical illness varies from severe (Reye's syndrome) to asymptomatic (mild glomerulonephritis). Diagnosis is usually made on clinical grounds and investigations aim to identify an associated infectious agent or evidence of recent infection. Immunological studies (Table 13.2) tend to have secondary, supportive diagnostic significance.

Treatment is directed against the responsible organism, if still present, but otherwise and more often is supportive. In general, specific prophylactic measures are not available, but once the disease has developed specific prophylaxis is often applicable (for example, penicillin to prevent recurrent

Table 13.1. *Infection-related illness*

Illness	Related infections (examples)
Rheumatic fever	*Strep. pyogenes* (pharyngitis)
Post-streptococcal glomerulonephritis	*Strep. pyogenes* (nephritogenic strains)
Henoch–Schönlein purpura	*Strep. pyogenes* (upper respiratory tract)
Erythema nodosum	*M. tuberculosis* *Strep. pyogenes* *Yersinia enterocolitica* *Chlamydia trachomatis*
Kawasaki disease	Streptococci? Rickettsiae?
Reye's syndrome	Influenza B Chickenpox
Erythema multiforme; Stevens–Johnson syndrome	Herpes simplex, orf *M. pneumoniae*
Post-viral fatigue syndrome; Myalgic encephalomyelitis; Chronic fatigue syndrome	Coxsackievirus B, Hepatitis A and B, Epstein–Barr virus
Reiter's syndrome (Chapter 13)	*Chlamydia trachomatis* *Ureaplasma urealyticum* *Salmonella*, *Campylobacter*
Encephalitis (post-infectious) (Chapter 6)	Mumps, measles, rubella, chickenpox
Guillain–Barré syndrome (Chapter 6)	Respiratory or gastrointestinal illness, cytomegalovirus, Epstein–Barr virus
Haemolytic uraemic syndrome	*E. coli* *Shig. dysenteriae*

Table 13.2. *Abnormal immunological responses in patients with illness after acute infection*

Raised serum immunoglobulins (IgM or IgG and IgM; infecting agent usually no longer detectable)
Auto-antibodies
Serum complement (high initially, then low)
Increased immune-complexes
Acute phase proteins (e.g. C-reactive protein)
Increased T-suppressor cells and reversal of T-helper: T-suppressor cell ratio

rheumatic fever). Fortunately, the outcome is usually favourable, although recovery is often delayed.

SPECIFIC SYNDROMES

Rheumatic fever

Aetiology and epidemiology

This disease is a complication of pharyngeal infection by group A streptococci (*Streptococcus pyogenes*) and affects 1–3 per cent of patients with streptococcal pharyngitis. Pharyngeal infection is an important factor since streptococcal skin infections do not usually result in rheumatic fever. The disease develops about two weeks after the pharyngitis when the antibody response is at its maximum. Children 5–15 years old are mainly affected, especially in poor and overcrowded living conditions. Distribution is world-wide with the highest incidence and severity now in the tropics.

Clinical presentation

The diagnostic features are classified as major and minor (Table 13.3). Recent *Strep. pyogenes* pharyngitis and two major, or one major and two minor criteria are necessary for diagnosis. The differential diagnosis includes infective endocarditis, gonococcal polyarthritis, rubella, parvovirus, Lyme disease, brucellosis, and rheumatoid arthritis.

Table 13.3. *Major and minor criteria for the diagnosis of rheumatic fever*

Major criteria:

- polyarthritis typically ‘flitting’ and especially large joints
- pancarditis; friction rubs are common but pericardial effusions are rare
- chorea
- erythema marginatum (see Fig. 13.1)
- subcutaneous nodules, especially on the extensor surfaces near large joints

Minor criteria:

- fever
- raised ESR or C-reactive protein
- prolonged P–R interval on ECG
- arthralgia
- past history of rheumatic fever or rheumatic heart disease

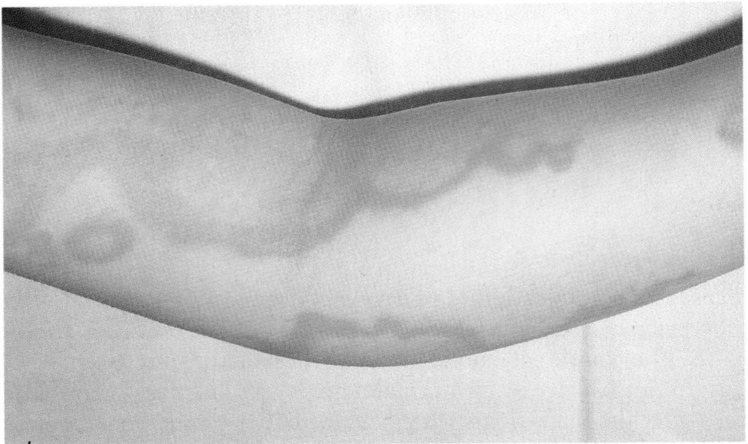

Fig. 13.1 Erythema marginatum.

Investigations

Evidence for *Strep. pyogenes* infection is sought by taking throat swabs for culture and sera for antistreptolysin O titres. If the latter are normal, other streptococcal antibody tests (antideoxyribonucleotidase B, antihyaluronidase, antistreptokinase, and antinicotinamide-adenine dinucleotidase) are performed and minor criteria sought. Rarely, biopsy material from affected tissues may be required: the Aschoff body is pathognomonic.

Management

Benzylpenicillin is given to remove any remaining organisms and high-dose aspirin to reduce fever and inflammation. Bed rest is advisable during treatment if there is cardiac involvement. Chorea can be difficult to manage even with sedation and tranquillizers, but fortunately is self-limiting. Resolution can be monitored by measuring the ESR.

Further attacks can and should be prevented. Factors associated with recurrence are a previous attack within five years, pre-existing rheumatic heart disease and a vigorous antistreptolysin O response. Prophylaxis by monthly intramuscular benzathine penicillin, or daily oral penicillin should continue during the school years or for at least five years. Thereafter, those with increased exposure to infection (attending further education, in contact with schoolchildren or working in hospitals) should continue prophylaxis. Attacks are much less common after 40 years of age.

Prognosis

Most attacks subside in three amonths, but 5 per cent last for more than six months. About a quarter recur, usually within five years. Some patients develop chronic cardiac disease, usually valvular.

Post-streptococcal glomerulonephritis

Aetiology and epidemiology

The illness usually starts 10–14 days after *Strep. pyogenes* infection of throat or skin (for example, streptococcal impetigo or infected scabies). It particularly affects young male children 1–5 years old and is uncommon in adults. Only a few M protein serotypes of streptococci (mainly type 12 in developed non-tropical countries) are nephritogenic and those that are have a variable effect. A few epidemics have been recorded and, like rheumatic fever, it is commoner in the tropics but occurs world-wide.

Clinical presentation

The disease may be asymptomatic, mild, or severe enough for admission to hospital. Onset is acute with malaise and pyrexia. Signs of nephritis are oedema (especially around eyes and eyelids), oliguria, proteinuria, and haematuria. In older children, blood pressure may rise and cause encephalopathy or cardiac failure. Oliguria usually remits in a few days with polyuria and progressive recovery. Complete recovery of kidney function may take a month or longer. A few patients develop chronic renal failure.

Investigation

Swabs from throat and from skin lesions, if present, should be cultured for *Strep. pyogenes*. Blood samples may show high or rising antistreptolysin O titres; if these are normal, other streptococcal antibodies (see section above on rheumatic fever) should be assessed. Immune complexes are increased and serum complement reduced. Urine examination can show proteinuria, granular casts, and haematuria. Creatinine clearance is impaired. Rarely, renal biopsy may be necessary to detect immune complexes deposited on glomerular basement membranes.

Treatment

Residual streptococcal infection is treated with penicillin. Electrolytes, fluid balance, and blood pressure are monitored and controlled until there is a natural remission, usually in a few days but sometimes up to a month. More severe renal failure may require dialysis. Because there are so few nephritogenic strains of streptococci, subsequent prophylactic antibiotics are unnecessary.

Prognosis

The long-term prognosis is good: 95 per cent of patients recover completely. During the acute illness about 1 per cent have severe disease with uraemia, encephalopathy, and cardiac failure. Of the 5 per cent who do not recover completely, many are asymptomatic for years before developing chronic nephritis or hypertension. Intercurrent infection (for example, varicella during convalescence) may provoke return of haematuria. About 2 per cent have a second attack with a different strain of *Streptococcus*.

Haemolytic uraemic syndrome

Aetiology and epidemiology

This syndrome of haemolytic anaemia, thrombocytopenia, and acute renal failure can occur in any age-group but is commonest in children under four years. The commonest identified cause is a verotoxin-producing strain of *E. coli* especially *E. coli* 0157. The syndrome has shown family clustering and association with pregnancy and oral contraceptive usage.

Clinical presentation

Classically, the child has initial upper respiratory tract or gastrointestinal infection. There may be persistent vomiting, followed after about a week by purpura, pallor, tiredness, lassitude, and reduced urine output. Hypertension is present in half of the patients. About a third have hepatosplenomegaly and a third anuria. There may be haemorrhagic colitis.

Investigation

A feature of the haemolytic anaemia is red-cell fragmentation from blood passing through damaged, small blood-vessels ('microangiopathic haemolytic anaemia'). Thrombocytopenia is usual, sometimes with fully developed disseminated intravascular coagulation. Diagnosis can usually be made from clinical features and the blood film, with raised serum urea and creatinine. *E. coli* 0157 may be found by stool culture and tested for toxin production.

Management

Treatment is supportive, particularly of acute renal failure, sometimes with blood transfusion. The effectiveness of heparin, fibrinolytic therapy, plasma transfusions, and vitamin E remains non-proven.

Prognosis

Renal function usually improves in a few weeks. Prolonged oliguria or hypertension are bad prognostic factors. The mortality rate exceeds 30 per cent in untreated cases; many survivors develop hypertension and renal impairment. With good supportive care, mortality is about 5 per cent.

Henoch-Schönlein purpura

Aetiology and epidemiology

This is usually but not always preceded by streptococcal sore throat or upper respiratory infection. The usual age-group affected is 2–20 years.

Clinical presentation

Typically, 1–3 weeks after a sore throat, the patient develops a macular or purpuric rash (Fig. 13.2) on limbs and buttocks (Fig. 13.3) due to vasculitis. The rash may be urticarial before becoming haemorrhagic. Other common features are fever and marked malaise, arthralgia and arthritis—especially of lower limbs. Glomerulonephritis affects a third of cases, with haematuria and albuminuria.

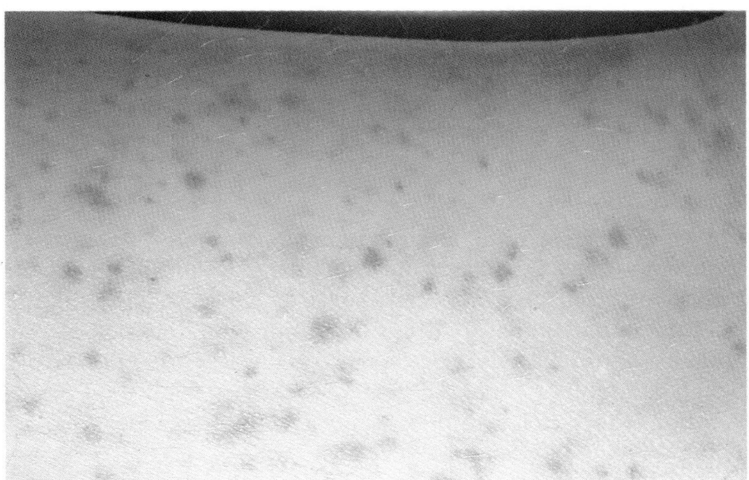

Fig. 13.2 Henoch–Schönlein purpura.

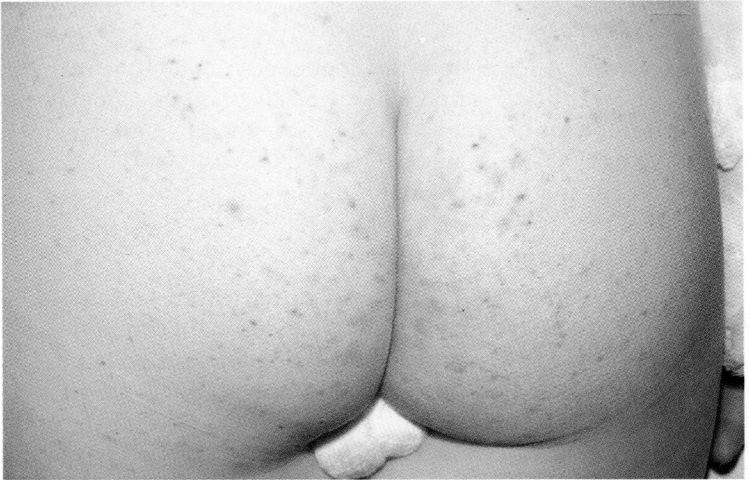

Fig. 13.3 Henoch–Schönlein purpura.

Abdominal pain is probably due to vasculitis and blood loss may cause anaemia. Intussusception is a rare complication in young children.

When all these features are present, diagnosis is not difficult. If rash is predominant, the differential diagnosis includes meningococcal infection, haemolytic uraemic syndrome, and battered baby syndrome; if renal lesions predominate, post-streptococcal glomerulonephritis, systemic lupus erythematosus, and polyarteritis nodosa must be considered.

Investigation

The bleeding abnormality is due to vasculitis with deposition of IgA immune complexes. Platelets and coagulation factors are normal. There is leucocytosis and a raised ESR. IgA levels are raised. Renal biopsy shows focal, proliferative glomerulonephritis.

Management and prognosis

Treatment is supportive until natural remission, usually within 1–6 weeks, perhaps longer in adults. Corticosteroids and other immunosuppressive drugs do not influence the renal prognosis but may alleviate gastrointestinal features. Analgesics and antiinflammatory drugs help joint symptoms.

The prognosis is usually good and depends on the degree of renal involvement. Rarely, dialysis may be required for acute renal failure, or surgery for intussusception. Illness relapses in 5–10 per cent and some, especially adults, develop chronic renal disease and hypertension.

Erythema nodosum

Aetiology and epidemiology

Streptococcus pyogenes infection is the commonest cause, but many infective agents (Table 13.1), drugs (especially penicillin, sulphonamides and oral contraceptives), sarcoidosis, and other diseases (such as inflammatory bowel disorders) may cause the illness. In the UK, half the cases have no demonstrable cause. Young adults and children are primarily affected.

Clinical presentation and investigation

Pyrexia and a characteristic skin eruption are usual features. Multiple, bilateral, very tender, shiny, dusky red nodules of various sizes appear on the shins (Fig. 13.4) and more rarely on anterior thighs (Fig. 13.5), forearm, and face. These may crop and diapedesis produces a bruised appearance which slowly fades. The lesions never ulcerate. Arthalgia and arthritis can occur. The diagnosis is usually made from clinical findings. Leucocytosis and raised ESR are common. Underlying conditions should be identified, especially streptococcal infection, tuberculosis, and sarcoidosis, or sometimes drugs.

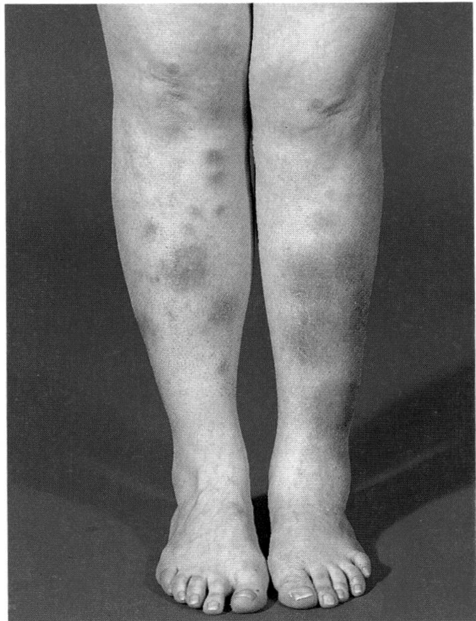

Fig. 13.4 Erythema nodosum.

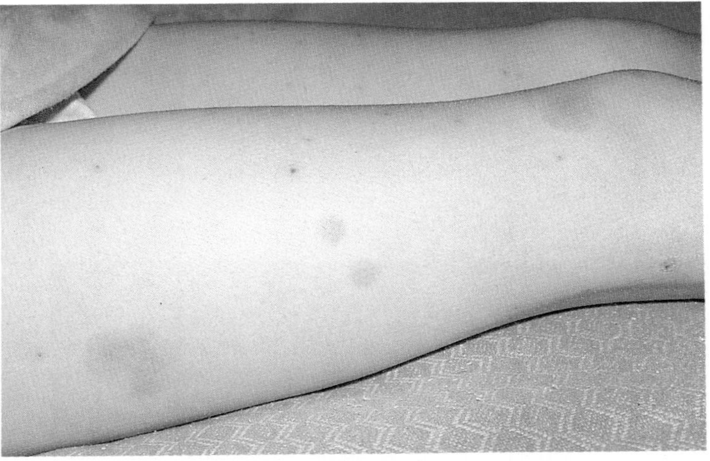

Fig. 13.5 Erythema nodosum.

Management and prognosis

Treatment or removal of the precipitating factor is necessary. Steroids may temporarily relieve symptoms but require caution in case underlying infection is activated. The rash may persist for many weeks but the long-term prognosis is usually good.

Kawasaki disease

Aetiology and epidemiology

The cause of this disease, also known as mucocutaneous lymph node syndrome, is unknown. It is thought to result from abnormal immunological responses to an antigen, such as bacteria, house-dust mites (there is association with recent carpet cleaning) viruses or rickettsiae. First described in Japan in 1967, cases have been reported world-wide, although the greatest prevalence is in Japan and Hawaii. It is endemic in Japan, yet epidemics have occurred in the USA and Finland. Young children are usually affected (median age two years) but adults can also be affected. There is male predominance.

Clinical presentation (Fig. 13.6a–e)

The main features are shown in Table 13.4. There may also be pneumonia, diarrhoea, jaundice, arthritis, aseptic meningitis, and most importantly cardiac

(a)

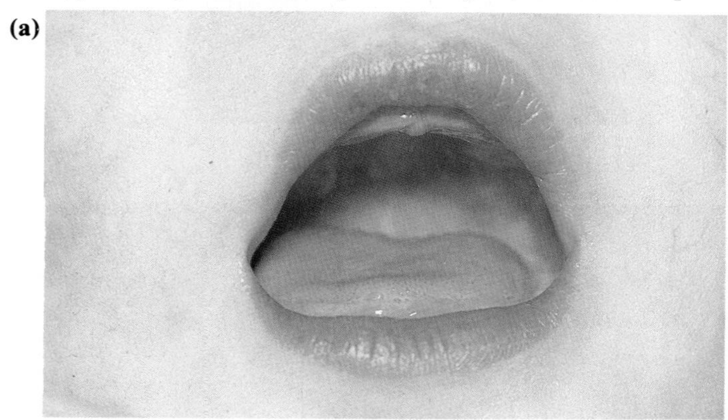

(b)

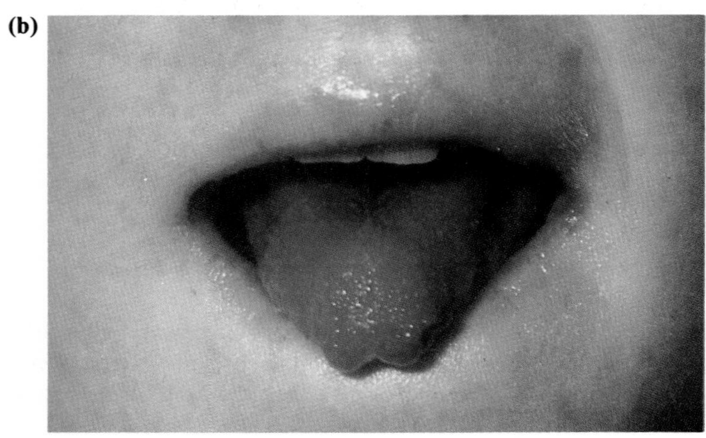

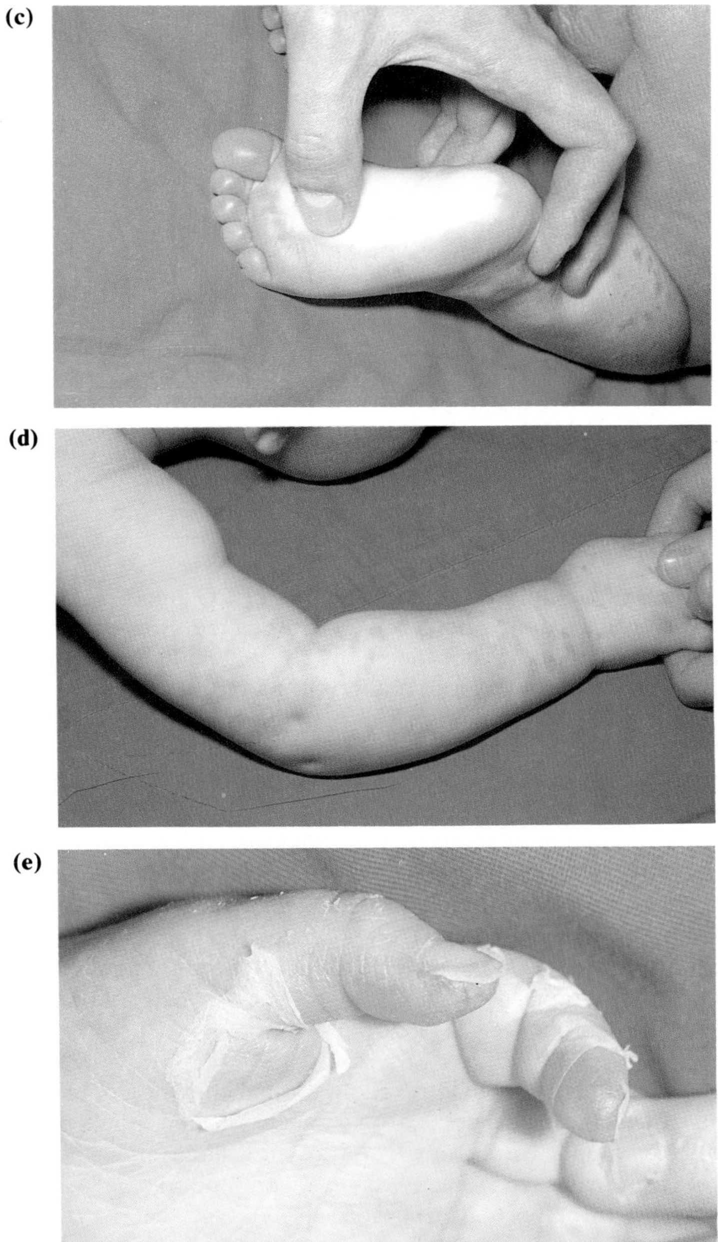

Fig. 13.6 Kawasaki disease: **(a)** red, fissured lips with normal intra-oral appearance; **(b)** strawberry tongue; **(c)** red soles—erythema blanched by pressure; **(d)** rash; **(e)** desquamation.

Table 13.4. *Main features of Kawasaki disease*

Fever (>5 days) unresponsive to antibiotics
Bilateral, non-purulent conjunctivitis
Changes around the mouth (at least one of the following):
red, fissured lips (see Fig. 13.6a)
strawberry tongue (see Fig. 13.6b)
pharyngitis
Changes in the extremities (at least one of the following):
red palms and soles (see Fig. 13.6c)
indurative oedema
desquamation (after 2–4 weeks; see Fig. 13.6e)
Diffuse erythema (Fig. 13.6d), without vesicles or crusts, especially on trunk
Non-suppurative cervical lymphadenopathy

abnormalities. Half the children have an abnormal ECG and there is substantial risk of left ventricular failure and coronary artery aneurysms developing from two weeks to several months after the onset. The differential diagnosis includes toxic shock syndrome, scarlet fever, and erythema multiforme.

Investigation

The diagnosis is clinical, since there are no pathognomic laboratory findings. White blood cells, C-reactive protein and ESR are raised. Anaemia and massive thrombocytosis are common. Pyuria occurs, but urine culture is usually negative. The CSF may show lymphocytic infiltrate but viral cultures are usually negative. Hepatic enzymes are mildly elevated. There is no evidence of auto-antibodies. Because of the risk of aneurysm, serial cross-sectional echocardiograms of aortic root and coronary arteries should be performed. Aneurysms of muscular arteries are not usually important.

Management and prognosis

There is no specific treatment but intravenous normal immune globulin with high dose aspirin reduces the incidence of coronary aneurysm and should be started as soon as possible. Supportive therapy by intravenous fluids may be required. In the third week aspirin dosage should be reduced, continuing for 6–8 weeks to help prevent coronary thrombosis. Corticosteroids increase the risk of coronary aneurysms and are contraindicated. Fortunately, most patients recover after 4–6 weeks. Mortality is about 1 per cent, mainly where fever and raised ESR persist for 10 days or more. Coronary aneurysms are responsible for late fatalities. Children should therefore be followed up for at least one year.

Reye's syndrome

Aetiology and epidemiology

The aetiology of this syndrome of acute encephalopathy associated with fatty degeneration of the viscera, especially liver, is unknown, but it is associated with the recovery phase of influenza B, varicella, and occasionally other infections. It primarily affects children under 16 years of age. Ingestion of aspirin and other salicylates may increase the risk of acquiring the disease and the incidence has fallen with decreased aspirin usage. Familial cases and localized outbreaks occur.

Clinical presentation

Usually after several days of mild upper respiratory tract infection, the more serious illness is heralded by sudden onset of vomiting. Central nervous system manifestations such as lethargy, drowsiness, irritability, behavioural changes, convulsions, and coma evolve rapidly over the next 24 hours. The liver is enlarged but jaundice may be absent or minimal. There are signs of raised intracranial pressure and cerebral oedema. The disease can be rapidly fatal, usually from respiratory failure.

Investigations

Cerebrospinal fluid should be examined to exclude meningitis. Herniation seldom follows lumbar puncture, possibly because the cerebral oedema is diffuse, but ventricular tap may be preferred. The CSF is usually normal apart from a low glucose level. Biochemical studies of blood show raised ammonia levels (especially significant), increased prothrombin time, hypoglycaemia, and abnormal liver function. Post-mortem shows fatty degeneration of the liver but no histological evidence of encephalitis.

Management and prognosis

Management is supportive and artificial respiration is often required. Cerebral oedema should be reduced, for example by mannitol, preferably with continuous intracranial pressure monitoring. Blood glucose should be assessed, but hypoglycaemia can resist treatment. If the patient survives the first 72 hours the outlook is greatly improved. The overall fatality rate is 30–40 per cent, and those with the most cerebral dysfunction have the worst prognosis. Nevertheless, the prognosis for Reye's syndrome appears to be improving, probably because of better supportive care. Recurrent cases have been described.

Erythema multiforme and Stevens–Johnson syndrome

Aetiology and epidemiology

With only skin involved this illness is known as erythema multiforme. The more severe Stevens–Johnson syndrome also involves mucous membranes. It may follow infections—for example herpes simplex or *Mycoplasma pneumoniae*—but also follows use of certain drugs (especially penicillins, sulphonamides, and other antibiotics). In about half the cases, no cause is found. Children and young adults are mainly affected and the disease is not infectious.

Clinical presentation (Fig. 13.7a–d)

The severity varies greatly but it often starts with 1–2 days of fever and malaise suggesting infection. The rash of erythema multiforme is an erythema of many shapes and sizes, often confluent, including 'target lesions'—raised annual red areas with a purple centre; the centre can change colour like a bruise, and may vesiculate. Extensive skin involvement leads to bullae which can break down and denude extensive areas. The lesions are characteristically symmetrical affecting palms and soles and extensor areas of limbs.

In Stevens–Johnson syndrome, mucous membranes of conjunctiva, genital areas and mouth are also involved with crusting of lips. Pneumonia is the commonest complication. Patients with conjunctivitis should be observed for the rare complication of corneal ulceration.

The disease may be confused with toxic epidermal necrolysis (scalded skin syndrome: Ritler's disease, Lyell's disease—see Chapter 7) in which lesions are within the epidermis, not under it, so that areas of skin tend to wrinkle and peel off.

Investigation and management

The clinical features are usually diagnostic and patients often look more ill than they feel. There is usually leucocytosis and an elevated ESR.

Most mild cases resolve in a few weeks. Where there is extensive skin involvement the patient needs to be kept warm, monitoring fluid and electrolytes and preventing secondary infection. Corticosteroids help in severe cases. The eyes, mouth, and skin require careful nursing.

Prognosis

The prognosis is usually good, with recovery of mild cases in 1–2 weeks. In severe cases deaths occur, especially among babies. There may be recurrences, especially if herpes simplex reactivation is the cause when prophylactic acyclovir may be helpful.

Post-viral fatigue syndrome

Aetiology and epidemiology

Common viruses associated with this syndrome are listed in Table 13.1, but many others may rarely be involved. The condition affects only a very few patients infected by any one of the associated viruses. The principal abnormality appears to be immunological, but various secondary defects have been described. Young

(a)

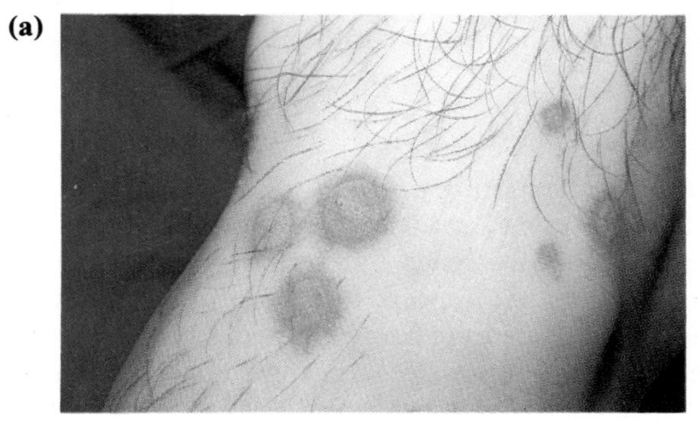

(b)

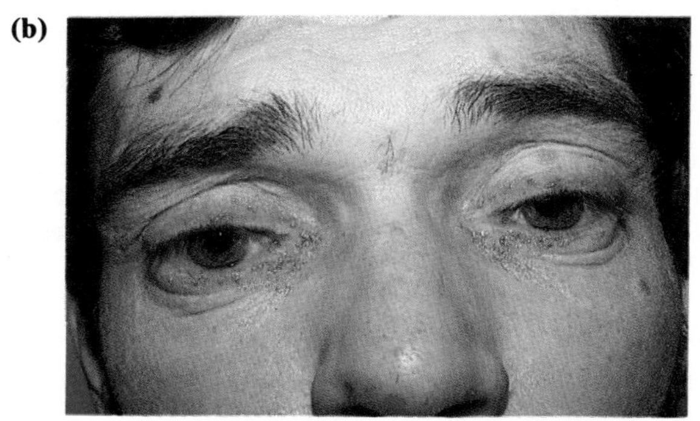

(c) (d)

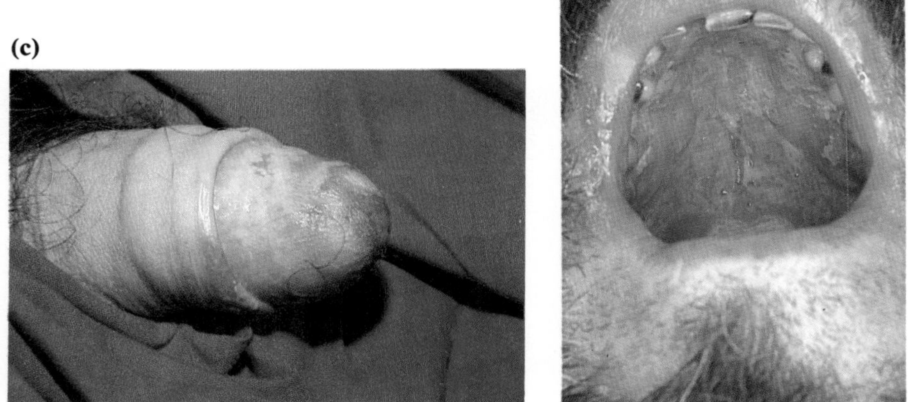

Fig. 13.7 Stevens–Johnson syndrome: **(a)** target lesion; **(b)** eye involvement; **(c)** genital involvement; **(d)** mouth involvement.

adults, especially those who are active and previously well, are predominantly affected.

Clinical presentation and investigation

There is often a history of initiating viral illness. Excessive fatigue and malaise are the main complaints, present for at least 3 months. Patients appear well but complain of great debility, worsened by exertion. Minimal exercise, especially compared to previous ability, appears to have prolonged effects. Tiredness is excessive and most patients require many hours of sleep. Symptoms can involve any system. Reduced concentration and/or memory and mood changes are common. Psychiatric referral is not unusual, but usually unproductive in early illness. A few may develop food allergies. A more serious and sometimes epidemic form of the illness is Myalgic Encephalomyelitis (formerly called 'Royal Free Disease').

Immunological examination in some patients shows increased T-suppressor cells and abnormalities of natural killer cells. Others may have metabolic changes in muscles which may explain tiredness. Some show suggestive serological evidence of continuing infection with coxsackie B viruses. Otherwise, investigations are usually remarkably normal. Diagnosis is clinical and requires exclusion of other causes of the symptoms.

Management and prognosis

There is no specific treatment. Rest is essential and attempts to exercise back to health should be resisted. The patient should be reassured that recovery will eventually occur but may be slow and even take years. Ideally, patients should change life-style to adjust activity to current reduced abilities. Most patients require much support and symptomatic treatment.

14

Imported infections

INTRODUCTION

Travel, particularly foreign travel, entails exposure both to organisms not endemic to the home country and to unfamiliar strains of familiar organisms, such as *Escherichia coli*, enterotoxin-producing strains of which cause much traveller's diarrhoea (see Chapter 4). More people now travel than ever before, and in 1989 30.8 million from Britain took holidays abroad. Most went on package holidays to Mediterranean countries, but more distant and tropical countries (with more risk of infections!) are increasingly offered for short holiday 'packages'. Others such as business people, air and sea crew, members of the armed forces, technical advisers and project workers, educators, missionaries, and voluntary workers, also travel abroad. The risks of infection vary with increased exposure in relation to some or all of the following:

(a) *Interpersonal contact* in crowded transport or accommodation, or by sexual activities.

(b) *Faecal–oral transmission of pathogens.* Kitchen hygiene may be inadequate (most travellers do not cook for themselves), water supplies less pure, and the visited population may have a higher prevalence of carriers of pathogens such as *Salmonella typhi* or *Entamoeba histolytica.* Foodstuffs may be contaminated by use of excreta as fertilizer or from flies.

(c) *Biting arthropods*, such as mosquitoes, tsetse flies, mites, or ticks can transmit a wide range of pathogens.

(d) *Diseases common in the country visited.* As the term 'tropical diseases' implies, many diseases, particularly infections, have a limited geographical range, although often not strictly within the tropics. The risk of contracting a particular infection is obviously higher where it is endemic, but varies even within endemic zones. Thus, except during an outbreak there is little risk in an urban area of being bitten by a mosquito carrying yellow fever virus.

(e) *Duration of stay.* Longer stays allow more opportunities for infection although degrees of resistance to local pathogens can develop. Intensity of transmission may vary through the year; thus malaria increases in rainy seasons when mosquito breeding sites are more abundant.

(f) *Individual life-style.* An overland traveller in the tropics eating and drinking local food and water is clearly at more risk of infection than one staying for a week in a four-star hotel in Europe. However, simple food, freshly cooked and eaten hot under village conditions may be safer than elaborate dishes prepared by 'food-handlers' in a large hotel! In general, inexperience, travel within rural areas, eating out, sexual promiscuity and intravenous drug abuse lead to more infections.

(g) *Occupation*. Some activities entail risk from specific infections. Thus veterinarians may be exposed to rabies in endemic areas, and midwives to hepatitis B where this is endemic with high carriage rates.

(h) *Personal precautions*. The risk of contracting some infections can be reduced by simple measures such as boiling water, avoiding insect bites, active or passive immunization, and chemoprophylaxis (Chapter 16).

Assessment of the size of the problem

Surveys of returning travellers

Surveys mainly of package holidaymakers, show that diarrhoeal and respiratory illnesses are commonest (Fig. 14.1). For areas frequently visited by British holidaymakers, risks are greatest along the North African coast, less for European Mediterranean countries, and less again for north European countries (Table 14.1).

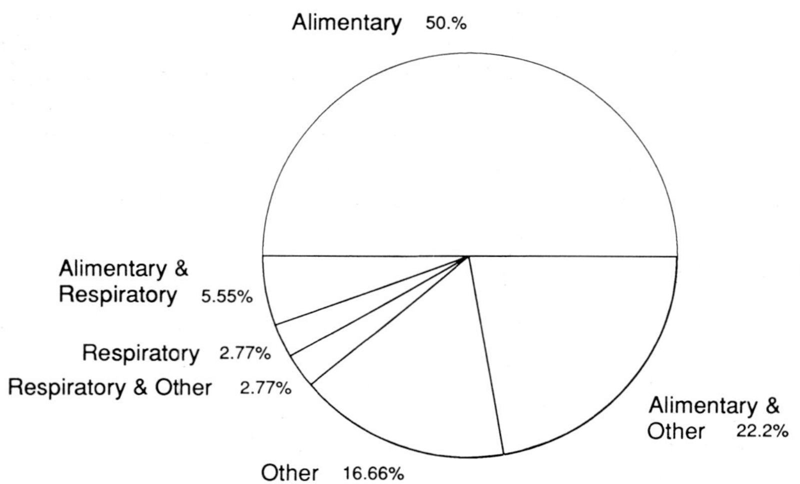

Fig. 14.1 Types of illness reported by the 36 per cent of 13 816 travellers in Scottish surveys 1977–85 who reported some illness. (J. H. Cossar *et al.*, *J. Infection*, 1990, **21**, 27–42.)

Table 14.1. *Percentage of surveyed travellers to different countries reporting illness*

Country	Percentage	Country	Percentage
Egypt	56	Greece	16
India/Nepal	50	Spain	16
Gambia	46	Yugoslavia	14
Turkey	41	Caribbean	14
Kenya	34	France	9
Tunisia	34	W. Germany	5
S.E. Asia	28	The Netherlands	3
Bulgaria	22		

From *Holiday Which?*, May 1987 (Consumers' Association).

Notifications

Some imported infections are compulsorily notifiable because of their severity or public health consequences, and their incidence is reflected in officially published reports.

Laboratory data

For other infections, diagnostic laboratory reports of identified pathogens may provide the only source of information. These show that gastrointestinal pathogens predominate (Table 14.2). However, not all illness is fully investigated, laboratory techniques are not fully sensitive, and the figures are heavily biased towards pathogens readily identified by standard laboratory procedures such as bacterial stool culture or blood film examination. Viruses, for instance, are hardly represented in Table 14.2 although they cause much of both respiratory and diarrhoeal illness; jaundice, however, often entails tests for hepatitis viruses.

The patient from abroad

Most patients returning to Britain with imported infections manifest symptoms of fever, diarrhoea, respiratory involvement, jaundice, or skin rashes, singly or in combination. The differential diagnosis and management of these problems are discussed below with emphasis on the more common and severe diseases. Some infectons which are rarely seen in developed countries but are important in the tropics, are described in Chapter 15.

'Have you been abroad?' is an important question to be asked in acute illness. The patient may not associate symptoms which develop after return with the travel, which may not be mentioned. Also, there may be delay before some

Table 14.2. *Percentage distribution of imported infections reported by consultants in public health medicine, Scotland 1988*

Disease	Percentage
Salmonellosis	50
Campylobacteriosis	16
Shigellosis	11
Malaria	6
Giardiasis	5
Hepatitis A and B	4.5
Typhoid/paratyphoid fever	1.5
Legionnaires' disease	0.5
Others	5.5

Data from Communicable Diseases (Scotland) Unit. The total number of infections reported was 746.

infections declare themselves; thus, vivax malaria (Table 14.3) and onchocerciasis often present a year or more after return.

When presenting symptoms are associated with travel, further details should be elicited: dates of departure and arrival, countries visited and stops *en route*, approximate life-style particularly as regards food and water, illness and treatment abroad, recognized illness in people or populations visited, and exposure to bites by insects (Fig. 14.2) or other animals. The immunization history and use of chemoprophylactic agents should be noted.

Table 14.3. *Time from return to UK to diagnosis of malaria, 1990*

Infectious agent	Time (months)			
	<1	1–5	6–11	>12
P. falciparum	651	100	3	2
P. vivax	150	214	163	29
P. ovale	12	54	17	13
P. malariae	7	5	—	—

Data from Malaria Reference Laboratory.

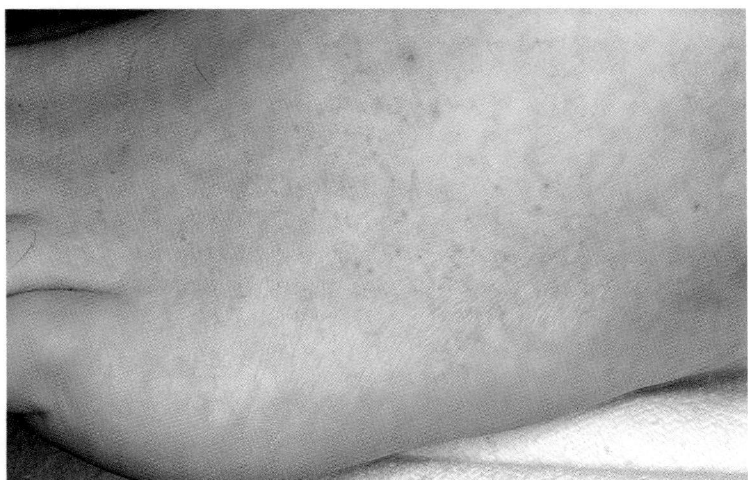

Fig. 14.2 Mosquito bites. Each small red dot represents one bite. Both ankles and feet were covered by such bites. This engineer had falciparum malaria.

FEVER

The diagnostic approach to fever is described in Chapter 9, but further consideration must be given to diseases endemic in the particular area visited by the patient.

Common causes of fever such as respiratory and urinary tract infections may also be 'imported'. Malaria, typhoid and paratyphoid fevers, and tuberculosis (Chapter 9) should be considered. Less common causes include amoebic liver abscess, Legionnaires' disease (Chapter 2), acute brucellosis (Chapter 9), filariasis, tick typhus (Fig. 14.3), visceral leishmaniasis (Chapter 15), dengue and African trypanosomiasis (Chapter 15). Viral haemorrhagic fevers are rarely imported but, when on epidemiological and clinical grounds they are considered, special precautions are required (see below). Katayama syndrome (fever, cough, wheeze, arthralgia, and eosinophilia) can develop some weeks after heavy exposure to schistosomes by swimming or bathing in endemic tropical areas, usually African.

Malaria

Aetiology

Four protozoal species cause malaria: *Plasmodium falciparum*, *P. malariae*, *P. ovale*, and *P. vivax*.

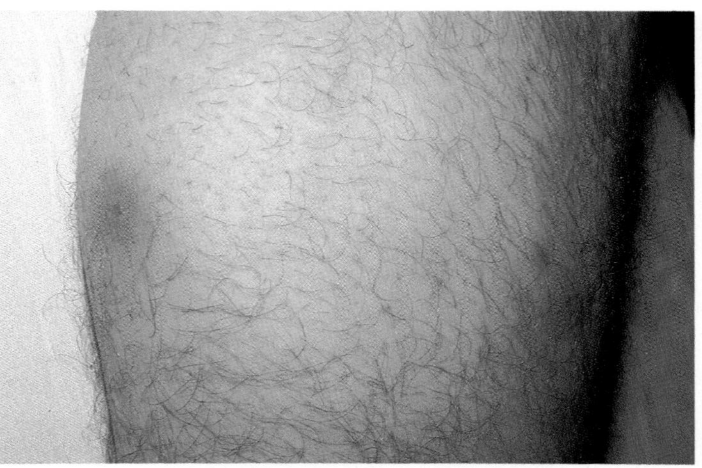

Fig. 14.3 Tick typhus. Pustular lesions with central necrosis ('eschars') at sites of infected tick bites on traveller to rural Zimbabwe.

Epidemiology and pathogenesis

Plasmodium falciparum is present in most endemic areas (Fig. 16.9). *Plasmodium vivax*, *P. malariae*, and *P. ovale* are more localized. Sub-Saharan Africa is the most intensely endemic focus of malaria. The protozoa must spend the sexual stage of their life-cycle in female anopheline mosquitoes (Fig. 16.8), reaching the sporozoite stage in salivary glands and then transmitted to man when the mosquito takes a blood meal. Injected sporozoites are taken up by the liver where they divide within hepatocytes during the incubation period. This varies from 12–30 days; that for *P. falciparum* is usually the shortest. Incubation periods may be prolonged, especially by chemoprophylaxis if parasites are partly resistant. Just before the end of the incubation period, merozoites are released from hepatocytes, invade erythrocytes, and grow as trophozoites (Fig. 14.4) into multinucleated schizonts which rupture the red cells after about 48 hours (72 hours with *P. malariae*) releasing more merozoites to re-invade erythrocytes and continue the cycle. A few merozoites develop within erythrocytes into male or female gametocytes which are available to be taken up by a mosquito and complete the life-cycle. Some of the original sporozoites of *P. vivax* and *P. ovale* may remain dormant within hepatocytes as 'hypnozoites' which can become active months or occasionally years later. Malaria can also be transmitted through sharing needles by intravenous drug misusers and by blood transfusion which, in highly malarious areas, should be accompanied by antimalarial therapy.

Fever occurs when merozoites are released and, if there is severe haemolysis, with falciparum malaria, anaemia and jaundice ensue. Falciparum malaria also causes capillary plugging, local haemorrhage, and oedema partly by increased

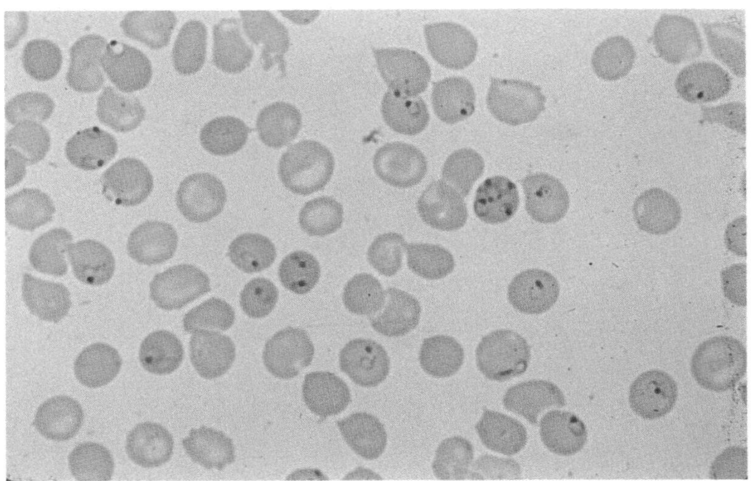

Fig. 14.4 Malaria. Trophozoites of falciparum malaria within red cells. Some red cells have been entered by more than one parasite.

stickiness of non-parasitized red cells. In the brain this leads to 'cerebral malaria'. Together with hypovolaemia and intravascular haemolysis, disseminated intravascular coagulation contributes to renal and pulmonary failure and shock. Immune complex deposition in glomeruli in chronic *P. malariae* infection can cause progressive glomerulonephritis.

Clinical presentation

Intermittent fever is characteristic. The rapid initial rise in temperature results in a rigor, accompanied by profound malaise, headache, nausea, and aching in limbs. After about 30 minutes, a temperature of around 40 °C or above is reached and the patient feels 'burning up'. After another hour or so the temperature falls precipitately with profuse sweating. In vivax and ovale malaria those febrile paroxysms usually recur ever 48 hours, whereas *P. malariae* has a 72-hour cycle. Fever in falciparum malaria is more irregular, often with daily spikes, presumably because populations of parasites have cycles at different stages.

In falciparum malaria, a few days of fever and rigors can lead to rapidly progressing, severe, life-threatening illness, especially in non-immune persons such as the average British traveller. Ominous signs are jaundice, drowsiness or confusion, and black urine (Fig. 14.5) suggesting haemoglobinuria ('Blackwater fever'). Splenomegaly is common (Fig. 14.6). Malaria should be excluded in anyone with fever who has recently been in an endemic area, however briefly, whatever else the symptomatology.

Table 14.4 *Rarer causes of imported fevers in Britain*

Disease	Aetiology	Diagnostic pointers	Laboratory findings
Amoebic liver abscess (Chapter 5)	*Entamoeba histolytica*	High, spiking, intermittent fever; drenching sweats; tender liver but *no* jaundice. Rapid response to metronidazole.	Leukocytosis with normal differential. ESR raised. Alkaline phosphatase increased. Transaminases normal. Amoebic serology positive. Sonolucent area on ultrasound.
Dengue fever	Dengue viruses 1–4 (transmitted by Aedes mosquitoes)	Sharp, flu-like illness with prominent headache and limb pains. Fever for 5–7 days. Maculopapular rash or generalized erythema after 3–4 days. Occasionally petechiae or shock syndrome. Travel mostly to Far East, also Indian subcontinent, S. America and Caribbean.	Leucopenia, thrombocytopenia. Virus isolation from blood. Specific antibody response.
Filariasis (usually Bancroftian or Brugian from Indian subcontinent or S.E. Asia—Chapter 19)	*Wuchereria bancrofti* *Brugia malayi*	Cough and wheeze (tropical pulmonary eosinophilia); fever, lymphangitis, and lymphadenopathy, especially in lower limbs and genitals. Chronic lymphoedema after repeated infections leading to elephantiasis.	Eosinophilia. Filarial serology positive. Microfilariae in midnight capillary blood or venous blood after provocation with diethylcarbamazine.
Rickettsioses			
(a) Tick typhus (in Old World)	*Rickettsia conorii*	Fever, headache, myalgia, malaise. Primary eschar (Fig. 14.3) at bite site; enlarged tender regional lymph nodes. Often maculopapular rash, may become haemorrhagic.	Serological reactions: Weil–Felix reaction; CF, MA, and IFA tests.*

(b) RMSF† (tick-borne) (in New World)	*R. ricketsii*	Tick exposure. Severe, sustained fever, rigors, headache, myalgia. Pink rash on limbs, often haemorrhagic.	do.
(c) Epidemic typhus (human louse-borne)	*R. prowazekii*	From crowded, low-hygiene conditions (war, refugees). Fever, headache, myalgia. Macular rash on trunk.	do.
(d) Murine typhus (rat flea-borne)	*R. mooseri*	From crowded, rat-infested conditions. Similar to (c) above but milder.	do.
(e) Scrub typhus (rodent mite-borne in central, eastern and south-eastern Asia, North Australia)	*R. tsutsugamushi*	Exposure in scrub or tropical forest clearing. Primary eschar at bite site. Fever, headache, rash on trunk then limbs.	do. (Weil–Felix test may be negative)
African trypanosomiasis (Chapter 15)	*Trypanosoma brucei*	Hard, painful nodule at site of bite (usually legs or head)—subsides. Recurrent febrile episodes up to one week each with headache, malaise, enlarged lymph-nodes. Often afebrile intervals of weeks. Circinate erythematous rash. Sleeping sickness a late manifestation.	Blood (thick film, buffy coat, and filtration), CSF, and lymph-node aspirate show trypanosomes. CSF IgM raised.
Visceral leishmaniasis (Chapter 15)	*Leishmania donovani*	Insidious onset, mild fever, cough, diarrhoea. Abdominal distension, hepatomegaly, gross splenomegaly.	Pancytopenia. Albumen lowered. Gammaglobulin raised. Histology (Fig. 15.10) and culture of bone marrow, splenic aspirates, or liver biopsy show leishmania. Serological response.

*CF = complement fixation; MA = microagglutination; IFA = indirect immunofluorescence.
†RMSF = Rocky Mountain spotted fever.

Fig. 14.5 Malaria. Haemoglobinuria due to rapid intravascular haemolysis.

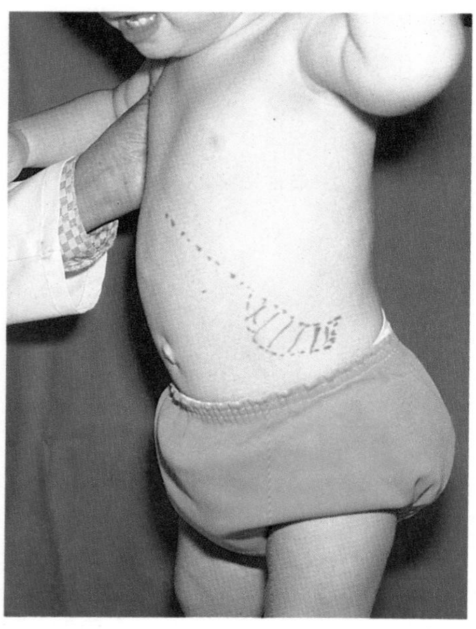

Fig. 14.6 Malaria. Splenomegaly is common in the non-immune and in children in endemic areas.

Investigation

Diagnosis is made by inspecting thick and thin blood films taken before therapy (Fig. 14.7). The stained thick film reveals white cells and parasites released from lysed red cells and allows a relatively large amount of blood to be screened quickly. Repeated examinations may be required. Species identification is

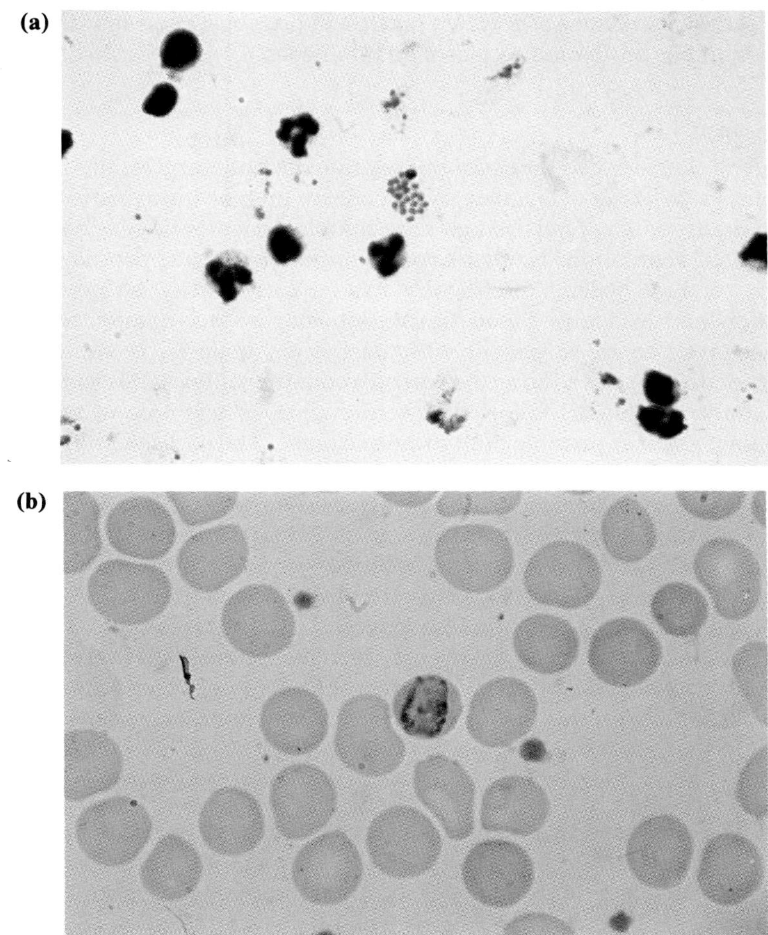

Fig. 14.7 Malaria. **(a)** Thick film showing stained white cells, trophozoites and schizonts of *P. vivax* released from lysed red cells; **(b)** thin film is best examined where red cells are just separated from each other. The trophozoite is of *P. malariae*. Platelets are also visible.

performed on the thin blood film where individual intact red blood cells containing trophozoites, schizonts, or gametocytes may be seen. Prolonged search may be necessary to find sufficient typical examples for the species to be distinguished. More than one species may be present. If the film shows falciparum parasites the percentage of red cells parasitized should be counted. Parasite numbers are reduced rapidly by appropriate treatment. Some common antibacterial agents, like cotrimoxazole, also have an antimalarial effect and can make diagnosis difficult. If antimalarials have been given and parasites cannot be found, serological evidence of recent malaria in previously non-immune patients can be found as antibodies to plasmodial antigens.

Management

Severely ill patients with falciparum malaria constitute urgent, life-threatening emergencies. Without delay, they should receive quinine intravenously if unable to take orally, with supportive and resuscitative measures such as renal dialysis and assisted ventilation. Neither heparin nor steroids have proved helpful for DIC or cerebral oedema. Refractory hypoglycaemia may be present. Fluid restriction and exchange blood transfusion may reduce complications. Oral quinine, given to those less ill with falciparum malaria, is substituted for intravenous quinine as soon as the patient's condition allows. It is usual to give a 7-day course of quinine, followed by tetracycline, or one dose of Fansidar, or mefloquine to cover possible multidrug resistance. The parasitaemia of patients with falciparum malaria is monitored for falling parasite counts. Asexual forms but not gametocytes (which do not cause disease) normally disappear within five days of starting effective treatment. It is important to continue effective prophylaxis after recovery until the patient has been away from the endemic area for 4 weeks in case further parasites are developing. Those with hepatocytes vivax, ovale, or malariae malaria should receive chloroquine orally. Hypnozoites of vivax and ovale malaria in the liver are not susceptible to the above drugs and are usually removed by a two-week course of primaquine. This drug should not be used in those with severe G6PD deficiency, in whom it causes haemolysis.

Prognosis

Delay in diagnosis and treatment is dangerous. Mortality is high in non-immune patients with falciparum malaria and parasitaemia of more than half their red cells or who have developed jaundice, cerebral symptoms, or haemoglobinuria. Otherwise, with treatment recovery is usually uneventful provided there is no drug resistance. In endemic areas falciparum malaria causes high mortality in children, particularly with poor nutrition or intercurrent infection. Survivors develop antibodies slowly over years and further attacks become less severe. In the other forms, even without treatment, symptoms settle over 3–4 weeks with clearance of vivax and ovale parasites from the blood. Repeated infections cause chronic debility and anaemia. *Plasmodium malariae* can cause chronic, asymptomatic parasitaemia. After treatment, recurrence may result from inadequate treatment, drug resistance by parasites, premature discontinuation of chemoprophylaxis, reinfection, or activation of *P. vivax* or *P. ovale* hypnozoites.

Prevention

No vaccine for general use is currently available. Non-immune travellers to malarious areas should usually take chemoprophylaxis, avoid mosquito bites and report promptly febrile symptoms even months after returning (Chapter 16). Chemoprophylaxis is usually impracticable for the populations of endemic areas, to whom drug treatment is given for symptoms. The number of mosquitoes can be reduced by spraying insecticides and reducing stagnant water sites suitable for egg-laying.

Typhoid and paratyphoid fevers

Aetiology

These fevers are caused by *Salmonella typhi* or *S. paratyphi A*, *B*, or *C*. The typical illness is also termed *enteric fever*.

Epidemiology and pathogenesis

These organisms are present worldwide but are not common in developed countries with higher standards of water supplies, sewage disposal, and catering. In these countries, such as Britain, indigenous chronic carriers become fewer with the passage of time. Spread is by faecal–oral or urine–oral routes and can become epidemic if sewage containing these pathogens contaminates drinking water. Local outbreaks and endemic transmission are more often due to contamination of food or milk by asymptomatic, chronic, faecal, or urinary carriers of the organisms. After ingestion many organisms are killed by gastric acid; but where the infecting dose is high, when organisms are within food boluses, if gastric emptying is rapid or gastric acidity reduced, viable organisms can enter the small intestine. They are then absorbed and multiply within macrophages during the incubation period to be released into the bloodstream. Symptoms relate mainly to septicaemia but organisms also enter the gastrointestinal tract (via bile) and the urinary tract after passing through liver and kidney, respectively. Intense immune response in lymphoid tissues of Peyer's patches may lead to ulceration, perforation, or haemorrhage.

Clinical presentation

Early symptoms are fever, headache, and dry cough. The fever is not usually associated with rigors, and peaks in the afternoon when headache is worst. Mild abdominal discomfort and anorexia are present. Constipation is commoner than diarrhoea, especially in adults. Confusion may be present and the patient may appear deaf with slow response to questions. Vivid and bizarre dreams, visual and auditory hallucinations are common. As illness progresses the patient becomes stuporose. Common signs are relative bradycardia (increasing by less than the normal 15 beats per minute for each centigrade degree rise), step-ladder rising fever (Fig. 14.8), splenomegaly (but not lymphadenopathy), and 'rose spots'. These are pink, macular spots, visible only on unpigmented skin, that blanch on pressure (Fig. 14.9); they may be few in number and often scattered over the lower thorax or upper abdomen, anteriorly or posteriorly. Rose spots

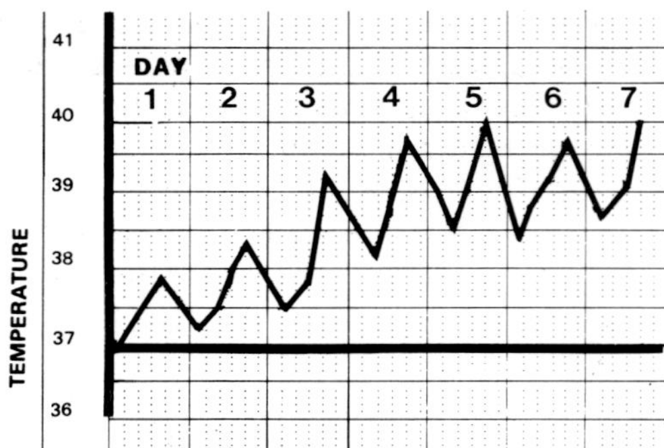

Fig. 14.8 Typhoid fever. 'Step ladder' rise of fever. The fall of temperature, spontaneous or following drug therapy, usually mirrors this.

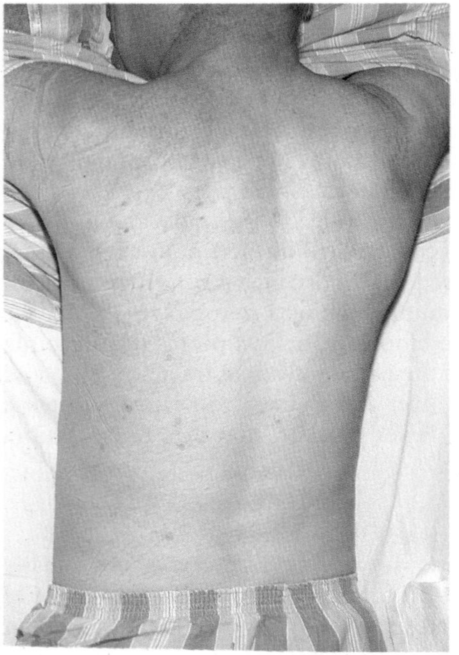

Fig. 14.9 Typhoid fever. Rose spots.

last only a day or two, may appear in crops and are usually more numerous in paratyphoid B.

Investigations

Blood culture is usually positive. Urine and stool should also be cultured, but isolates from these sources may be due to chronic carriage (usually asymptomatic). Slight leucopenia is usual in acute infection. The Widal test identifies antibodies to O (somatic), H (flagellar), and Vi (somatic) antigens of the organisms but must be interpreted with care where there has been past exposure or immunization. High agglutination titres (over 640) especially with O antigens are strongly suggestive, and a fourfold rise in titre is diagnostic. Vi antibodies are present in about 75 per cent of chronic carriers.

The differential diagnosis includes conditions listed in Chapter 9, especially malaria and tuberculosis which also presents with normal or low white-blood-cell counts. When fever and confusion co-exist cerebral malaria must be urgently excluded (see above).

Management

About 60 per cent of patients with typhoid fever died in pre-antibiotic days. Antibiotic treatment markedly reduced mortality and morbidity. If there is no drug resistance, chloramphenicol, amoxycillin, co-trimoxazole, mecillinam, trimethoprim, and ciprofloxacin are equally effective given orally unless the patient is very ill. Treatment should be given for 14 days to lessen the chance of relapse. The temperature falls only slowly, often in step-ladder fashion over four or five days. Resistance to drugs may be present or develop; multiresistant strains are common in the Indian subcontinent. Organisms sensitive by *in vitro* tests do not always respond *in vivo*. If there is no response, a different drug should be tried.

Complications

The most severe complications, perforation, or haemorrhage secondary to reaction at Peyer's patches, are the main contributors to overall mortality of about 3 per cent. Haemorrhage may require blood transfusion. Perforation is treated conservatively by intravenous infusion and nasogastric suction (unless peritonitis becomes generalized) because friable bowel wall is difficult to oversew. Cases presenting late in the illness may be in a 'typhoid state'—a stuporose condition, difficult to reverse and probably caused by endotoxaemia. Infection secondary to septicaemia may localize at any body site, causing meningitis, osteomyelitis, cholecystitis, prostatitis, arteritis, or other conditions. Even after optimal treatment about 10–20 per cent of cases relapse, commonly 10–14 days after stopping treatment, usually with mild illness. One to 2 per cent suffer a second relapse. Chronic carriage of the organism in 3 per cent of cases usually causes faecal rather than urinary excretion in a ratio of 100 to 1. Where urinary schistosomiasis is prevalent, as in Egypt, urinary carriage is commoner. Clearance may be attempted by ampicillin or ciprofloxacin for three months but sensible carriers in developed countries present little risk to others provided they wash hands after defaecation or urination and before preparing food. Faecal carriage is difficult to clear in the presence of gallstones; if clearance is imperative—for example, in the case of a cook—cholecystectomy is advised.

Prophylaxis

Killed *S. typhi* vaccine is widely used and is 70–90 per cent effective for 3–7 years. Three doses of oral attenuated vaccine Ty21a and single-dose purified parenteral

Vi polysaccharide vaccine have similar efficacy. Just as after natural infection, protection may not withstand exposure to large infecting doses. Killed paratyphoid vaccines are ineffective and no longer generally available in Britain. Good hygiene and sanitation are the essential measures to interrupt transmission.

Viral haemorrhagic fevers (VHF)

Many viruses cause febrile illnesses with haemorrhagic rash, and the most important are listed in Table 14.5. Most are zoonotic and endemic in limited tropical areas (Fig. 14.10), with little potential for spread within the community after importation into Britain. The risk of transmission to nursing and medical staff is also low, and cross-infection is rare but can cause severe illness and death. The main risk is from viraemic blood, which can contaminate venepuncture needles. Most of the causal viruses (Table 14.5) are categorized in Britain by the Advisory Committee on Dangerous Pathogens as 'group 4 pathogens' presenting maximal danger. Most British attention focused on Lassa fever after outbreaks with 40 per cent mortality in Nigeria, although later serological surveys showed that non-fatal infections are widespread in endemic areas. However, Congo/Crimean haemorrhagic fever, Marburg, and Ebola viruses can also spread, mainly by blood contact, and entail a similar risk for hospital personnel. Dengue haemorrhagic fever (DHF) or dengue shock syndrome follow immune enhancement with second or subsequent exposure to dengue viruses, primary infections causing less severe dengue fever. DHF is therefore rare in travellers and develops

Table 14.5. *Viruses that can cause severe haemorrhagic disease in man*

Source	Disease
Mosquito-borne	Dengue, types 1–4 Rift Valley fever Yellow fever (see Chapter 15)
Tick-borne	Congo/Crimean haemorrhagic fever* Kyasanur Forest disease* Omsk haemorrhagic fever*
Rodent-borne	Argentinian haemorrhagic fever (Junin)* Bolivian haemorrhagic fever (Machupo)* Lassa fever* Haemorrhagic fever with renal syndrome (Hantavirus)
Unknown source	Marburg virus* Ebola virus*

*Group 4 pathogens.

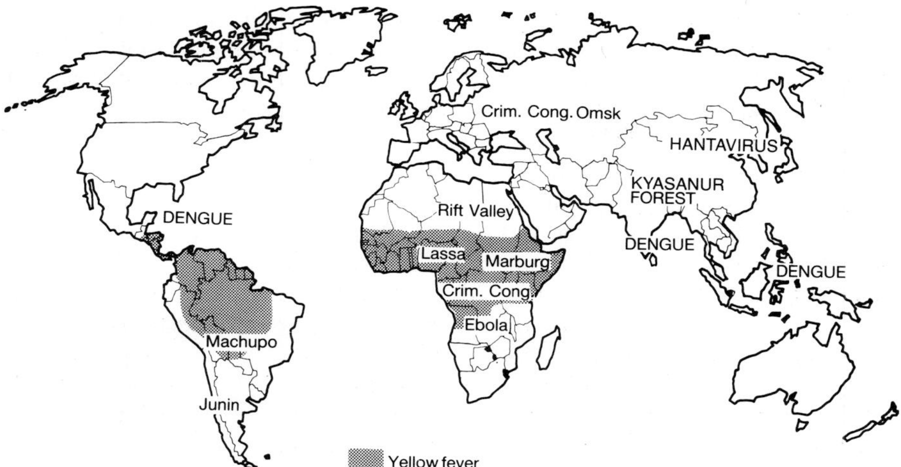

Fig. 14.10 Viral haemorrhagic fevers. Approximate areas of endemicity superimposed.

mainly in southern Asia, but there have been recent epidemics in South America and the Caribbean.

Clinical presentation

Lassa fever usually has an insidious onset with painful ulcerated throat, backache, and extreme lassitude. The onset of most other types of VHF is usually sudden with severe flu-like symptoms. Jaundice is present in about a third of yellow fever cases (Chapter 15); a maculopapular rash appears about the fifth day in Marburg and Ebola fevers. Gastrointestinal upset is common in all with haemorrhage, circulatory failure, DIC, haemorrhagic rashes, and shock in severe cases. Progression to renal failure is characteristic in some patients with Hantavirus infection.

Diagnosis and management

Viral haemorrhagic fevers should be considered whenever fever develops within 21 days of leaving an endemic area (for Britain this is usually mainland sub-Saharan Africa), especially after living or working in non-urban areas, in villages, or on safari. As the initial symptoms of fever, malaise, lethargy, myalgia, headache, and sore throat are relatively non-specific, attention is paid to epidemiological factors, including occupations such as medicine and nursing, and whether malarial prophylaxis has been taken (if not, this makes malaria more likely). If VHF cannot be excluded after consultation with the infectious disease physician, the local Consultant in Public Health Medicine should be contacted, and the patient isolated where he or she is until experienced assessment has been made. In the light of this assessment, transfer may be arranged to a secure isolation unit with virus-filtered ventilation or high-security isolation in a plastic isolator (Fig. 14.11). A high-security laboratory is required to process dangerous specimens. Most such patients have malaria (Table 14.6). Ribavirin intravenously is indicated in Lassa fever. Administration of convalescent immune serum, if available, may be helpful. Virus may be excreted for several weeks in urine, also in semen in the case of Marburg virus.

Low-grade contacts of proven cases, such as fellow passengers in buses or aeroplanes, are not usually followed up, but closer contacts (household or those in contact with secretions or excretions or body fluids) require daily surveillance for 21 days.

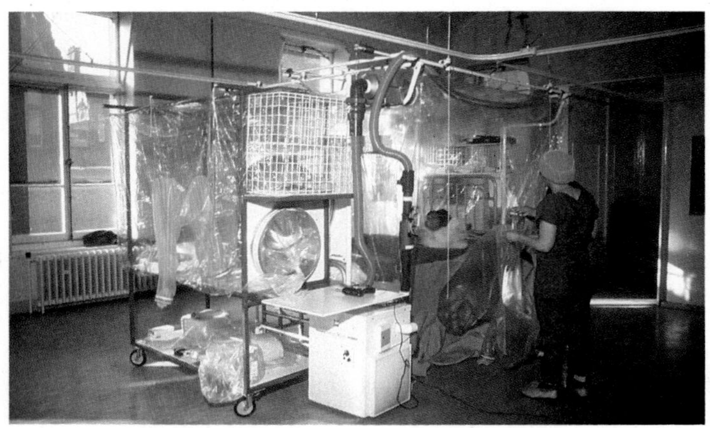

Fig. 14.11 Viral haemorrhagic fevers. Trexler isolation unit at Ruchill Hospital.

Table 14.6. *Final diagnosis of patients with suspected viral haemorrhagic fever, 1982–90 (England and Wales)*

Final diagnosis	Number of patients
Falciparum malaria	79
Lassa fever	5
Dengue haemorrhagic fever	1
Other infections (including typhoid fever, tick typhus, hepatitis A, dysentery, respiratory infections, urinary tract infections, viral pericarditis)	86
Undiagnosed	12
Total	183

Data from B. Bannister, personal communication (Coppetts Wood Hospital, London).

Rickettsioses (Table 14.4)

Rickettsiae are widespread in the world as zoonotic infections of mainly small mammals, transmitted by parasitic vectors (mainly ticks or mites) which may also bite and infect humans. *R. prowazekii* is adapted to humans, louse-borne, and capable of latent persistence after recovery, recurring years later as the mild fever 'Brill's disease'. Rickettsiae cause focal vasculitis, often producing rashes which, in the tick-borne 'spotted fevers' of America may become haemorrhagic like meningococcal septicaemia. Many of the illnesses are severe and life-threatening. Specific diagnosis is serological, preferably by modern tests more specific than the original Weil–Felix reaction. In treatment supportive measures are important, and a prompt response usually follows therapy by chloramphenicol or a tetracycline such as doxycycline.

DIARRHOEA

Most imported bowel infections (Table 14.2) are also endemic in Britain and are managed no differently, although the relative frequency of some, like giardiasis and *E. coli* enteritis, is increased in travellers. It must be remembered that infection may precipitate exacerbation of chronic inflammatory disease or irritable bowel syndrome, or may induce disaccharidase deficiency or tropical sprue. This last condition is typified by chronic malabsorption after visiting certain tropical areas, especially after prolonged stay under insanitary conditions in the Caribbean, India, or south-east Asia. Sprue may respond to long-term tetracycline. Amoebiasis and cholera are discussed below. Common intestinal parasites like ascaris or hookworm (Chapter 4) do not usually cause diarrhoea, but this may be caused by trichuriasis (whipworm disease, a nematode infection) and strongyloidiasis (a helminthic infection of the duodenum and upper jejunum).

Amoebiasis

Aetiology and epidemiology

This infection by the protozoon *Entamoeba hystolytica* spreads by the faecal–oral route usually via water or food contaminated by an ill or often asymptomatic cyst passer. The disease is especially common in countries with poor sanitation, but can be particularly dangerous for those who develop it in areas of low prevalence where the diagnosis may be delayed or missed. Because acute amoebic dysentery can mimic ulcerative colitis and can be difficult to confirm, many advocate a course of anti-amoebic treatment before steroid therapy (which can activate or exacerbate amoebiasis) or surgery in patients with ulcerative colitis from areas of high prevalence.

Clinical presentation

The incubation period varies from a few days to several months. Symptoms may present acutely with fever, pain, and diarrhoea with blood and mucus. More often fever and systemic symptoms are absent and the patient has relapsing diarrhoea with blood and mucus (Fig. 4.4) over many months. Later, perhaps even without clinically recognized primary infection, complications such as liver abscess (Chapter 5), intestinal strictures, large bowel ulceration, or 'amoeboma' can occur. When the liver is involved it is usually tender and enlarged but jaundice is uncommon. Amoeboma (localized granulomatous infiltration of the bowel-wall containing amoebae) usually presents as a mass in the right iliac fossa.

Investigations

In acute amoebic dysentery, motile amoebae containing red cells are diagnostic. They are most readily seen in mucus passed per rectum or scraped from an ulcer during sigmoidoscopy. Microscopy must be performed immediately, because on cooling amoebae lose motility and round up like leucocytes. Sigmoidoscopic and radiological appearances of amoebic dysentery are occasionally indistinguishable from those of ulcerative colitis and can resemble infection with other agents such as *Salmonella*, *Campylobacter*, and *Cl. difficile*. The distinction can sometimes be made histologically. Concurrent intestinal infection is found in only 50 per cent of amoebic liver abscesses. Antibodies to amoebae are present in over 90 per cent of cases of liver abscess but only in severe amoebic dysentery, i.e. after tissue invasion. Radiography and scanning can help to detect an abscess or amoeboma. Cysts may be found in the faeces of symptomatic or symptom-free patients.

Management and prognosis

Metronidazole is effective in amoebic dysentery and liver abscess. If fever does not resolve rapidly after starting metronidazole in liver abscess the diagnosis must be suspect. Metronidazole resistance may be present and if dysenteric symptoms do not subside tinidazole can be tried. Abscesses may resolve with drug therapy alone but pus must be aspirated if rupture seems imminent (Fig. 5.11). Occasionally, dilatation or resection of affected parts of bowel is required. Amoebiasis can be followed by the irritable bowel syndrome. For clearing intestinal cysts from asymptomatic carriers, metronidazole alone is less effective than diloxanide furoate.

Cholera

Aetiology and epidemiology

Vibrio cholerae O group 1 exists in two biotypes, classical and El-Tor. El-Tor strains now cause most human cases and most outbreaks. The disease is primarily due to an enterotoxin elaborated by the organism. Non-O1 strains are generally non-toxigenic but occasionally cause illness which is usually sporadic rather than epidemic. Several pandemics during the nineteenth century originated in the Indian subcontinent, and since the 1960s the El-Tor biotype has spread from Indonesia to many parts of the world, including Europe, Africa, or South America. Faecal–oral transmission in epidemics is usually through infected water supplies, but infection can also be contracted from convalescent carriers and through infected seafood. The organisms are not known to cause infection except in humans although they may survive in coastal saline water.

Clinical features

The incubation period can be from a few hours up to four or five days. Sudden onset of profuse, very watery ('rice-water') stools without fever or systemic upset is typical of severe illness. Since up to 30 litres of stool can be passed in 24 hours, rapid dehydration can be severe and fatal within a few hours if rehydration is delayed. However, mild and asymptomatic infections are commoner.

Diagnosis

The motile *Vibrio* can be visualized by dark ground microscopy in a drop of liquid faeces hanging from the underside of the coverslip. Culture is required for confirmation.

Management

The mainstay of treatment is replacement of fluid and electrolytes, usually by oral solutions containing electrolytes and glucose to encourage sodium absorption. Those most severely affected require initial intravenous rehydration. If vomiting is present, oral rehydration can still be achieved by a Ryle's tube passed into the jejunum. Tetracyclines or furazolidine can be useful when the strain is sensitive.

Prevention

This depends mainly on hygiene and protection and purification of water supplies. The killed vaccine gives only limited protection, is not effective for contacts or epidemic control and is no longer recommended by the World Health Organization for tourists on standard itineraries. Tetracycline chemoprophylaxis is useful for close contacts. Both clinical illness and asymptomatic infection can be followed by a carrier state lasting several months. First or isolated cases in an area should be reported to the World Health Organisation.

OTHER SYNDROMES

Jaundice

Hepatitis A (Chapter 5) is commonly imported, because of increased faecal–oral transmission in many areas of the world. Less well recognized are importations of enterically transmitted non-A non-B hepatitis (hepatitis E). Hepatitis B and C can also be imported after infection, especially from casual sexual contact or intravenous drug abuse in countries with high carriage rates such as China and south-east Asia.

Hepatitis is managed as described in Chapter 5. It is important to note that the patient is afebrile when jaundice develops. If fever is present, falciparum malaria must be excluded and other common causes of hepatitis sought; for example, Epstein–Barr virus and cytomegalovirus. Yellow fever (Chapter 15) presents with fever and jaundice but is very rare in travellers returning from endemic zones and is preventable by the vaccine.

Rashes and skin infections

Since dust, humidity, and heat encourage infection of minor cuts and bites, secondary skin infection is very common in the tropics, often with suppuration or spreading cellulitis usually due to *Staph, aureus* or *Strep. pyogenes*, respectively. Hot and humid environments can cause *prickly heat* and this too may become infected. *Impetigo* is common (Fig. 14.12). Suntan emphasizes pityriasis versicolor. Urticaria may develop in prodromal hepatitis B and is commonly a feature of systemic helminthic infection such as filariasis. 'Swimmers' itch', schistosomal dermatitis, develops within a few hours of massive exposure to the cercariae of various schistosomes which may be present in surface waters of the tropics.

Leprosy (Chapter 15) should be considered when there is hypopigmented skin, especially if anaesthetic; local, thickened, cutaneous nerves should be sought.

Fig. 14.12 Impetigo in traveller from Zimbabwe. Lesions had spread over body untreated over few weeks. Silvery sheen is due to calamine ointment.

Occasionally, particularly in Indians, leprosy may present as polyneuritis in the absence of skin lesions (see Fig. 15.3). The diagnosis may be confirmed by biopsy of the patch or of a thickened superficial nerve.

Any chronic skin ulcer should be specifically diagnosed. Biopsy of the edge may be needed to distinguish the many causes; for example, cutaneous leishmaniasis (Fig. 14.13) or mycobacterial ulcer.

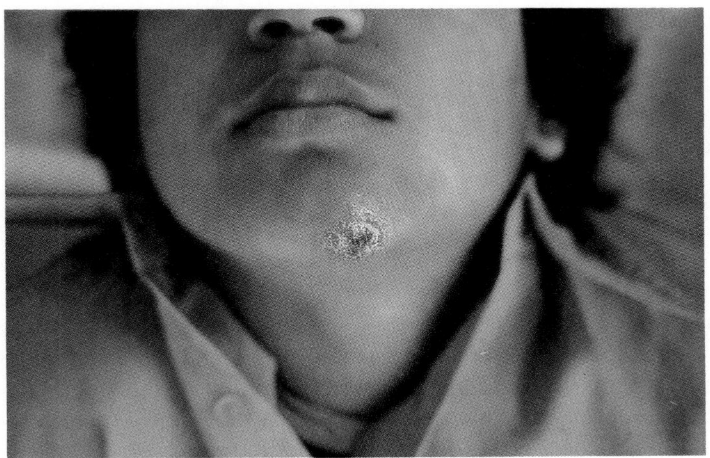

Fig. 14.13 Cutaneous leishmaniasis. Indolent ulcer with crusting for several weeks in patient from Middle East.

Respiratory infections

Increased human contact during travel make respiratory infections very common. Most are of viral aetiology. *Legionnaires' disease* must always be considered in those with severe pneumonia, and *tuberculosis* in those with chronic cough. Cough and wheeze associated with fever, arthralgia and eosinophilia, are common early sensitivity responses to systemic helminthic infection such as schistosomiasis, ascariasis, filariasis, or hookworm disease. Pharyngitis with exudate spreading beyond the tonsils must be treated as diphtheria while laboratory confirmation is sought.

Animal bites: rabies

Although contact with strange domestic and wild animals should be avoided in rabies-endemic areas, many travellers get bitten (Fig. 14.14) or licked. Immunization before travel is not usually appropriate (Chapter 16). Suspect wounds should be washed and scrubbed with copious soap and water and then alcohol. Local medical advice should be sought as both specific passive and

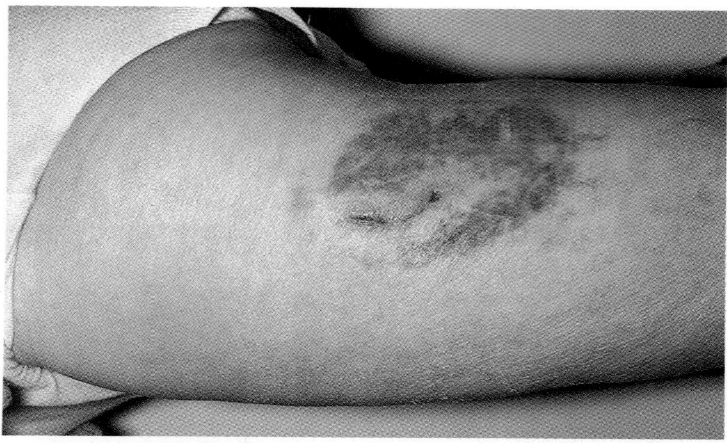

Fig. 14.14 Rabies. Dog-bite on thigh.

active rabies immunization may be necessary. Local police should be informed. When travellers return from endemic areas to Britain and report animal bites or licks, urgent attempts should be made by telephone or through the travel company to trace the responsible animal: if still healthy 10 days after the bite then it was not rabid. If not, active immunization should be started promptly or continued and passive immunization with hyperimmune globulin considered (Chapter 16).

The incubation period of rabies is usually 2–8 weeks, especially in severe exposures and bites to the head, but exceeds three months in 17 per cent of cases. Early symptoms are headache, fever, and perhaps pain or tingling at the site of the original bite. Stupor ('dumb') or mania ('fury') may develop followed by hydrophobia with intense spasms on being offered foods or fluids, and despite intensive care, eventual death. Diagnosis is by immunofluorescence studies of corneal imprints, snips of hairy skin (for example, nape), or post-mortem brain; virus may be cultured from saliva or brain. Negri bodies are typically seen on histology of brain. Serological tests may be diagnostic but difficult to interpret when vaccine has been given.

Genito-urinary infections

Sexual encounters abroad may transmit a wider range of pathogens than in UK. These organisms are more often resistant to standard treatment regimes; for example, penicillinase-producing gonococcus. Infections with hepatitis B virus or HIV, prevalent world-wide are also transmitted sexually and are more likely in areas of high carriage such as south-east Asia and Central Africa, respectively. Those who have had casual sexual intercourse abroad or been treated abroad for sexually transmitted disease should attend their local genito-urinary medicine clinic for check-up on return. Macroscopic and microscopic haematuria may be due to urinary schistosomiasis (Chapter 15).

SPECIAL CONSIDERATIONS

Immigrants

This chapter has largely concentrated on infections in travellers from temperate to warmer climates. One should remember that those who come as immigrants, migrant workers, or students from these warmer and often less developed countries are commonly heavily affected by 'exotic' infections and often by those which are, or were formerly, prevalent in temperate zones. Circumstances or choice often impose crowded living conditions, with poor hygiene and facilities, on recently arrived immigrants trying to establish themselves in the receiving country. Latent infections such as tuberculosis may be reactivated and find opportunities for spread within crowded groups, especially among children.

Up to 75 per cent of some immigrant groups may have infection identifiable on screening, mostly with intestinal helminths, intestinal bacterial pathogens, or *M. tuberculosis*. They may not complain of illness because of language or cultural differences—recurrent low-grade fevers or the passage of intestinal worms may be considered normal.

Screening of stools, of blood for haemoglobin, malaria parasites, eosinophils, and HBsAg, chest X-ray, and tuberculin tests may be helpful.

Some additional important tropical infections are described in Chapter 15.

Screening the traveller

Routine screening of the returned British traveller rarely detects infection in those who are symptomless, with no clinically detectable abnormality. Should the patient require further reassurance one can examine faeces for ova, cysts, and parasites and a blood film for eosinophils. Eosinophils may be increased slightly with bowel helminths, but very high eosinophil counts suggest tissue invasion by certain helminths, for example, filaria, stronglyoides, or larva migrans. Treatment is often offered for parasites found, but a light, symptomless helminth load is unlikely to have serious consequences. Some serological tests—for example, for schistosomiasis and filariasis—can be helpful, as may chest radiography. The possibility of HIV infection requires consideration and perhaps counselling.

15

Infection in the tropics

Many infections of great importance in a world context are rarely seen in countries such as Britain. Climate, other environmental conditions, and poverty contribute to the tropical and subtropical regions being major reservoirs of such infections. Even in developed countries, doctors require some knowledge about them since travellers may be exposed and symptoms may also affect visitors or immigrants from the tropics.

Some of these infections are discussed in other chapters. For example, Chapter 14 covers the infections most frequently imported into developed countries; poliomyelitis (Chapter 6) is described as a classical example of CNS infection; intestinal helminth infections are discussed in Chapter 4. Also numerous infections seen commonly in temperate regions, such as influenza, are world-wide in distribution, being no less common in tropical than in temperate regions.

The World Health Organization (WHO) in conjunction with the World Bank has a special programme for research and training in tropical diseases. It also supports an 'Expanded Programme for Immunisation' intended to provide basic childhood immunization for all. Of the diseases covered by these programmes, those not discussed elsewhere in this book are considered here as examples of major world problems being tackled through international co-operative efforts.

LEPROSY

Aetiology and epidemiology

Leprosy is due to chronic infection with *Mycobacterium leprae*. There is no intermediate host. Infection is spread through respiratory droplets from the nasal mucosae of patients with multibacillary disease (see below). The route of entry into the body, whether respiratory or by skin or both, is unclear. It is now primarily a disease of tropical and subtropical parts of the world, more because of poor socio-economic conditions and ease of spread than predilection for warm climates. In highly endemic areas asymptomatic infection is common and less than 5 per cent of those infected develop clinical features.

Clinical presentation

The organism has a predilection for cooler parts of the body, hence the skin, nasal mucosae, superficial nerves, eyes, and testes are particularly affected. Clinical features range from *lepromatous* disease with widespread lesions containing vast numbers of lepra bacilli (multi-bacillary), through *borderline* to *tuberculoid* disease where lesions are few, often solitary and in which lepra bacilli are very rarely seen. Less than 5 per cent of patients present with lepromatous disease. *Indeterminate* leprosy describes cases which do not fit into these categories, usually in early disease, which may resolve or progress towards other forms.

The incubation period is thought to be between about 3 and 15 years with lepromatous disease taking longer to appear than tuberculoid leprosy.

Lepromatous leprosy. In the early stages, skin lesions are diffuse and poorly defined. There may be hypopigmentation and, in pale skins, erythema. Eventually the skin becomes thickened, macules, papules, and nodules develop and there is hair loss including the eyebrows. These lesions do not characteristically lose sensation as is the case in tuberculoid disease. Superficial nerves are involved later, usually symmetrically, and sensation is reduced giving the features of peripheral neuropathy. Facial nerve involvement can lead to lagophthalmos. Resorption of phalanges may progress to complete loss of digits, mainly due to repeated trauma and physical stresses on the peripheries unrecognized as harmful by the patient because of the anaesthesia (Fig. 15.1). Testicular involvement leads to atrophy and gynaecomastia (see Fig. 15.4). Nasal mucosae can be destroyed with ulceration and loss of inter-nasal septum. As mentioned above, nasal involvement is important as the primary source of infective organisms.

Tuberculoid leprosy. The hallmark of tuberculoid disease is the anaesthetic, hypopigmented or erythematous, sharply demarcated skin patch (Fig. 15.2).

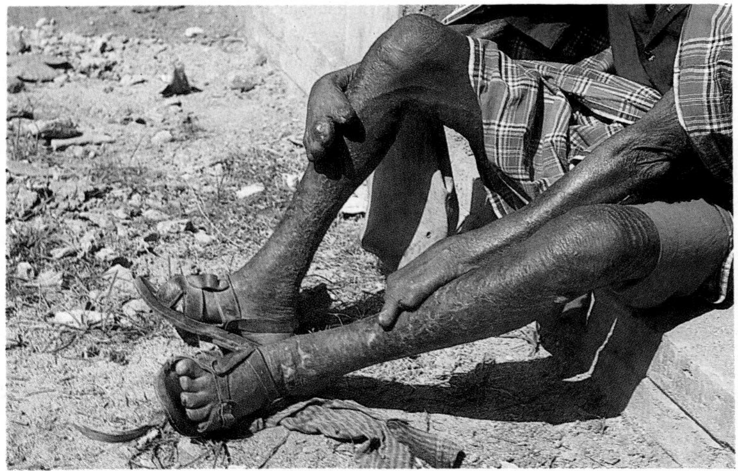

Fig. 15.1 Loss of digits in lepromatous leprosy.

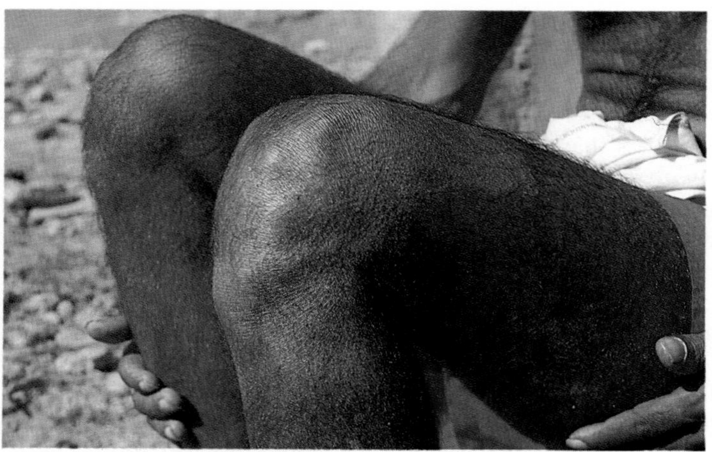

Fig. 15.2 A depigmented, anaesthetic skin patch in tuberculoid leprosy.

This may have a raised edge with some central healing and loss of sweating. Asymmetrical nerve involvement with palpable thickening usually affects a single nerve trunk close to the skin lesion. This can lead to ulnar or radial nerve palsy, or foot drop. Sometimes there is nerve involvement with no skin lesions (neural leprosy).

Borderline leprosy. This term describes cases which have features of both the lepromatous and tuberculoid forms. Whereas cell-mediated immunity is marked in tuberculoid disease, and typically absent from lepromatous disease, borderline cases vary immunologically and can evolve towards either end of the clinical spectrum.

Diagnosis

Multibacillary disease is detected by the 'skin smear' technique performed on any involved part of the body, such as the ear-lobe. A little intracutaneous fluid is removed with a scapel, spread on a slide, and stained by a modified Ziehl–Neelsen technique. Lepra bacilli are numerous in lepromatous and scanty or absent in tuberculoid disease. After successful treatment bacilli degenerate and appear 'beaded'. Organisms can also be obtained from the nasal mucosa or expelled nasal mucus.

In tuberculoid disease the diagnosis is usually obvious but biopsy of the edge of a lesion shows typical histological changes in skin and superficial nerves.

Lepromin test. Killed lepra bacilli are injected intradermally. If after 48 hours an area of erythema appears, delayed hypersensitivity to the bacillus is present. If a nodule more than 5 mm diameter is present at three weeks the patient is capable of developing cell-mediated immunity. Neither test can diagnose leprosy, but in conjunction with clinical features they can help monitor prognosis.

Management

Relatively short courses of rifampicin monthly plus dapsone and clofazimine daily appear to be effective in all forms of disease, and may probably replace lifelong treatment by dapsone or clofazimine alone formerly given to lepromatous patients. Rifampicin is rapidly bactericidal and appears to shorten infectivity markedly when cost allows.

Surgery can be important, for example to correct deformities by releasing contractures or to increase mobility by tendon transplants. Dividing the Schwann sheath can relieve pressure on a nerve and prevent permanent damage, if done early in paralytic disease (Fig. 15.3).

Avoiding burns, trauma, and physical stress on anaesthetic limbs is important to prevent deformities; for example, by protective gloves and footwear.

Erythema nodosum leprosum. This usually develops during treatment and is probably a reaction to immune complexes. Features are inflamed maculopapular skin lesions (Fig. 15.4), arthritis, nephritis, iritis, and accentuation of neuropathy. The reactions may be persistent and recurrent but are generally self-limiting. Erythema nodosum leprosum can be associated with progression towards the lepromatous end of the clinical spectrum. Rest plus aspirin and/or clofazimine (if not already in use) can be helpful for their anti-inflammatory effect. Corticosteriods quickly control symptoms but their side-effects must be carefully considered because they may be difficult to withdraw.

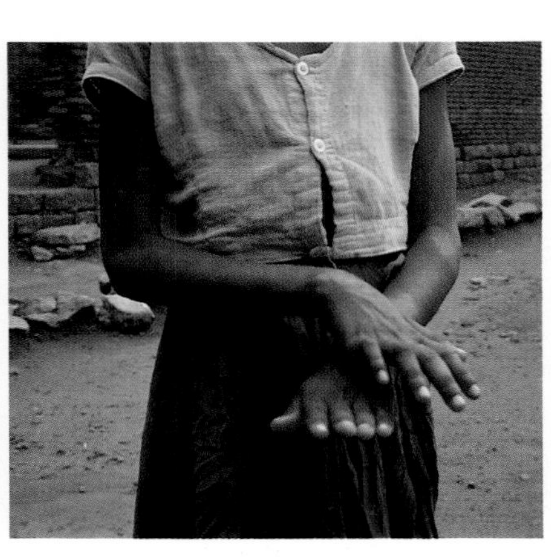

(a)

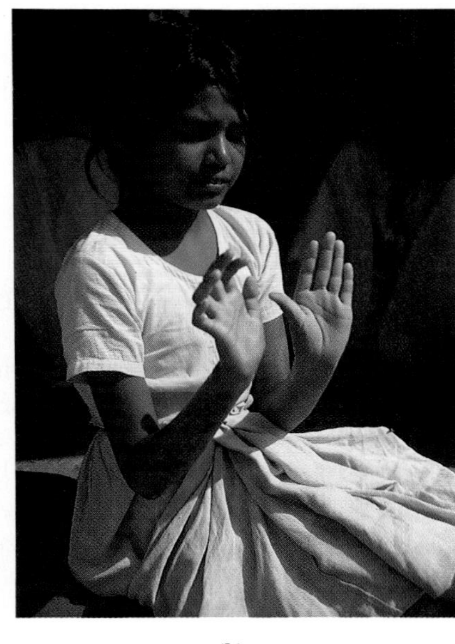

(b)

Fig. 15.3 Leprosy: radial palsy **(a)** before surgery; **(b)** after nerve decompression.

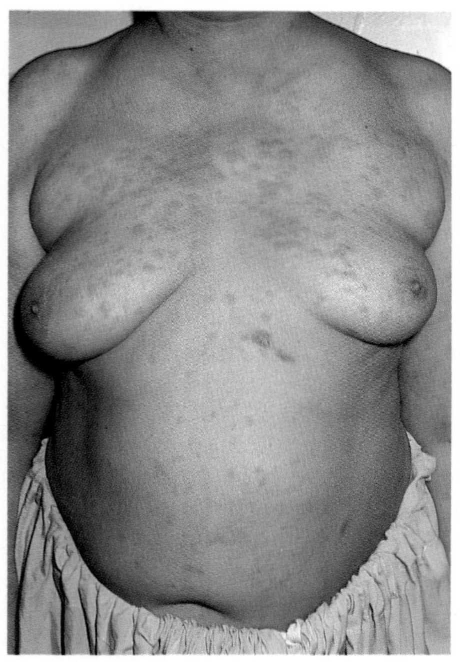

Fig. 15.4 The rash of erythema nodosum, also showing gynaecomastia from testicular atrophy.

Prevention

Leprosy is associated with poor socio-economic conditions but early detection and effective treatment of multibacillary disease is important for control. No leprosy vaccine is yet available. Some trials suggested that BCG vaccine as used against *Mycobacterium tuberculosis* gives some protection, but overall the evidence is inconclusive.

FILARIAL INFECTION

Nematode worms of the family Filarioidea which affect man are summarized in Table 15.1. They are important causes of morbidity in tropical and subtropical regions. Two of the most common and important, *Wuchereria bancrofti* and *Onchocerca volvulus*, are considered below.

Wuchereria bancrofti

Aetiology and epidemiology

Larvae enter through the bite of an infected mosquito and then settle in lymphatics, principally of the groins and axilla. The worms mature and females

Table 15.1. *The commonest filarial infections in man*

Nematode	Distribution	Transmission
Wuchereria bancrofti	Asia, Africa, S. America, Oceania	Bite of mosquito (culicine and anopheline)
Brugia malayi	Asia, especially Far East	Bite of mosquito (mainly culicine)
Loa loa	West and Central Africa	Bite of deer fly (chrysops)
Onchocerca volvulus	Africa, Central America, Yemen	Bite of the black fly (simulium)
Dracunculus medinensis (Guinea worm)	Africa, Asia, Middle East, S. America	Drinking infected water

after fertilization release microfilariae into the blood, from where they can be ingested by another feeding mosquito. *Wuchereria bancrofti* infection is confined to man. It is widespread in Asia and also found in Africa, South America, Indonesia, and the South Pacific Islands. Being spread by various species of mosquito it is found wherever breeding sites are available in both urban and rural areas. *Brugia malayi* and *B. timori* cause similar infections in areas of SE Asia.

Clinical presentation

Many infections are asymptomatic and microfilariae may be found coincidentally in the blood. Several months after initial infection intermittent attacks of lymphadenitis can occur with fever and inflammation over associated lymphatic vessels. Sometimes precipitated by physical exertion, these attacks are most common among young adults. The commonest sites for lymphadenitis are the groin and axilla and there may be transient lymphoedema. The breast and spermatic cord may be involved. Microfilaraemia is not the cause of these attacks which are often due to secondary infection particularly with *Streptococcus pyogenes*. Abscesses can form leaving a scar. Hydrocoeles are common. With repeated attacks associated lymphoedema tends to resolve more slowly as lymphatics fibrose. Eventually gross deformity and disability can result. Severe swelling most commonly affects the legs or scrotum but also sometimes the arms, breasts, or vulva. Lymphatic varices can cause leakage of lymph through skin or intra-abdominal organs leading for example to chyluria.

Tropical eosinophilia. This is characterized by marked blood eosinophilia, sometimes accounting for 50 per cent of the white blood cells and total counts of 20×10^9/litre or more. Often there are associated pulmonary features such as paroxysmal nocturnal cough and bronchospasm. Pulmonary infiltration may be present and seen on chest X-ray. This condition is widespread in Asia and thought to be an allergic response to *W. bancrofti* (or *B. malayi*). Children and young adults are most often affected. Marked eosinophilia without chest symptoms also may present in onchocerciasis.

Diagnosis

Although microfilariae may be seen at any time of day, *W. bancrofti* shows marked nocturnal periodicity, and most microfilaria enter the blood between 2200 and 0100 hours. The periodicity is related to the biting habits of the mosquito vectors and *B. malayi* shows diurnal periodicity. Microfilariae can be stimulated to enter the blood during the day by a dose of 50 mg of diethylcarbamazine, but this may provoke allergic reactions.

Microfilariae can sometimes be seen live in a drop of fresh blood on a microscope slide or can be stained, for example, with Giemsa. Concentration methods include filtration, or centrifugation and examination of the buffy coat of the blood (Fig. 15.5). They are rarely seen in tropical eosinophilia. Complement fixation, fluorescent antibody, and ELISA tests for antibodies can be helpful when microfilaria are not seen but false positives and negatives are common from cross reactions with other (including animal) nematodes.

Treatment

Diethylcarbamazine rapidly kills microfilariae and, less reliably, adult worms also if given in repeated courses. 'Herxheimer reactions', due to antigens from dead or dying microfilariae, are common soon after treatment starts, and cause fever, malaise, and sometimes vomiting. Anti-inflammtory drugs such as aspirin and antihistamines help to control these reactions, as can starting treatment with small doses. Scrotal elephantiasis can be treated surgically by excision of affected tissue but chronic oedema of the limbs is difficult to improve. Although microfilariae are rarely seen in tropical eosinophilia, response to diethylcarbamazine is usually rapid but recurrence is common.

Prevention

This depends upon eradicating mosquito breeding sites, preventing mosquito bites, early and regular treatment of patients and sometimes mass treatment to reduce the 'human pool of infection'. Repeated streptococcal infections can be prevented by prophylactic penicillin.

Onchocerca volvulus

Aetiology and epidemiology

Onchocerciasis ('River blindness') is mainly found in Equatorial Africa, occasionally in south and central America and the Yemen. It is spread by a small blackfly (*Simulium damnosum*) which breeds in swiftly flowing rivers. The bites are usually painful. The parasite affects only man, and adult worms live in subcutaneous tissues or nodules for many years, releasing microfilariae mainly into skin rather than blood or lymphatics.

(a)

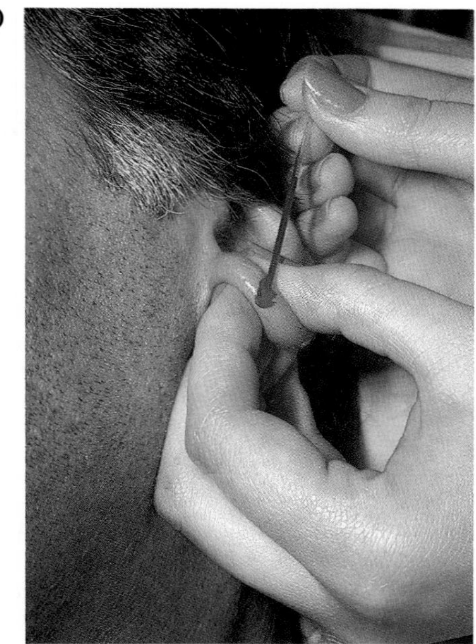

(b)

Fig. 15.5 (a) and **(b)** Live filariae may be seen in the 'buffy coat' of centrifuged capillary blood.

Clinical presentation

Symptoms are primarily due to microfilariae, adult worms usually causing no symptoms even when migrating. Dead microfilariae in skin cause inflammation and pruritus, leading to excoriation and secondary infection. Hypopigmentation and inguinal or femoral lymphadenopathy may be present. Skin changes are most prominent on the lower half of the body where most bites occur. There may be lymphoedema, especially in the groins or genitalia, but not to the same extent as in *W. bancrofti* infection (see above). Microfilariae can invade the cornea, and eye involvement with visual impairment is common in those living in endemic areas. Punctate keratitis, vascularization and scarring of the cornea, iritis and less often retinal damage and optic atrophy cause visual field defects. Serious damage is usually due to repeated infections which short-term visitors to endemic areas generally escape.

Diagnosis

Eosinophilia is usual and microfilariae may be visualized in 'skin snips', the cornea or the anterior chamber of the eye. Serological tests (for instance, ELISA) are useful. There are cross-reactions with other nematode infections. Aggravation of itch and rash after a dose of diethylcarbamazine is suggestive of the diagnosis (Mazzotti reaction).

Treatment

Treatment is with invermectin, which can initially aggravate symptoms but less so than diethylcarbamazine, used formerly. With heavy infection this can cause increased rash and itch, lymphadenitis, arthropathy, postural hypotension, and most seriously iritis which warrants steroid therapy. Mobilization of microfilariae and formation of immune complexes are responsible. These drugs do not kill adult worms so alone are not curative. They can be followed by a 'macrofilaricidal' drug, suramin. Suramin given intravenously, usually weekly for five weeks, can cause febrile toxic illness with renal damage and optic atrophy. The dose must be carefully regulated. For these reasons it is rarely used and preference is given to controlling microfilarias with an annual dose of ivermectin. For those continually exposed this treatment may need to be indefinite, but for those who have left endemic areas 5–10 years treatment (the life-span of the adult worms) is usually sufficient. Obvious nodules containing adult worms can be surgically removed.

Prevention

Unlike the mosquito the *Simulium* fly travels long distances making control difficult; also breeding sites are provided by many irrigation projects. Bites of blackflies should be avoided by clothing and insect repellents. Insecticides may control the larval stages although they are difficult to use because the fly breeds in running water.

SCHISTOSOMIASIS (BILHARZIASIS)

Aetiology and epidemiology

Schistosomiasis is due to infection with a blood fluke (trematode). Adult worms live for many years in mesenteric or vesicular veins and produce eggs which cause granulomas and scarring in tissues. *Schistosoma mansoni* and *S. japonicum* mainly affect the intestine and liver; *S. haematobium*, the urinary tract, especially the bladder. Endemic areas overlap to some extent and some other species are locally important. *S. mansoni* and *S. haematobium* usually infect only man but *S. japonicum* infects other mammals such as dogs and pigs. For all types, certain freshwater snails are necessary intermediate hosts and are penetrated by larvae (miracidia) which hatch from ova excreted in human stools and/or urine. After several weeks' development in the snail, cercariae are released and can penetrate intact human skin during activities such as washing, swimming, or paddling. Occasionally the site of entry is buccal mucosa, from drinking water. The parasites, now transformed into schistosomula, then enter the blood stream, proceed to lung and liver; After maturation adult worms migrate to intra-abdominal veins. Schistosomiasis is widespread in Africa, South America, and the Middle and Far East (Fig. 15.6). Irrigation projects often increase exposure and in some communities most people are infected, most having symptoms.

Clinical presentation

Initial infection may be asymptomatic, but two to six weeks after exposure features such as fever, diarrhoea, cough, hepatospenomegaly, and urticarial skin rash can develop. These symptoms are commoner in visitors from non-endemic areas and are unusual in *S. haematobium* infection. Later features depend on the site of tissue damage by the presence of ova.

Schistosoma haematobium. Dysuria and haematuria, often terminal, are usually present. Granulomas with fibrosis and calcification form in the urinary bladder mucosa and may present as papillomas or simulate neoplasia. Ureters can be obstructed with resulting hydronephrosis and renal failure. Carcinoma of the

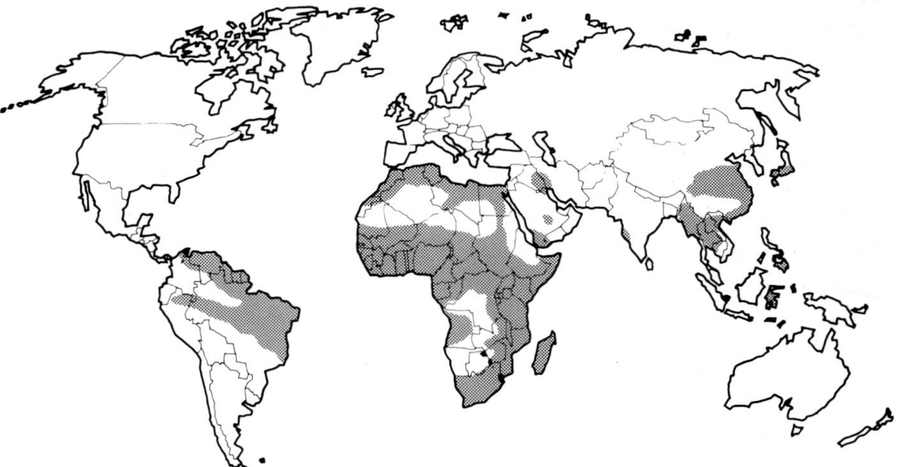

Fig. 15.6 The areas of the world where most transmission of schistosomiasis occurs.

bladder is a complication, and pulmonary hypertension can result from ova entering the systemic circulation.

Schistosoma mansoni. Infection with this species usually presents with anaemia due to blood loss and sometimes frank rectal bleeding. There may be abdominal pain, diarrhoea, or liver enlargement. Migration of ova to the liver can cause portal hypertension with splenomegaly. Intestinal granulomas can present as papillomas but have no proven link with carcinoma. In long-standing liver disease, deposition of ova in lungs causes pulmonary hypertension as with *S. haematobium*. *Schistosoma japonicum* infection is similar to *S. mansoni* but tends to be more severe.

Especially in heavy infection, adult worms can enter other parts of the body such as spinal cord, causing compression, or the brain, when tumour can be mimicked. Other complications include the nephrotic syndrome and persistent salmonella carriage, especially in the damaged urinary tract.

Investigations

Ova can usually be found on microscopy of urinary deposit in urinary tract disease (Fig. 15.7), but repeated examination may be necessary. Concentration methods are necessary for detecting ova in stools (Fig.15.8), and they can also be

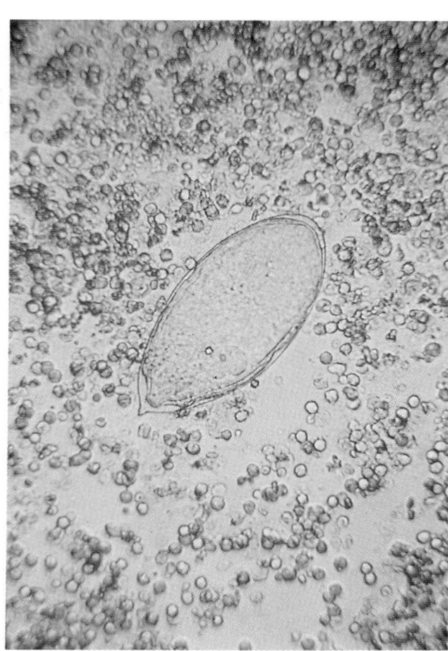

Fig. 15.7 Schistosomiasis. *Schistosoma haematobium* ovum in urine containing many red cells.

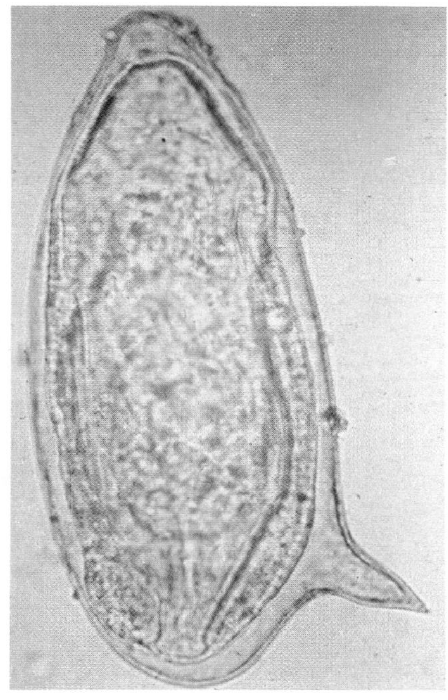

Fig. 15.8 Schistosomiasis. Ovum of *S. mansoni*.

found in biopsy specimens of rectal mucosa. Serological tests using complement fixation, fluorescent antibody techniques, or ELISA are available but false positives and negatives occur. Eosinophilia may be marked but can also be due to other invasive intestinal helminths such as *Strongyloides stercoralis* and *Trichinella spiralis*.

Treatment and prevention

Treatment can destroy living adults worms and, if reinfection does not occur, inflammation and granulomas may resolve. The new drug praziquantel given orally in a single dose has revolutionized treatment, replacing more toxic and parenteral alternatives. When whole communities are treated, transmission may be interrupted and prevalence reduced.

Other control methods involve attempts to control faecal and urinary contamination of fresh water, destroying snail intermediate hosts, and avoiding unnecessary exposure (most appropriate advice for travellers). No vaccine is yet available.

LEISHMANIASIS

Visceral leishmaniasis (kala-azar)

Aetiology and epidemiology

Kala-azar is caused by the protozoon *Leishmania donovani* throughout tropical Africa, in parts of South America, in both Central and East Asia, and around the Mediterranean (Fig. 15.9). Infection is transmitted from animal reservoir hosts,

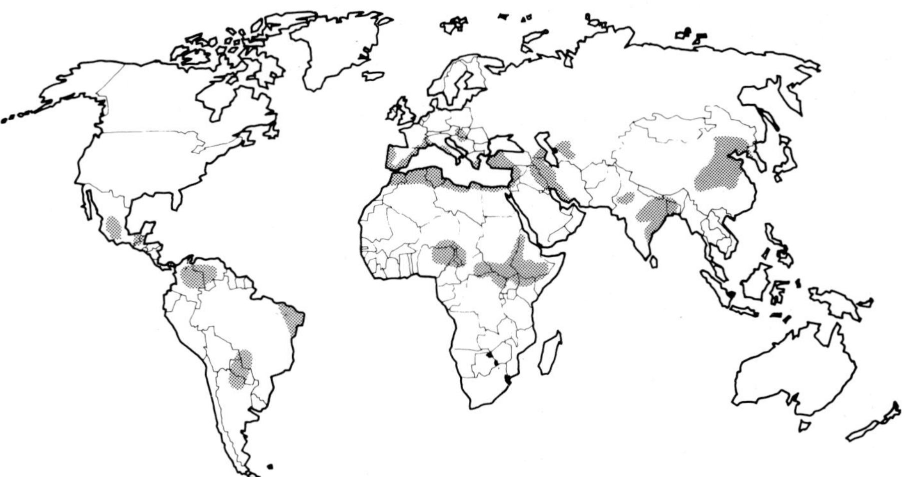

Fig. 15.9 The areas of the world where most transmission of *L. donovani* occurs.

such as dogs, rodents, or man by the bite of an infected sandfly (genus *Phlebotomus*). It has been transmitted by blood transfusion and congenitally. In the Mediterranean regions, North Africa, and Central Asia, kala-azar is usually a sporadic disease of children and infants; in the Indian subcontinent it causes epidemics which affect mostly older children and young adults; in Africa it is endemic, with both local foci of infection and superimposed epidemics most often affecting young male adults. Different subspecies are involved.

Clinical presentation

The incubation period of kala-azar varies from a few weeks to many months. Fever may be of insidious onset and intermittent, with sweating and malaise. Lymphadenopathy is more prominent in African and Mediterranean types; hepatomegaly and splenomegaly which may be marked, develop with anorexia, weight loss, and, if untreated, usually death. Some infections can be mild or subclinical, especially during epidemics. It may present with secondary infection, such as pneumonia or diarrhoea.

Investigations

Normochromic normocytic anaemia, neutropenia, and thrombocytopenia secondary to hypersplenism are typical. Changes in liver-function tests occur and albuminuria is common.

Splenic puncture (when the spleen is enlarged) usually reveals the parasite which may be found, less consistently, in lymph nodes, bone-marrow or the buffy coat of centrifuged peripheral blood (Fig. 15.10). Indirect fluorescent antibody tests and ELISA techniques can be helpful but may give cross-reactions with other protozoal infections such as malaria and toxoplasmosis. The Leishmanin

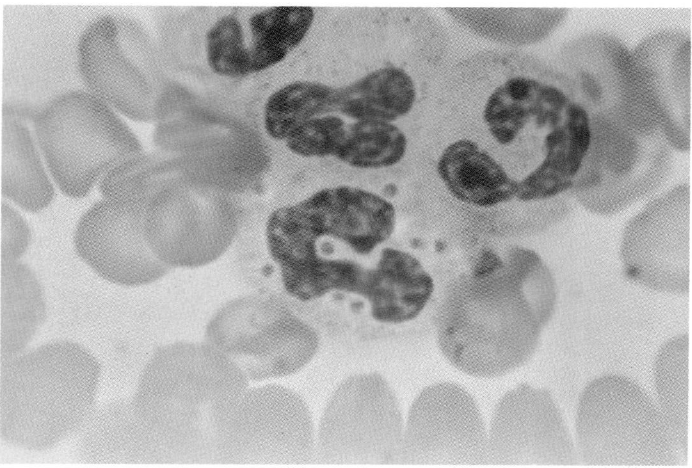

Fig. 15.10 Visceral Leishmaniasis. Amastigotes of *L. donovani* are occasionally seen in circulating white blood cells.

skin test for delayed hypersensitivity is useful in prevalence studies but usually negative in active infection.

Treatment

In those seriously ill, treatment may have to be started without laboratory confirmation of the diagnosis. Initial treatment is usually with parenteral pentavalent organo-antimony drugs and various regimes have been devised. Indian strains are generally more sensitive than African strains which usually require a more prolonged course. In resistant cases a diamadine drug such as pentamidine, may be necessary, but side-effects such as vomiting, hypotension, and diabetes are common. Occasionally, splenectomy combined with drug treatment is helpful, preferably followed by prophylactic penicillin and preceded by pneumococcal vaccination to prevent post-splenectomy pneumococcal infection.

Cutaneous leishmaniasis

This is caused by other species of leishmania and characterized by chronic ulcerating skin lesion(s) appearing up to several months after the sandfly bite, for example, Oriental sore in the Middle East, Central Asia, and Africa; espundia (mucocutaneous) in South America. Widespread in Asia, Africa, the Middle East, Central and South America, it contrasts with kala-azar in not usually causing systemic illness (Fig. 15.11). Drug treatment requires potentially toxic

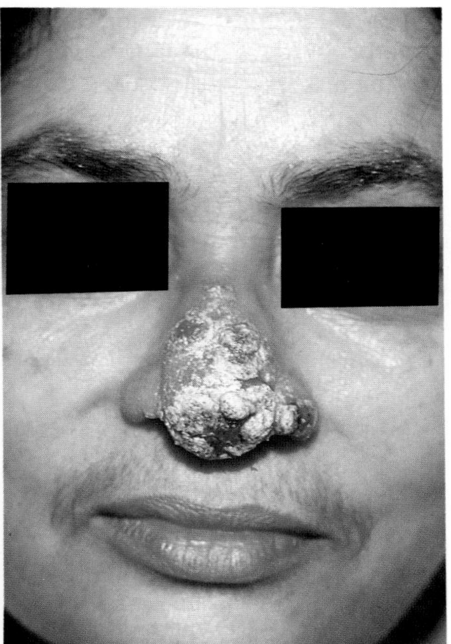

Fig. 15.11 Cutaneous leishmaniasis.

drugs such as pentavalent antimony for Asian and Mediterranean forms, or pentamidine for South American forms. Small lesions may heal spontaneously or be removed surgically without drug treatment.

Control

The most effective measures of control vary with local circumstances and include prompt treatment of human cases, destruction of sick dogs or other locally incriminated reservoir hosts and the use of insecticide against the sandfly. Sandflies generally bite at dawn and dusk and do not fly far from their breeding sites, although they may be blown further afield by wind currents. Avoiding bites with suitable clothing and netting is helpful.

TRYPANOSOMIASIS

Aetiology and epidemiology

The organisms causing African trypanosomiasis are conventionally designated *Trypanosoma brucei gambiense* and *T. b. rhodesiense*. Modern identification techniques however suggest that this is an oversimplification. Infection is found within the distribution of the tsetse fly (various *Glossina* species) which transmits infection while taking a blood meal. In West Africa infections are usually of *T. b. gambiense* type, mainly transmitted man to man by tsetse flies found close to running water. *Trypanosoma brucei rhodesiense*, found mostly in East and Central Africa, has an important reservoir in cattle and wild game and its distribution is more widespread, being transmitted by tsetse flies of dry savannas (Fig.15.12). In Mexico, Central and South America, *Trypanosma cruzi* causes

Fig. 15.12 The setting where *T. b. rhodesiense* occurs. Note the absence of cattle.

Chagas' disease. It is spread from the faeces of blood-sucking bugs (Reduviid) infected from human, domestic, or wild animal hosts. Sites of entry can be conjunctiva, mucous membranes, or skin wounds, usually bug-bites themselves.

Clinical presentation

In *T. b. gambiense* and *T. b. rhodesiense* infections, at the site of the initial *Glossina* bite which may be painful, a 'trypanosomal chancre' can develop, but this initial infection may be asymptomatic. After two to three weeks or longer there is blood invasion, with systemic features which tend to be more severe with *T. b. rhodesiense*, characteristically causing acute illness, contrasting with the more insidious onset of *T. b. gambiense*. Fever and lymphadenopathy, particularly in the posterior triangle of the neck, are early features, and this is when trypanosomes are most likely to be found in blood. Symptoms may then temporarily resolve, but fever can persist and splenomegaly develop with perhaps pleuritis, erythematous rashes, and arthropathy. After several months to two years or more, changes in behaviour and features of dementia can appear, with typical drowsiness during the day and wakefulness at night ('sleeping sickness'). Eventually other neurological features develop with tremors, spasticity, ataxia, and upper motor neurone signs. Response to pain is delayed. After many months, death commonly occurs from malnutrition, intercurrent infection, or coma. Spontaneous recovery has been observed during early illness but is rare once neurological features have developed. *Trypanosoma brucei rhodesiense* infection is generally more virulent. The patient may die before neurological features develop. Cardiac involvement can cause fatal arrhythmias, pleural and pericardial effusions.

In Chagas' disease, death may result from cardiac involvement during the initial acute toxic illness which presents with lymph node, liver, and spleen enlargement. Usually, a prolonged, largely asymptomatic period intervenes before complications such as cardiomyopathy, megaoesophagus, or megacolon present with cardiac failure, usually fatal, aspiration pneumonia, or intestinal obstruction.

Investigations

Trypanosomes can be visualized in blood films in early disease or concentrated by various methods, including centrifugation, when they are seen in the buffy white cell layer. Later they may be found in aspirates from lymph nodes or bone marrow, sometimes in cerebrospinal fluid, occasionally in the original skin lesion. Serological tests are helpful but not specific, which limits their value in persons continually exposed. There is commonly normochromic anaemia, very high serum IgM level, and raised ESR. In neurological disease, cerebrospinal fluid can show increased white cells, protein and IgM, but this test should not be done until treatment has been started in case infection is introduced into the CSF by blood on the lumbar puncture needle. In Chagas' disease, blood culture has been successful and serology is generally more useful.

Treatment

African varieties of trypanosomiasis are treated with toxic parenteral drugs. Pentamidine or suramin (for the more fulminant *T. b rhodesiense* infection) is used in early illness, followed by melarsoprol if there is nervous system involvement. This is an arsenical drug which itself causes encephalopathy and treatment must be carefully monitored. For Chagas' disease no consistently reliable treatment is available, but in the early stages nifurtimox and benzinidazole have been successful. Late complications can only be treated symptomatically or surgically.

Prevention

This depends on control of vectors, early treatment of infectious patients, and efforts to avoid bites in endemic areas. Since blood transfusions can transmit infection, care to select healthy donors is important and serological tests to exclude infected donors are advisable where Chagas' disease is endemic. No vaccination is available for humans. Pentamidine prophylaxis has been used for *T. b. gambiense*, but this may inadequately treat early disease and may mask neurological involvement.

YELLOW FEVER

Aetiology and epidemiology

Due to a mosquito-borne flavivirus, yellow fever is endemic in African and South American monkeys but strangely absent from Asia. It usually infects man when he enters the jungle environment and is bitten by mosquitoes normally feeding on monkeys. Subsequent spread from man-to-man is by man's own mosquito vectors, especially *Aedes* spp. which may breed in urban locations.

Clinical presentation

After an incubation period of 3–6 days, a severe pyrexial illness starts. This is followed by jaundice and extensive haemorrhagic manifestations leading on to coma and death in severe cases—a typical 'viral haemorrhagic fever' (Chapter 14) with prominent liver damage. In mild cases proteinuria and epistaxis are common.

Investigation and management

Investigation is done in specially designated laboratories. The virus (a dangerous pathogen) may be isolated from blood or rising titres of antibodies detected in serum. Liver biopsy shows characteristic necrosis with inclusion bodies. Management is supportive. Effective immunization is achieved by a live vaccine.

Prognosis

The prognosis is better in indigenous populations in non-epidemic periods, although there is a significant mortality (about 5 per cent). In epidemic periods and among visitors to the area, mortality in the un-vaccinated may be as high as 50 per cent.

16

Prevention of infection: Immunization and advice to travellers

INTRODUCTION

Communicable diseases can be prevented, in principle, by interrupting transmission of the causal organisms even when the causal organism is not yet known or no vaccine or antibiotic is available. Thus in AIDS, epidemiological clues provided a rational basis for reducing spread through changes in sexual practice and blood handling even before HIV had been incriminated. Good personal hygiene is important, and at community, national, and international levels, transmission of many infections is reduced by public health measures and legislation directed to ensure pure food and water, and sanitary disposal of excreta and other waste products; also, but rarely, by quarantine measures for unimmunized persons arriving in a country (for example, India) free of a feared infection (yellow fever) from another country where it exists (Africa).

Methods of reducing transmission of infections such as salmonellosis, rabies, and malaria can depend on control of *reservoirs* of infection in the environment by measures directed at animal hosts (for example, intensively reared poultry and cattle; foxes) or *vectors* (for example, mosquitoes) or by reducing human *exposure* by avoiding dangerous environments or wearing appropriate protective clothing.

These principles of control also apply to the infectious patient at home or in hospital, and to infectious materials in the laboratory. 'High isolation' units using plastic isolators, controlled ventilation with sterile air filtration, sterilization of excreta and other materials from the infected area, and limitation of the involved staff to a well-trained and supervised team, are facilities available at a limited number of designated centres for dealing with the most dangerous infections. Most infections, however, are managed satisfactorily with simple isolation facilities and common sense procedures, if these are carried out meticulously. The aim is to avoid cross-infection between patients and unnecessary contamination of staff. Hands are the most important vehicles of transfer of infections and should be thoroughly washed between procedures. Further details of measures to minimize hospital infection are given in Chapter 17.

Specific chemoprophylaxis

Chemoprophylaxis is the administration of a specific antimicrobial substance to prevent an infection from becoming established or progressing to manifest disease. The drug should be one to which the infectious agent is usually or known to be sensitive. Risks of possible toxic side-effects of the drug should be considered. Situations in which chemoprophylaxis may be appropriate are shown in Table 16.1.

Specific immunoprophylaxis

Immunoprophylaxis is the establishment of protection against infection through specific immunological mechanisms. It can be achieved *actively* by stimulating the body's own immune system by appropriate antigens, or *passively* by injecting pre-formed antibodies.

Table 16.1. *Situations where chemoprophylaxis may be helful*

Organism	Disease	Commonly used drug
Strep. pyogenes	Rheumatic fever	Penicillin or erythromycin
Gram-negative organisms (e.g. *E. coli*)	Urinary tract infection, especially in children	Co-trimoxazole
Strep. viridans	Infective endocarditis following tooth extraction	Penicillin or amoxycillin
Staph. aureus	Pneumonia in fibrocystic disease	Flucloxacillin
M. tuberculosis	Mantoux-positive disease-free children	Isoniazid (sometimes + rifampicin)
N. meningitidis	Meningococcal meningitis/ septicaemia	Rifampicin or minocycline
C. diphtheriae	Diphtheria in non-immune contacts	Erythromycin
Herpes simplex	Recurrent genital herpes	Acyclovir
Influenza A	Severe and complicated influenza in 'at risk' persons	Amantadine
Cytomegalovirus	(in AIDS)	Ganglicovir
Pn. carinii	Pneumonia in the immunosuppressed, e.g. AIDS	Co-trimoxazole
P. falciparum	Malaria	Chloroquine etc. (see below)
Toxoplasma	(in AIDS)	Fansidar

Active immunization

This was originally carried out by infecting the recipient with a live but less virulent variety of the organism against which protection was required—by 'vaccinia' (*vacca*, a cow), the cowpox virus used by Jenner to 'vaccinate' against the antigenically similar smallpox infection. Several useful live attenuated vaccines are currently available (Table 16.2). Current research is directed to genetic manipulation of these agents to produce non-virulent vaccines unable to regain pathogenic properties during growth in the vaccinee.

Further developments entailed culturing the agents in quantity and then inactivating them with formalin or other chemical or physical treatments. After suitable purification, these preparations of antigens could be injected to stimulate an immune response; for example, against typhoid. Current work on non-living vaccines (Table 16.2) is directed to purifying antigens more completely, possibly by separating appropriate antigenic components from disrupted organisms, and by 'genetic engineering' techniques. Thus the essential surface antigen of hepatitis B virus is synthesized on a commercial scale by modified yeast cells. It is becoming possible to produce far more potent vaccines with minimal side-effects by these techniques which also make it easier to prepare combined vaccines against several infections.

The objective of active immunization is to stimulate humoral and cellular responses and 'memory cells' to ensure rapid specific immune response to the infective agent when encountered even years later. The use of vaccines has dramatically reduced the incidence of infections in many parts of the world, currently with considerable success through the Expanded Programme of Immunization (EPI) of the World Health Organization concentrated on children

Table 16.2. *Some vaccines in current use**

Living attenuated vaccines	Non-living vaccines	
Poliomyelitis	Poliomyelitis	Tick-borne encephalitis
Measles†	Influenza	Japanese B encephalitis
Rubella†	Hepatitis A	Pneumococcal
Mumps†	Hepatitis B	Meningococcal
Yellow fever	Typhoid	Diphtheria (toxoid)
BCG (against tuberculosis)	Pertussis	Tetanus (toxoid)
Typhoid	Cholera	Anthrax
	H. influenzae type b	Plague
	Rabies	

*All given parenterally except live poliovaccine and live attenuated typhoid vaccine.
†Usually combined as 'MMR' vaccine.

in developing countres (Fig. 6.1). This programme aims to immunize against diphtheria, pertussis, tetanus, poliomyelitis, measles, tuberculosis and more recently hepatitis B. Where neonatal tetanus causes severe mortality, efforts are made to immunize women who then transfer passive immunity to their newborn infants. Vaccination remains important in developed countries both for protecting individual recipients (for example, tetanus which does not spread from person to person) and in order to achieve adequate herd (population) immunity to prevent outbreaks. A high level of herd immunity particularly benefits vulnerable unprotected groups such as newborn babies or the immunosuppressed, who are then less likely to be exposed to, for instance, measles, or pertussis. Except for smallpox, control of an infectious disease by immunization has not yet ensured its total and permanent eradication—through this is a possible goal on a more limited scale, for example, eradication of indigenous poliomyelitis from the USA and projected for several other countries.

The decision whether or not to immunize depends on balancing the dangers and costs to the individual and the community of the infection against those of the

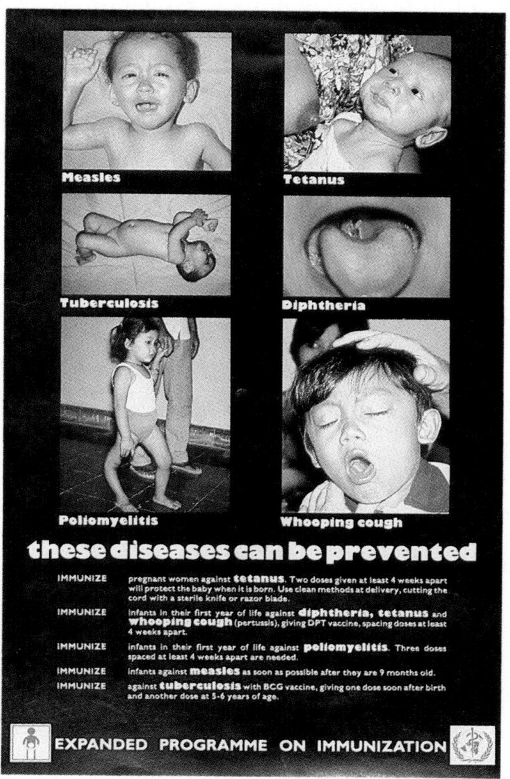

Fig. 16.1 The World Health Organization's Expanded Programme of Immunization aims to protect children throughout the world.

vaccine. An ideal vaccine is cheap, easy to administer, effective, and has no side-effects.

Passive immunization

This was originally carried out with antibody-containing sera from animals, commonly horses, immunized by vaccination with appropriate organisms. In humans, these foreign proteins caused reactions ('serum sickness'), were rapidly metabolized and thus effective for only a short time. Sensitized recipients became liable to violent allergic reactions to future injections of 'antiserum' prepared from the same animal species. Later it was found possible to give better protection, effective for longer periods and with minimal side-effects, by the use of human serum prepared from the blood of persons who had recently recovered from the infection (for example, measles 'convalesent serum'). Pooled human immunoglobulin from routine blood donations may contain useful quantities of antibody to common infective agents; for example, hepatitis A. Care must be taken, by selecting donors carefully and by virus-killing procedures during manufacture, to avoid transmitting in these preparations any infectious agents present in the donations, such as hepatitis B virus, HIV, or cytomegalovirus. Even better preparations are the separated immunoglobulins from human sera which have been either selected by tests as having useful amounts of the antibody either during convalescence from the relevant illness or after immunizing volunteers. A growing range of such materials is now available (Table 16.3) and in future may include monoclonal antibodies produced by cultured human B lymphocytes.

Passive immunization is most effective if given shortly before or very soon after exposure to infection, before the infectious agent has established itself. Given a few days after exposure it is unlikely to prevent illness but may reduce its severity as with measles or chickenpox. By the time the disease is clinically manifest it is rarely possible to reduce its severity by passive immunization. Passive immunization may reinforce other measures in severe infections of immunocompromised patients but may sometimes provoke immune complex disease.

Table 16.3. *Immunoglobulin preparations for prophylaxis*

Human immunoglobulin preparation	Uses in prophylaxis
Normal	Hepatitis A Measles Coxsackievirus (neonates)
Specific	Varicella-zoster Hepatitis B Tetanus Rabies Vaccinia

Antibodies to toxins (antitoxins) sometimes of animal origin, but preferably derived from immune humans, are used for *treatment*, not prevention, of diphtheria, botulism, tetanus, and envenomation by snake bites. Management of Gram-negative sepsis may be improved by using specific antibodies to immune system triggers or mediators; for instance, against endotoxin or tumour necrosis factor.

ACTIVE IMMUNIZATION PROCEDURES

Live vaccines

In order to retain activity these are usually freeze-dried and stored at low temperatures. After reconstitution for use they are inactivated quickly by exposure to light or heat, and the viable period for these vaccines after reconstitution is usually about one hour. This leads to difficulties especially in hot countries or where there are no refrigeration facilities. Organization of a reliable 'cold chain' from manufacture to the point of use has been essential for the success of the WHO EPI (Table 16.5) in developing countries. All who use vaccines even in temperate climates must be aware that a break in the 'cold chain' of refrigeration may lead to administration of an ineffective product (Fig. 16.2). If care is not taken, the last doses of multi-dose vials used at immunization clinics may be ineffective.

Live vaccines induce a subclinical or mild form of the disease and usually give lasting or even lifelong immunity after only one dose. Exceptions are as follows:

(a) Oral poliomyelitis virus vaccine requires several doses to induce adequate response to all three poliovirus types. 'Boosting' doses in later life may not

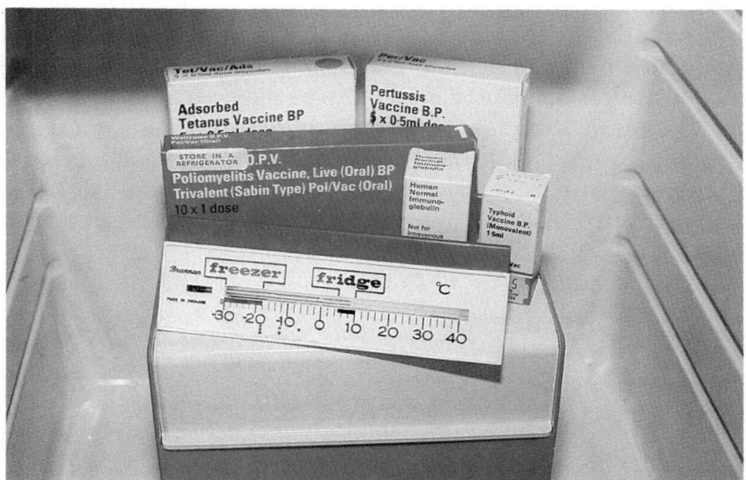

Fig. 16.2 Most vaccines must be stored cool (around 4 °C) but not frozen.

increase existing antibody levels but ensure that any of the component types which failed to 'take' previously has the chance to do so.

(b) Bacillus Calmette–Guérin (BCG) antituberculosis vaccine is the only widely used live *bacterial* vaccine and, like tuberculosis itself, does not give complete lifelong immunity.

(c) Ty21a (live attenuated oral typhoid vaccine), given in three or four doses, should be boosted by further three doses annually.

Live vaccines given to someone already immune confer little or no benefit but generally are not harmful. Giving live vaccines to neonates who have passively transferred maternal antibody which may neutralize the antigen is likely to be ineffective. Immune globulin (for instance against hepatitis A) given around the same time as a live vaccine (such as oral poliomyelitis vaccine) may reduce the latter's effectiveness. This is less likely to cause a problem with booster doses, but ideally live vaccines should be given at least 3 weeks before or 3 months after a dose of immunoglobulin if this is likely to contain the relevant antibodies.

When more than one live virus is required at one time, for example, before overseas travel, doses are best given simultaneously as with the combined measles–mumps–rubella (MMR) vaccine, or at least three weeks apart to prevent interferon response to the first immunization reducing response to the second. Live vaccines currently available are shown in Table 16.2 and vaccines for varicella and several other infections are under development.

Non-living (inactivated) vaccines

These suspensions of killed organisms, extracted antigenic components or modifed toxins should not be frozen. After injection they do not cause illness similar to the natural infection, and reactions are different from those to live vaccines. Except for the polysaccharide vaccines to meningococci, pneumococci and typhoid Vi antigen, more than one dose of inactivated vaccine is usually required. Booster doses in later life may also be required at intervals as short as 6–12 months (influenza vaccine) or ten years or more (tetanus toxoid). First dose of *toxoids* generally produce little or no detectable antibody and spacing between doses is important since a second dose given too early may induce little secondary response. Inactivated vaccines and toxoids currently available are shown in Table 16.2.

Administration, suitability, contraindications, and records

Routes of administration

Vaccines are given orally (for example live poliomyelitis), intramuscularly (for example hepatitis B), subcutaneously (for instance diphtheria/pertussis/tetanus (triple)) or intradermally (BCG, Fig. 16.3). The recommended route and dose

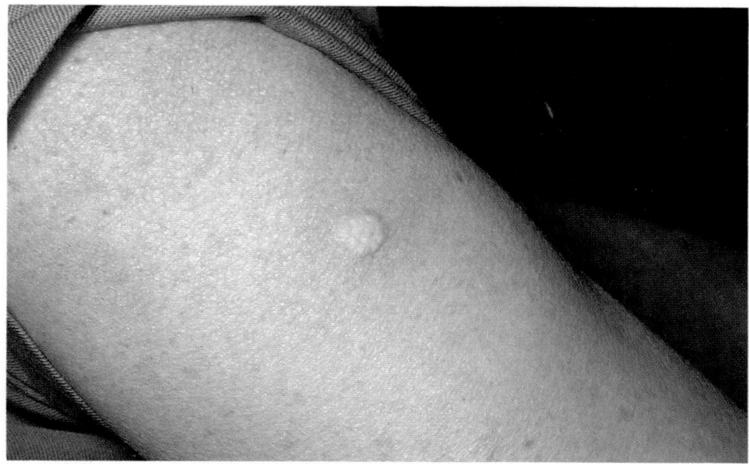

Fig. 16.3 Vaccines are given by various routes. Intradermal injection of BCG (against tuberculosis).

should be used—triple vaccine given superficially can cause painful swelling; BCG subcutaneously can cause an abscess and/or poor immune response.

Who should receive active immunization?

Recommendations are available from the World Health Organization and most countries have their owm recommended schedules. The British schedule is shown in Table 16.4; the EPI schedule of WHO for developing countries in Table 16.5.

There are often subtle but important differences in schedules used in different parts of the world. For example, where diphtheria and poliomyelitis are rare, primary immunization of infants is usually delayed for several months in order to allow a better immune response. However, where these infections are common, administration can start as early as one month of age but often with an additional dose during the second year of life.

Means of persuading people to accept vaccine vary, from a voluntary educational approach to legal compulsion, and immunization certificates are sometimes required for entry into certain countries or before attending college or school.

Contraindications

It is usually wise to postpone immunizing someone suffering from serious or febrile infection since it may be difficult to assess whether any 'reaction' is due to vaccine or concurrent infection. Milder illnesses such as nasal catarrh or slight persistent cough, however, are not contraindications.

After serious local or systemic reaction to a previous dose, the need for further doses must be carefully assessed in relation to the risks involved. Too many doses

Table 16.4. *British recommented schedule of immunization*

Age	Vaccine	Regime
During the first 6 months of life	Diphtheria, pertussis, tetanus (DPT) and oral polio (OPV)	1st dose at 2 months of age. 3 doses at monthly intervals completed by 6 months of age.
Around 15 months of age	Measles–mumps–Rubella (MMR)	One dose only. Previous history of any of the natural infections is not a contraindication.
At school or nursery school entry	Diphtheria, tetanus and OPV: booster MMR	Preferably 3 years after primary course. If not given previously.
Between 10th and 14th birthdays	BCG (tuberculosis)	Given to tuberculin negative subjects. Can be given from birth to contacts.
	Rubella (girls only)	Previous history of possible rubella is not a contraindication. (May be discontinued if infant MMR administration proves to give lasting immunity to rubella.)
On leaving school, starting employment or entering further education	OPV and tetanus	
Adult life	Booster doses (to those at risk every 10 years) of polio and tetanus and primary courses for those not previously immunized. Rubella for women not pregnant (e.g. immediately postpartum) and of child-bearing age if seronegative. Special requirements for the immunocompromized, for certain occupations and for overseas travellers. Influenza and hepatitis B for those in high-risk groups.	

of vaccine can produce hypersensitivity as with tetanus toxoid given repeatedly after injuries.

Some vaccines contain traces of allergens such as egg protein or antibiotics and should not normally be given to those with a history of serious allergy to these components such as anaphylaxis or angioneurotic oedema.

When parenteral vaccines are being administered adrenaline and other facilities for treating anaphylaxis should be available and the recipient observed for a short time after the injection.

Live vaccines should not usually be administered during pregnancy and are contraindicated in the immunocompromised (virulence may be expressed as disease or fetal damage).

Table 16.5. *Expanded Programme of Immunization (EPI) recommended by World Health Organization—especially applicable to developing countries*

Age	Vaccines
Birth	BCG + OPV + Hepatitis B
6 weeks	DPT + OPV
10 weeks	DPT + OPV
14 weeks	DPT + OPV
9 months*	Measles*
Women of child-bearing age	Tetanus toxoid

*At 6 months with high titred Edmonston–Zagreb vaccine if measles mortality is significant below age 9 months

Reporting adverse reactions

In Britain severe reactions should be reported to the Committee on Safety of Medicines and to the manufacturers. The manufacturers should be informed if a particular batch is associated with more than the usual 'normal side-effects' (Fig. 16.4).

Records

It is important to keep records of all vaccines given. These are often mislaid by the recipient. Some health authorities maintain records by liaison with primary care

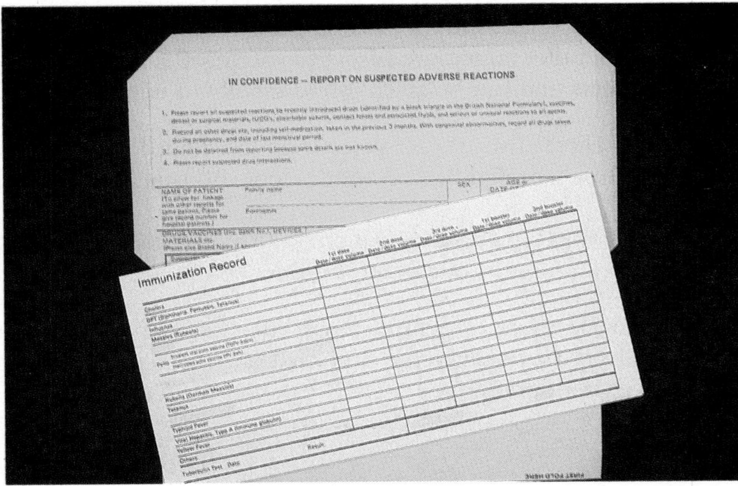

Fig. 16.4 Vaccination histories should be carefully recorded and side-effects reported to designated national authorities and/or manufacturers.

physicians and health visitors. These help is assessing the immune status of communities, following-up non-recipients and, if linked with disease registers, estimating the effectiveness of recommended immunization regimes. Immune status of population groups can also be determined by antibody surveys carried out on blood samples.

NOTES ON SOME INDIVIDUAL VACCINES COMMONLY USED IN BRITAIN

Diphtheria vaccine

Formaldehyde-treated diphtheria toxin ('toxoid', adsorbed to aluminium phosphate or hydroxide), by intramuscular (IM) or deep subcutaneous (SC) injection.

Primary course: Three injections, usually combined with pertussis and tetanus vaccines (DPT) or tetanus alone (DT). Boosters to high-risk groups every 10 years.

Notes: For persons aged 10 years or more the special low-dose vaccine for adults must be used for primary or booster immunization. This avoids possible severe reactions in immune recipients and makes prior Schick testing unnecessary.

Tetanus vaccine

Formaldehyde-treated tetanus toxin ('toxoid', adsorbed or plain). For a rapid response, for example after injury, use adsorbed vaccine. Plain vaccine should be used only for boosters and can be given intradermally to minimize reactions.

Primary course: Three IM or SC injections, combined for infants with diphtheria and pertussis, or diphtheria vaccine alone. Boosters every 10 years for those at risk.

Frequent boosters are unnecessary, can provoke hypersensitivity, and should not be given within 12 months of a previous dose.

Notes: Systemic side-effects are rare: local inflammation and nodules are common, especially if adsorbed vaccine is injected superficially. Previous clinical tetanus may not confer immunity.

Pertussis vaccine

Suspension of killed *B. pertussis* organisms including three pertussis serotypes (adsorbed or plain).

Primary course: Three IM or SC injections. Boosters not usually needed since disease is severe only in young children. Usually combined with diphtheria and tetanus vaccines; pertussis vaccine alone (monovalent) is available.

Notes: Transient fever, malaise, and painful local reactions at the injection site can occur. Convulsions and encephalopathy are occasionally associated with both the disease and with vaccination. Severe neurological reactions may follow about 1 in 310 000 injections of vaccine, but are not necessarily caused by vaccine. Contraindications attempt to exclude from vaccination those predisposed to serious reactions.

Contraindications: The British Joint Committee on Vaccination and Immunisation, 1990, advise that the vaccine is contraindicated where there is (a) history of any severe local or general reaction (including neurological reaction) to a preceding dose; or (b) an evolving neurological condition. Special consideration is necessary for (a) children whose parents or siblings have a history of idiopathic epilepsy; (b) children who had convulsions; (c) children with cerebral damage in the neonatal period.

Poliomyelitis vaccine

Attenuated live strains of polioviruses (types 1, 2, and 3) given orally. Inactivated poliovaccine (given by IM or SC injection) is an alternative, for example, when live vaccine is contraindicated for instance, by immunocompromise.

Primary course: Three doses. Reinforcing doses advised at school entry and on leaving school; also for travellers to endemic areas (10-yearly).

Cases of poliomyelitis are very rarely reported in association with live vaccine, usually in immunodeficient children or non-immune adult contacts. Some are coincidental, but possible renewed virulence of vaccine virus after passage through the intestinal tract may warrant vaccinating at the same time any inadequately immunized family members of children having primary immunization. Enhanced potency inactivated poliovaccine combined with bacterial antigens is used in some countries.

Measles–mumps–rubella (MMR) vaccine

Attenuated live viruses (freeze-dried). Single IM or SC injection about 15 months of age. Given safely after natural illnesses. Measles vaccine alone is given in some tropical countries at a younger age; since maternal antibody may then have been present, a boosting dose should ideally follow in the second year of life.

Notes: Up to 20 per cent of vaccinees develop mild 'measles' illness 5–10 days after immunization sometimes with febrile convulsions or rash (Fig. 16.5). Headache, fever, and mild aseptic meningitis from the mumps component has

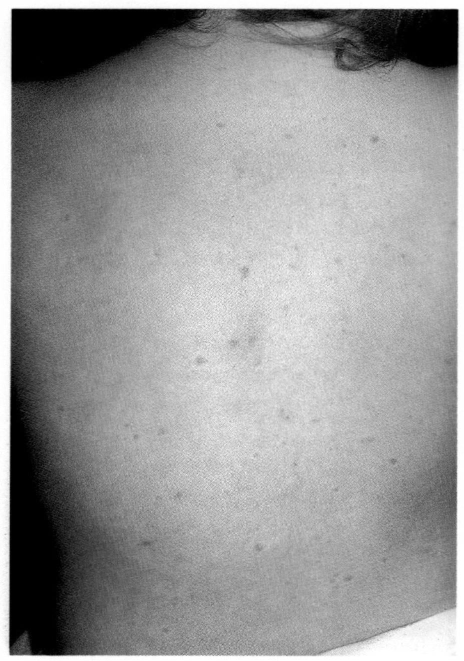

Fig. 16.5 A mild 'measles' rash following measles vaccination.

been reported. Complications are much less frequent than with natural infection. The vaccines can be given separately.

Contraindications: Since mumps and measles vaccines are made from chick embryo cell cultures, *anaphylaxis* or *severe* urticaria to egg ingestion is a contraindication, but not minor egg allergies or intolerance such as vomiting.

Rubella vaccine (currently used alone for teenagers and adults)

Attenuated live rubella virus (freeze-dried). Single IM or SC injection. Can be given after previous rubella-like illness. For use, see Chapter 11.

Contraindicated in pregnancy, but not proved to harm the fetus. May cause mild non-infectious rubella-like illness including arthralgia one to three weeks after immunization. Women should avoid pregnancy for three months after rubella vaccination.

Bacillus Calmette–Guérin (BCG) vaccine (against tuberculosis)

Bacillus Calmette–Guérin, living (freeze-dried). Single *intradermal* injection. Boosters can be given to those at high risk if exposure to infection continues and if skin tests for hypersensitivity to tuberculoprotein are negative (Chapter 9), usually not more often than 10-yearly.

Notes: Local BCG reaction lasts several weeks, occasionally ulcerating and associated with local lymphadenopathy, finally leaving a small scar. Severe ulceration or abscess formation can be treated with Rifampicin and/or isoniazid. Isoniazid 'powder' can be used for small chronic ulcers. An accelerated reaction within a few days similar to a Mantoux reaction follows vaccination of those already hypersensitive. Except in infants, prior PPD skin testing is advisable (see Fig. 9.11).

Haemophilus influenza b vaccine

Protein-conjugated polysaccharide vaccines have been developed which appear safe and effective even in infancy, superseding unconjugated capsular vaccines to which those under 2 years of age fail to respond effectively. They were licensed in the UK in 1992 for administration in the normal childhood schedule given at the same time as DPT (triple) vaccine.

Influenza vaccine

Inactivated 'whole virus', 'split virus', or 'surface antigen' vaccines containing antigens of current influenza A and B virus strains. Composition reviewed annually. Live, attenuated vaccines for intra-nasal administration are difficult to standardize and not generally acceptable. Single injection confers about 70 per cent protection against corresponding influenza strains for about one year. Recommended for individual protection of 'at risk' persons such as the elderly (especially in institutions where outbreaks can spread rapidly) and those with chronic heart, lung, kidney, or other debilitating disease.

Notes: Mild local reactions to vaccine are common, also fever lasting 1–2 days. Guillain–Barré syndrome occasionally associated with vaccination.

Hepatitis B vaccine

Suspension of inactivated adsorbed and purified hepatitis B surface antigen particles derived from plasma of human carriers or from DNA-recombinant yeast cultures. Primary course of three IM or SC injections protects around 90 per cent of recipients. Frequency of booster doses required to maintain immunity currently assessed by antibody testing, usually not more often than 3–5 yearly. Recently included in WHO recommended schedule for infants (Table 16.5).

Groups for which vaccination against hepatitis B is specially important include:

(a) Health care personnel directly exposed to blood, including those in contact with haemophiliacs on maintenance therapy and patients receiving renal dialysis.

(b) Patients likely to be exposed, such as new admissions to long-stay institutions for the mentally handicapped where hepatitis B is prevalent.

(c) Patients receiving repeated concentrated blood products, renal dialysis or about to undergo major elective surgery requiring massive blood transfusions, especially if outside Britain.

(d) Sexual partners of carriers.

(e) Newborn infants of carrier mothers (a special problem in the Far East, Oceania and Africa).

(f) Prostitutes.

(g) Homosexuals.

(h) Drug addicts.

(i) Police, custodial and emergency service staff.

(j) Travellers to countries with high carrier prevalence, especially if staying for prolonged periods.

PASSIVE IMMUNIZATION

Immunoglobulins (Table 16.3) are given by intramuscular injection in dosages specified by the manufacturer (Blood Transfusion Service, Protein Fractionation Unit, or commercial supplier). They give temporary protection when used before or immediately after exposure.

Normal human immunoglobulin

This is useful for preventing hepatitis A in travellers to endemic areas (see below) or protecting vulnerable children temporarily against measles; for example, after accidental admission of a measles case to a ward containing non-immune patients.

Human specific immunoglobulin

This is more potent and useful when immunosuppressed patients have been exposed to varicella. Hepatitis B immunoglobulin can be given after accidental skin puncture or mucosal contamination with infective material and also soon after the birth of the baby whose mother is a carrier of hepatitis B (Chapter 11). Human rabies and tetanus specific immunoglobulins are used in conjunction with the corresponding vaccine (given at a different site) for post-exposure prophylaxis.

SPECIAL ADVICE FOR TRAVELLERS

Deficiencies of primary courses of immunization as for the home country should be made good and booster doses of tetanus or poliomyelitis vaccine given if none have been received in the past 10 years. Unless there is a special risk factor, for example related to life-style or occupation, no further immunizations are normally required for travellers from Britain to Europe, North America, Japan, Australia, or New Zealand. For other regions and especially if there will be high risk of faecal–oral infection (for example, rural living, travelling 'rough', eating and drinking in low-hygiene circumstances—Fig. 16.6), typhoid and hepatitis A protection should be considered.

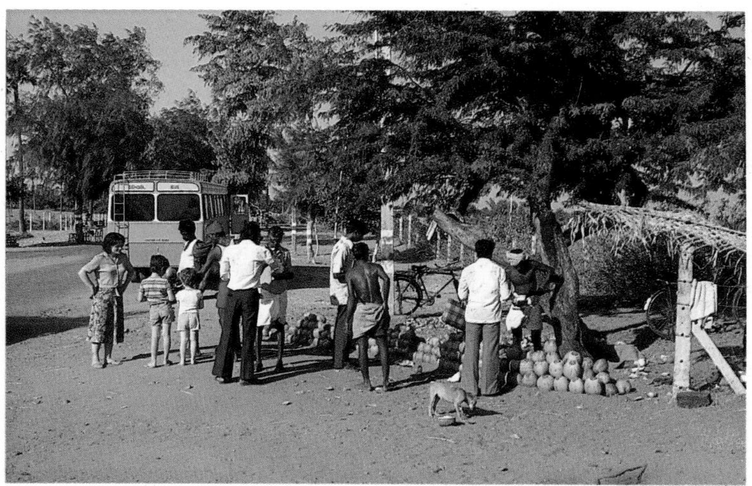

Fig. 16.6 Travellers eating and drinking in unplanned situations are especially at risk of faecal/oral infection.

Immunizations for special situations

Anthrax

Vaccination is advised only for those at occupational risk, e.g. through contact with infected hides, wool, hair, bone, feeding stuffs, and carcasses. Protective clothing should also be worn.

Cholera

Those at risk are people travelling to infected areas who are unlikely to be able to ensure hygienic water and food supplies. Cholera vaccine gives only partial and short-lived protection and is not effective in preventing spread of disease. No country now requires vaccination certificates as a condition of entry.

Diphtheria

Vaccination is advised for travellers to endemic areas likely to mix with the local population (e.g. health care workers or teachers). They should receive ten-yearly boosters of the adult (low dose) vaccine. This is in addition to following the normal British schedule for primary immunization.

Hepatitis A

Short-lived passive protection provided by immunoglobulin is being replaced by active vaccination with a killed vaccine providing 5–10 years' protection after three doses. It is indicated especially for adults travelling to or living in countries with poor food and water hygiene.

Hepatitis B

Advised for those at occupational risk (e.g. health care personnel) and also for long stay expatriates in countries with a high carriage rate in the local population.

Japanese 'B' encephalitis

Advised for prolonged stays, or repeated visits, travel during outbreaks, and when avoiding mosquito bites is impractical in Asia, east of and including the Indian subcontinent. Children are at special risk of serious disease.

Meningococcal A and C

Advised for those visiting countries in Sub-Saharan Africa when epidemics are likely, i.e. during the dry season or for prolonged stays. Children are at special risk. Spread is through close contact (e.g. within families, schools, dormitories) hence risk is usually small for organized package tourists.

Plague

Recommended for those going to areas where plague is prevalent and if living in poor conditions likely to bring them in contact with wild rodents.

Rabies

Pre-exposure vaccination can be given to those travelling to countries where rabies is present and who intend to either:

(a) Have regular contact with animals (e.g. Veterinarians).
(b) Be more than 24 hours away from a source of post-exposure vaccine.

Extra boosters should be obtained as soon as possible after a suspect bite.

Tick-borne encephalitis

Only those exposed to tick bites are at risk, e.g. long-term local residents, ramblers, scouts and those occupationally exposed such as foresters. Endemic areas include Eastern Europe, USSR and forested regions of Austria, Sweden, and Germany.

Tuberculosis

Vaccination is advised for travellers to countries where tuberculosis is common and mixing with the local population is likely, e.g. those going for long stays, and health care workers.

Vaccination is especially useful for young children and can given from birth.

Yellow fever

Vaccination is compulsory for travellers to those countries which require a certificate as a condition of entry. It is recommended for persons aged 9 months or over travelling through or living in infected areas in South America and Sub-Saharan Africa.

Detailed advice on risk areas, current outbreaks and indications is available from World Health Organization publications and specialist centres. The above information is based upon the database (TRAVAX) available to general practitioners in Britain (Fig. 16.7).

Malaria prevention

Effective prevention depends upon some or all of the following: (1) controlling the anopheline mosquito vector, (2) avoiding mosquito bites, (3) using prophylactic drugs, (4) receiving prompt treatment to prevent fatal illness. Anopheline mosquitos are most active in the evening and at night, when clothes should cover exposed skin as far as possible. Insect repellants containing diethyltoluamide for exposed skin (face and hands) and netting over windows and beds are helpful (Fig.16.8).

Prophylactic drugs

The non-immune traveller entering a malarious area (see Fig. 16.9) is generally advised to take antimalarial drugs principally to prevent the dangerous *Plasmodium falciparum* infection. This applies to all ages, including babies and during pregnancy. One or two doses prior to exposure help to confirm tolerance and allow time to achieve adequate blood levels. Medication must be taken regularly during exposure and continued for at least four, preferably six, weeks after leaving the malarious area. This is because the parasites first develop in the liver and only later cause illness after re-entering the blood where most prophylactic drugs take effect. If doses are missed and infection becomes established, the dosage used for prophylaxis is unlikely to be effective for cure. In highly endemic areas such as most of Africa, prophylaxis should normally be advised for travellers. In less endemic areas such as much of the Far East prompt treatment and avoiding mosquito bites may be preferred to risking side-effects of prophylactic drugs.

(a)

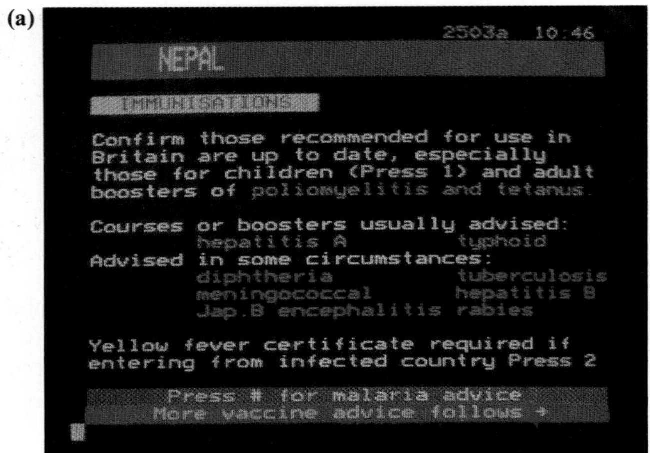

(b)

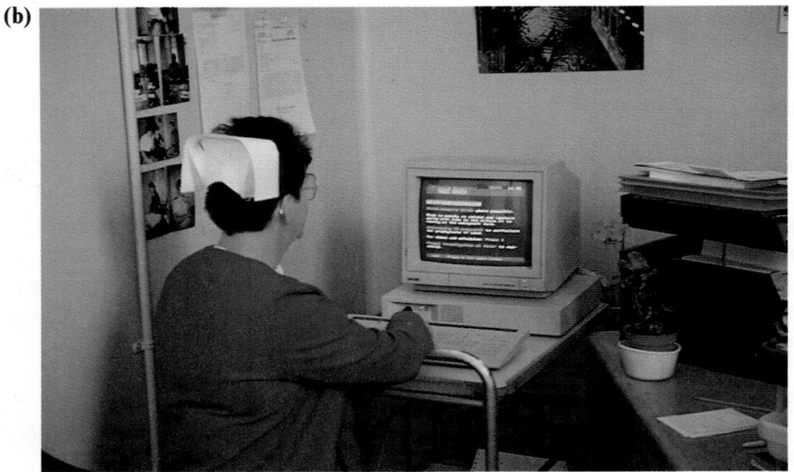

Fig. 16.7 (a) and **(b)** The TRAVAX database provided by the Communicable Diseases (Scotland) Unit, Ruchill Hospital.

Choosing the most appropriate drug

The leaflet *Travellers' Guide to Health* (TI) issued by the Department of Health in Britain lists countries where malaria is present. There is no perfect antimalarial drug or combination of drugs, only a reasoned choice balancing the likelihood of serious infection against possible side-effects of tablets. Knowledge of current resistance patterns in the country to be visited and the likelihood of exposure is needed. Advice can be sought by the advising doctor from national reference centres, and be confirmed by the long-stay traveller at his destination.

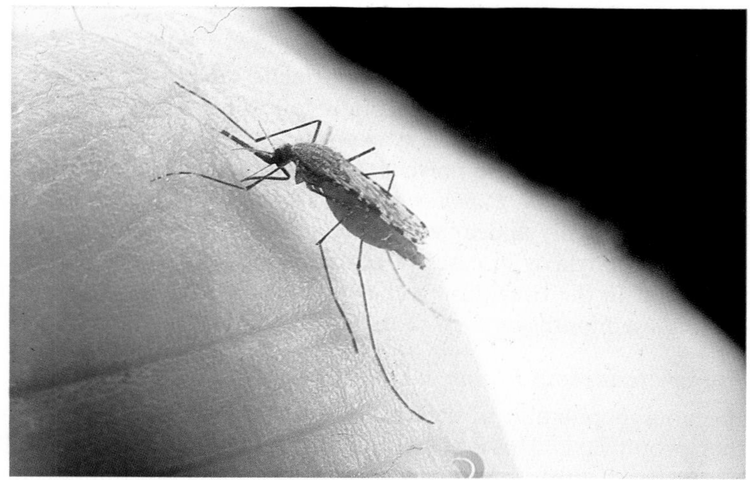

Fig. 16.8 The female anopheline mosquito transmits malaria. (Courtesy of the World Health Organization.)

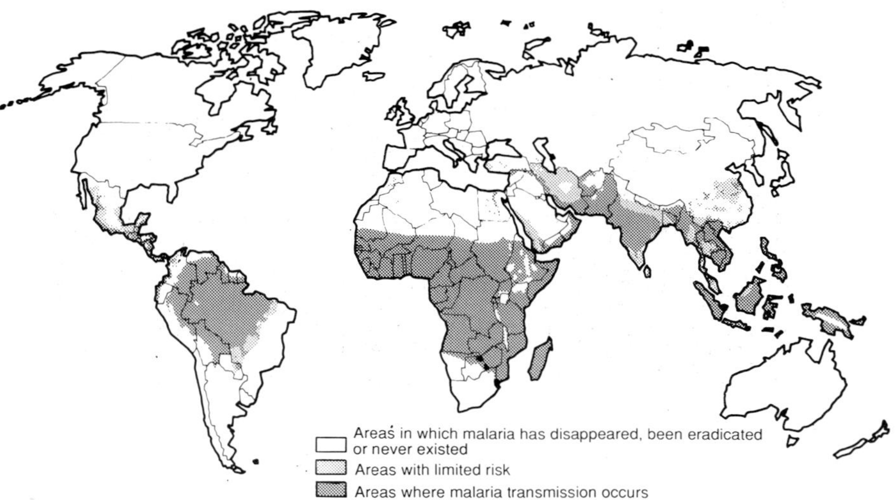

Fig. 16.9 The areas of the world where most malaria transmission occurs (Courtesy of the World Health Organization.)

The usual first-line drug is either chloroquine or proguanil. Where there is widespread chloroquine-resistant falciparum malaria, both drugs should be taken or alternatively a newer preparation such as mefloquine. Chloroquine plus proguanil has a reputation for minimal side-effects and this combination is especially useful for babies and during pregnancy. Chloroquine resistance is currently widespread in parts of East Asia, South America, and throughout Africa. It is spreading in the Indian subcontinent.

If malaria develops despite taking prophylactic tablets, this may be due to irregular ingestion, inadequate absorption following gastrointestinal illness, parasite resistance or especially heavy exposure.

Prophylactic tablets can also prevent the less immediately dangerous forms of malaria, *Plasmodium vivax*, *P. malariae*, and *P. ovale*. Unlike *P. falciparum* with which symptoms usually appear within six weeks of infection (or six weeks of discontinuing prophylaxis in cases of partial resistance), other plasmodia can remain dormant in the liver during chemoprophylaxis to become re-activated and cause disease months or even years later.

The temporary resident in Britain

The indigenous population in malarious areas usually relies upon treatment rather than prophylaxis. This is not without risk and is not advisable for the non-immune traveller. Visitors to Britain from endemic areas lose acquired immunity, sometimes within as little as one or two years. Before returning they should be warned of the need for immediate treatment of renewed malarial fevers. A short period of regular prophylaxis after return can be helpful in these circumstances since immunity to early developmental forms of the parasite may be boosted while tablets are being taken.

17

Epidemiology in action

INTRODUCTION

The diagnosis and treatment of infections requires vigilance that extends beyond the surgery or hospital bed. Knowing which infections can spread, in what circumstances, and when to take appropriate action is all part of effective management. Doctors, patients, their relatives, and many others such as health visitors, environmental health officers, other paramedical staff, veterinarians, and employers can be involved.

Sometimes immediate response is required, as for an outbreak of food poisoning in hospital, an imported case of viral haemorrhagic fever, or one of meningococcal meningitis. It may be necessary to mount an investigation (for example, in response to a cluster of cases of acute hepatitis B in a long-stay institution or of Legionnaires' disease in hospital). Detecting disease early in contacts, who may still be asymptomatic (for example, tuberculosis), can be life-saving. The methods are those of '*clinical epidemiology*' (Fig. 17.1), where action depends upon understanding the vulnerability of those exposed and the routes whereby infection could spread.

The microorganisms prevalent in a community are constantly changing. New infections or different presentations of old ones continue to be recognized (Chapter 1). Those living in countries with well-established public health systems may not realize the consequences of failing to maintain clean water supplies, effective sewage disposal or immunization programmes, especially when the infections thereby controlled have as a result become rare.

Monitoring patterns of infection (Fig. 17.2), introducing preventive measures and monitoring their effectiveness, all require ongoing collection of information about the current 'state of play', an organized activity called *surveillance*.

SURVEILLANCE

No one method is complete in itself and the most suitable vary with the objectives; for example, responding appropriately to a case or outbreak of infection, recording disease incidence, monitoring the effectiveness of an immunization campaign, defining the type of illness caused by an organism or the range of infectious agents causing a syndrome. Each method provides

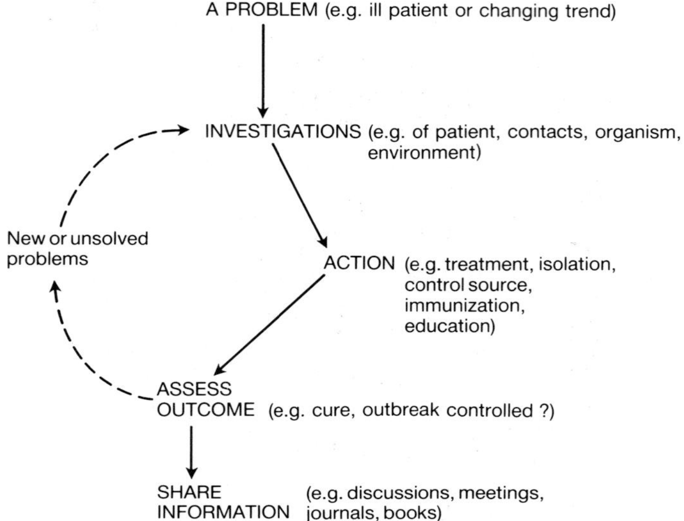

Fig. 17.1 The work of the clinical epidemiologist.

information which reflects a constantly changing picture. These methods are also basic to documentation of the usual (and unusual) symptoms and signs of disease, making diagnosis and rational management possible.

Facilities available affect the methods used. Thus, in some countries the most appropriate way of studying the prevalence of poliomyelitis may be analysis of laboratory isolates or antibody surveys, but in others the recording of acquired paralysis in children. Some examples of different methods are discussed below.

Methods of surveillance

Notifications

Official notifications of disease to central authority require cooperation by those making a diagnosis. Only a small proportion (often 1–10 per cent) of legally notifiable cases are usually reported, yet fluctuations over time often reflect important trends. Only selected infections are notifiable (Fig. 17.3). Notifications can show where and when cases and epidemics are occurring; they are more complete for some infections than others (for example, malaria reports in Britain are fairly complete, those for measles are not).

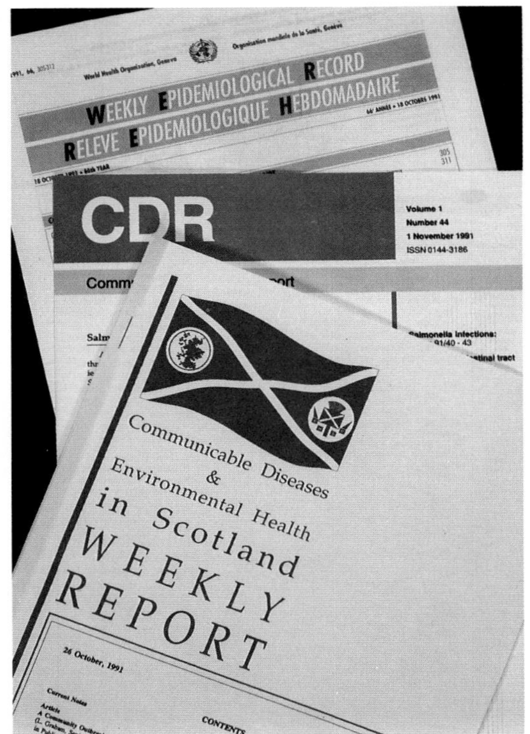

Fig. 17.2 Weekly reports of **(a)** World Health Organization; **(b)** Public Health Laboratory Service (England, Wales, and Northern Ireland); **(c)** Communicable Diseases and Environmental Health (Scotland) Units.

Laboratory evidence of infection

This reflects use of the laboratory by clinicians. This in turn depends on such factors as the need for confirmed diagnosis, a department's special interests, and availability of particular techniques in the laboratory. Laboratory reports can be studied collectively, either locally in a hospital or group of hospitals, or nationally as at the Communicable Disease Surveillance Centre (for England, Wales, and Northern Ireland) and the Communicable Diseases (Scotland) Unit for Scotland. Reports need careful interpretation, but clustering can give the first clue to an epidemic (Legionnaires' disease or meningococcal meningitis). Sometimes unusual or unexpected isolates are discovered and retrospective associations with symptoms made. Pooled information helps to guide management (for example, antibiotic sensitivities or choice of strains to include in the current influenza vaccine).

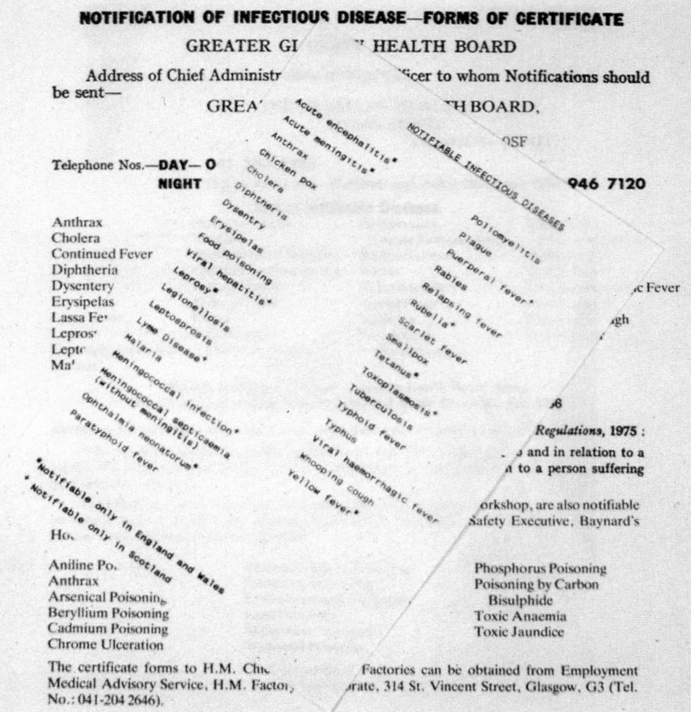

NOTIFICATION OF INFECTIOUS DISEASE—FORMS OF CERTIFICATE

GREATER GI... HEALTH BOARD

Address of Chief Administr... ...cer to whom Notifications should be sent—

GREA... ...H BOARD,

Telephone Nos.—**DAY— 0** ... **946 7120**
NIGHT

Anthrax
Cholera
Continued Fever
Diphtheria
Dysentery
Erysipelas
Lassa Fe...
Lepros...
Lept...
Ma...

...c Fever
...gh

...6

Regulations, **1975** :
... and in relation to a
... to a person suffering

...orkshop, are also notifiable
...Safety Executive, Baynard's

Ho...

Aniline Po...
Anthrax
Arsenical Poisonin...
Beryllium Poisoning
Cadmium Poisoning
Chrome Ulceration

Phosphorus Poisoning
Poisoning by Carbon Bisulphide
Toxic Anaemia
Toxic Jaundice

The certificate forms to H.M. Ch... Factories can be obtained from Employment Medical Advisory Service, H.M. Facto... ...rate, 314 St. Vincent Street, Glasgow, G3 (Tel. No.: 041-204 2646).

NOTIFIABLE INFECTIOUS DISEASES
Acute encephalitis*
Acute meningitis*
Anthrax
Chicken pox
Cholera
Diphtheria
Dysentry
Erysipelas
Food poisoning
Viral hepatitis*
Leprosy•
Legionellosis
Leptospirosis
Lyme Disease•
Malaria
Meningococcal infection•
Meningococcal septicaemia (without meningitis)
Ophthalmia neonatorum*
Paratyphoid fever
Poliomyelitis
Plague
Puerperal fever*
Rabies
Relapsing fever
Rubella•
Scarlet fever
Smallpox
Tetanus•
Toxoplasmosis•
Tuberculosis
Typhoid fever
Typhus
Viral haemorrhagic fever
Whooping cough
Yellow fever•
*Notifiable only in England and Wales
• Notifiable only in Scotland

Fig. 17.3 Certain infectious diseases are officially notifiable.

Diagnostic records

These sources of data also have limitations. In *general practice* infection has been given special attention in some studies. Much infection is not, however, brought to general practitioners, even though causing much morbidity. Surveys of illness at home are necessary to determine the incidence of, for example, common colds or sore throats. *Hospital* admission diagnoses commonly reflect symptoms, not aetiologies (for example, diarrhoea or fever). Discharge diagnosis when hospital stay is short, as is common with many infections, may be registered before investigations are complete. Many viral infections are not fully investigated, especially by those not specializing in infection.

Causes of death

Although still used to guide much health service planning, death certifications are largely unable to reflect the prevalence of infection since modern antimicrobials and other methods of management make death from infection unusual in countries such as Britain. Non-infectious diseases may be given prominence on certificates because they are more easily diagnosed.

Serological surveys

Availability of stored serum specimens (for example, taken for diagnostic purposes or from blood donations) enables useful serological studies of infection patterns to be carried out. These can reveal evidence of past infection and present immunity and susceptibility in different groups.

Immunization records

Careful documentation of the uptake of immunizations helps to monitor their distribution and identify 'at risk' groups. Together with studies of disease incidence or serological surveys, the effectiveness of immunization programmes can be assessed.

Sickness certificates

Gross fluctuations in numbers of persons off work from sickness are generally due to infection. One of the first indicators of an influenza epidemic is increase, sometimes dramatic, in sickness absence from work. This can be observed nationally from doctors' certificates or more locally, and rapidly, by firms keeping careful checks on their employees. In general, however, diagnoses on sickness certificates are too vague or preliminary to provide much accurate and useful information on prevailing infections.

Specific studies

A method which facilitates *a quick response* is reporting by general practitioner 'spotters' of new cases of infections such as influenza-like illness on a daily or weekly basis (Fig. 17.4). This usually gives early warning, and allows contingency planning for increased general practice and hospital work-load.

For '*new*' or rapidly changing *diseases* (for example, acquired immunodeficiency syndrome: AIDS), clinicians and laboratories, especially those with particular interest in infection, can be specially asked to supply relevant information. Prevalence studies of HIV infection now include anonymous antenatal and neonatal screening.

To help answer a particular question, a study can be mounted nationally. Thus, the National Childhood Encephalopathy Study was set up to seek any association between neurological disease in childhood and whooping cough vaccination.

Reference laboratories

Infections with potentially serious clinical or public health implications generally involve reference laboratories to provide information quickly about new isolates. Examples are reference laboratories for meningococci, legionellosis, malaria, viral hepatitis, HIV infection, mycology, mycobacteria and tuberculosis, enteric, and *Salmonella* infections.

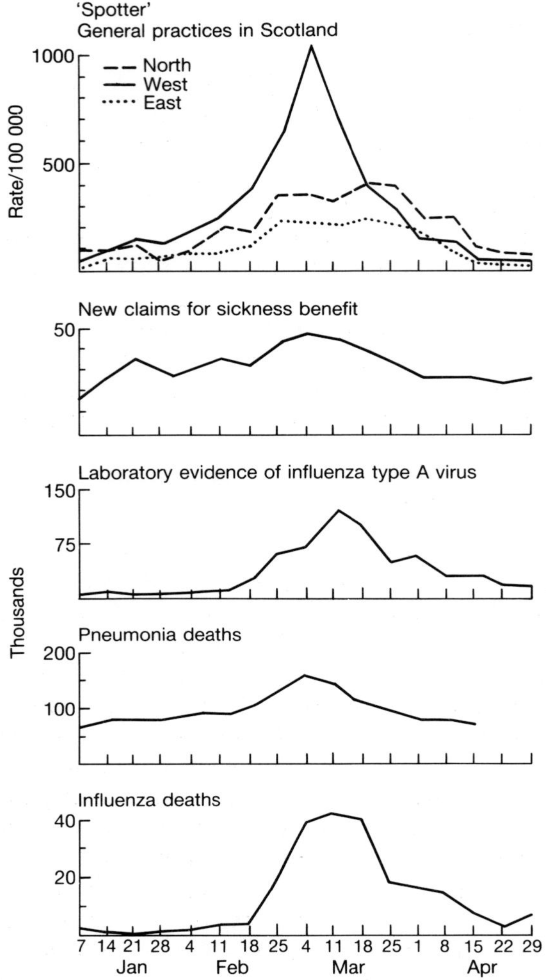

Fig. 17.4 Early warning of an influenza epidemic can come from 'spotter' general practitioners.

SOME CASE STUDIES

The following examples from home, school, general practice, the general community, and hospital (where cross-infection is a special hazard) demonstrate this broad approach to managing problems of infection.

Case 1: A family problem

The problem: A schoolboy developed persistent diarrhoea, nausea, and anorexia. His sister had milder but similar symptoms.

Investigation: After three weeks without improvement, examination of the boy's stool showed *Giardia lamblia* cysts (see Chapter 4). A course of treatment with metronidazole gave only temporary relief. Because asymptomatic giardial infection is possible and cysts are often transmitted between close family members, the stools of other family members were examined. The sister and mother were both positive.

Action and outcome: The whole family was treated and the need for personal hygiene emphasized. Follow-up confirmed recovery. The case was discussed over coffee with other doctors at the health centre so that the diagnosis could be considered in other symptomatic contacts (for example, school colleagues).

Case 2: Tuberculosis in a school

The problem: A 6-year-old boy at primary school developed fever and, a week later a squint. He was found to have tuberculous meningitis.

Investigation: Because 'open' tuberculosis (see Chapter 9) has become unusual in Britain, a source of infection among his close contacts was likely. Parents and other close relatives, including grandparents, and then school staff were interviewed and examined by chest X-ray. The school caretaker was found to have asymptomatic tuberculosis with organisms in his sputum. As a result all the other schoolchildren were examined and given chest X-rays and tuberculin skin tests. Two had evidence of asymptomatic primary pulmonary tuberculosis, one of tuberculous bronchopneumonia; 6 per cent had positive skin tests (a substantially higher level of positives than expected).

Action and outcome: The caretaker was isolated and treated. Children with active disease were also treated and those with positive skin tests given chemoprophylaxis and carefully followed up. Negative skin tests were repeated two months later. Proven cases were notified as required by statute. Resulting publicity led the local education authority to pursue more actively regular health checks for school staff in close contact with pupils.

Case 3: Influenza and the elderly

The problem: Earlier in winter than usual, general practitioner 'spotter' practices (see above) showed marked increase in 'flu-like' illness.

Action: Arrangements were made through the general practitioner responsible for the local old peoples' homes to offer the current recommended influenza vaccine to residents, since this age-group is especially at risk from influenza complications, including death. Other general practitioners were informed so that they could consider vaccine for at-risk individuals (for example, those with chronic pulmonary disease). The local hospital was warned that increased admissions were anticipated.

Outcome: During the ensuing influenza epidemic, no outbreak occurred in the residential homes. Two recipients of vaccine had marked febrile reactions which were reported to the manufacturers. Increased admissions to hospital necessitated temporary use of a surgical ward for medical cases. Chemoprophylaxis by amantadine was not offered because laboratories reported type B influenza, not type A, as the prevalent strain.

Case 4: An imported infection

The problem: A man who had returned two days previously from a working trip to an African refugee camp developed fever, painful cervical lymphadenopathy and sore throat. Lassa fever was considered unlikely after discussion with local infectious disease consultants—but he was admitted to hospital for isolation and investigations. Throat swabs showed organisms on staining resembling *Corynebacterium diphtheriae*. The patient had not received diphtheria vaccine since childhood.

Investigations: Close contacts, including family, general practitioner, and travelling companions were examined, throat swabs taken and records of diphtheria immunization obtained. A toxigenic strain of *Corynebacterium diphtheriae* was isolated from throat swabs from the patient and also from his 4-month-old non-immunized child with fever.

Action: Treatment was instituted for father and child. Prophylactic antibiotics were given to close contacts considered to have doubtful immune status (see Chapter 16). Primary courses of vaccine or boosters were offered, recognizing that the short incubation period of diphtheria limited their value. Follow-up was arranged.

Outcome: No further cases occurred. Repeat throat swabs were negative. Cases were notified as legally required. The importance of adequate immunization for travellers likely to be exposed was publicized.

Case 5: Milk-borne salmonellosis

The problem: Three babies from the same locality were admitted with diarrhoea during the same week to an infectious diseases ward. *Salmonella typhimurium* was isolated from the stools of all three. None had been weaned but all were receiving bottled cows' milk.

Investigation: Enquiries showed that all the famlies had purchased their milk from the same source—a single farm (producer/retailer)—and that it was not pasteurized. Reports of salmonella isolations from the same area during the previous month, recorded in weekly returns from the Salmonella Reference Laboratory, linked 10 other *S. typhimurium* infections to the same milk supply.

Further study showed that 14 outbreaks associated with unpasteurized milk had occurred in Scotland during the previous four years, 1980–83, with several deaths.

Action and outcome: Retail sales of milk from the farm involved were suspended. Investigations and control action by the Environmental Health Officer and veterinary colleagues were instituted. In view of the national importance of the continuing problem, legislation made pasteurization of cows' milk on sale to the public mandatory in Scotland from 1 August 1983. Since then public outbreaks ceased, but cases continued in farm employees who could receive raw milk free for their families until September 1986 (Fig. 17.5).

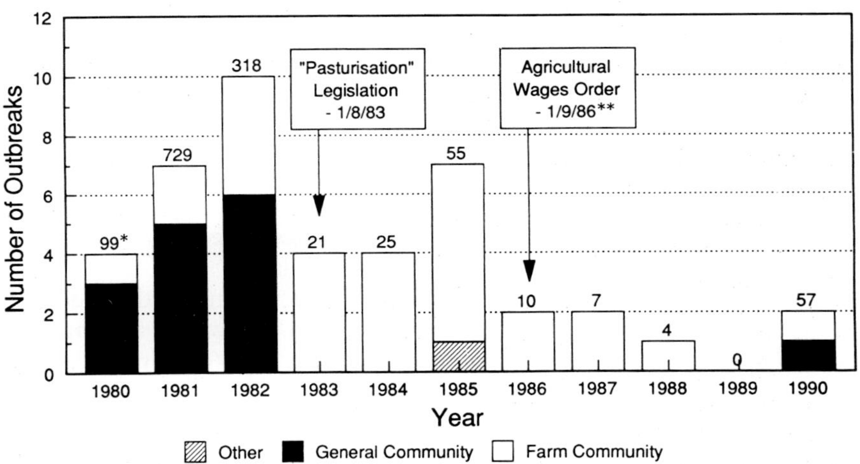

Fig. 17.5 Milk-borne infection in Scotland, 1980–90. *The number denotes the total persons affected. **This order allows only heat-treated milk as wage-benefit.

Case 6: An outbreak of 'food poisoning'

The problem: Many of those who attended a ball were attacked by diarrhoea and vomiting within 48 hours. The suspected source was the cold buffet, which had provided a wide choice including chicken, pate, mayonnaise, rice, and raw oysters.

Investigations: Food histories were taken from over one hundred persons who attended the function. Association between illness and consumption of particular foods was significiant only for oysters: 35 of the 37 who had eaten oysters were ill, whereas only 5 of 81 who had not eaten them were ill. Inspection showed a high standard of hygiene and cleanliness, but the oysters had been received frozen from abroad, then thawed and rinsed in chilled water before serving in dishes on ice. It was not understood that the foreign-language instructions were to 'cook at high temperature' 'after thawing', intending that the oysters should not be served uncooked.

Microbiological examinations showed some contamination of many foodstuffs with *E. coli*, suggesting cross-contamination between foods. Faecal samples were negative bacteriologically. Electron microscopy showed some 'small round structured viruses' in two oysters and four stool specimens.

Action: Caterers were reminded of food hygiene regulations and made aware of the need to cook raw oysters, and attention was drawn to the need for instructions on imported foods to be labelled in a language intelligible to the caterers.

Case 7: An outbreak of legionellosis

The problem: Several patients undergoing cardiac investigations in the same ward in hospital developed lobar pneumonia. The duration of their stay in hospital suggested that the illnesses had been acquired in hospital. Legionnaires' disease was diagnosed.

Investigation: Water in the hospital's air cooling system was found to contain *Legionella pneumophila* of the same serotype as caused the illnesses. The ventilation system in affected wards had become contaminated. Other patients developed the disease over the next few days.

Action and outcome: Early consideration of the diagnosis allowed appropriate treatment. Despite this, several patients died. The ward was temporarily closed and the incriminated equipment sterilized. New routines with more frequent checks on the cooling plant of the ventilation system were established, including regular cleaning, the use of biocides and bacteriological checks. Because similar outbreaks have been recognized in other hospitals, legislation has been introduced to regulate maintenance of cooling and ventilation systems in Health Service premises.

Case 8: Hepatitis B infection in a long-stay hospital

The problem: Two cases of acute hepatitis B involved staff members of a hospital for the mentally handicapped. No drug abuse or sexual association was determined.

Investigation: Selected screening showed a higher proportion than expected of patients and staff with antibodies to hepatitis B. Two 'high risk' carriers (*e*-antigen positive) were identified among patients known at times to scratch and bite attendants.

Action and outcome: Awareness of the likely source of the infections allowed special precautions to be taken while caring for those patients involved. Vaccination was arranged for those who did not have antibodies and were likely to be exposed in the future. An education campaign at the hospital on the natural history and modes of transmission of the disease did much to relieve staff anxieties.

CROSS-INFECTION IN HOSPITAL

Sometimes formal isolation in hospital of those with infections is necessary for the protection of the general public. More often infectious patients are admitted for their own special medical or nursing needs and the risks are then to hospital staff and other patients. The hospital environment itself predisposes to hospital-acquired (nosocomial) infection which is an increasing problem in many centres. The presence of numerous staff, immunocompromised patients, and invasive equipment that easily becomes infected (Fig. 17.6), provide pathogenic micro-organisms with unusual opportunities to spread and cause severe problems. The use of antibiotics may encourage resistant organisms to appear, and spread to vulnerable patients if control of infection techniques are poor, as shown by the recent emergence of methicillin-resistant staphylococci capable of causing deep seated infections such as septicaemia and endocarditis. Avoiding serious cross-infection may require not only special awareness and expertise but also a high proportion of cubicle accommodation.

Prevention of cross-infection

Procedures

With infectious patients, the necessary procedures vary with the mode of transmission of the particular infection. Contaminated hands are by far the

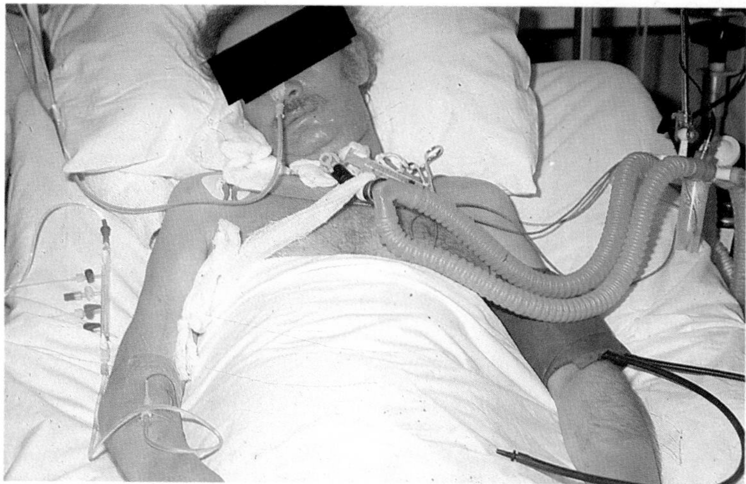

Fig. 17.6 Invasive hospital procedures provide sites of entry for 'opportunistic' infection.

commonest vehicle of spread. Special care with excreta is required, avoiding unnecessary close exposure, contamination of clothing, or spread from infected surfaces. Visiting may be restricted, gowns worn, and special care taken with washing before leaving the patient (Fig. 17.7). When relatives are admitted to accompany children, cross-infection may take place in shared bathrooms and toilets. Children engender special risks during diarrhoeal illnesses especially if toileting and nappy changing are done by staff who subsequently give feeds.

Some infections of skin may spread by transfer of dust from cubicle to cubicle, for example coccal infections; cubicles should be cleaned by methods that do not stir up dust, such as vacuum cleaning or 'damping down' before sweeping or dusting. After discharging a patient, time is necessary to clean thoroughly cubicles and reusable equipment such as mattresses, pillows, and toys, prior to a new admission. Wooden spatulas should be broken to prevent re-use. Instruments such as stethoscopes or thermometers may be reserved for the use of one patient or otherwise disinfected between patients. Patients suffering from the same infection can usually be nursed together. This may be useful; for example, during a whooping cough epidemic when illness may be prolonged and continuous observation then requires fewer staff. However, some infections are not confirmed microbiologically until late in the illness or after discharge.

Specimens

Collection and transport of specimens for laboratory investigations also require care related to the nature of the risk. Infectious urine, sputum, and stool samples

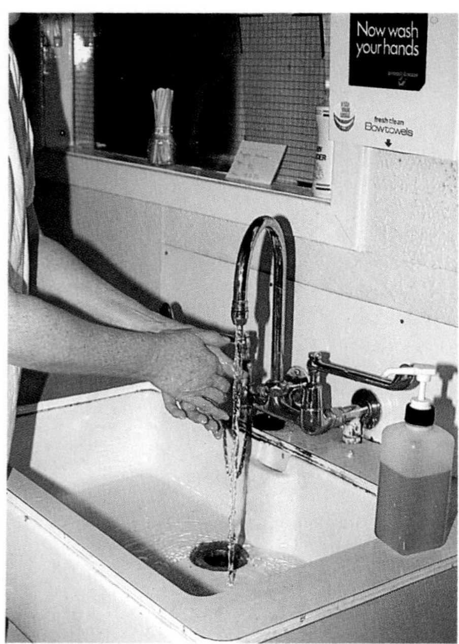

Fig. 17.7 Washing hands is most important in preventing cross-infection.

must be taken carefully and hygienically. Containers for transporting specimens must close securely, avoiding breakable glass where possible. Sometimes before submitting specimens it is necessary to inform the laboratory that they carry a particular risk (for example, suspected viral haemorrhagic fever, HIV infection or hepatitis B) (Fig. 17.8). However, it is more rational to adopt precautions for all specimens since asymptomatic carriers cannot reliably be recognized.

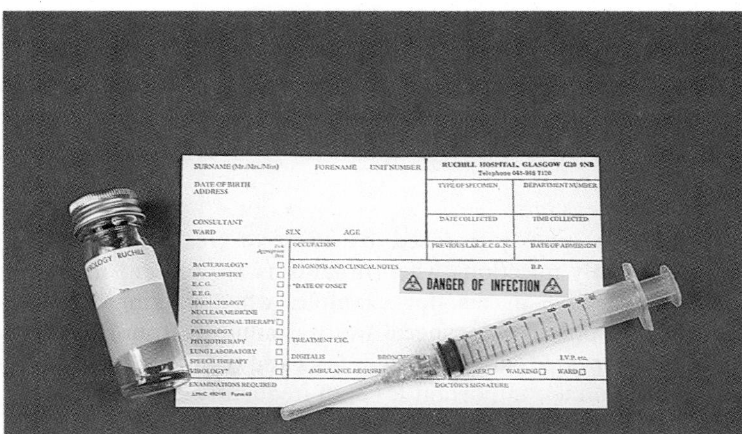

Fig. 17.8 Infectious specimens must be labelled carefully to warn the laboratory.

Staff

When staff are not constantly dealing with infection, the chance of cross-infection may be greater because of lack of awareness of risks, necessary procedures or available facilities. Specially trained staff (for example, infection-control nursing officers) can monitor infection daily in general wards with special attention to new admissions and vulnerable patients. They can scrutinize microbiological laboratory reports and be available to advise and discuss transfer to isolation facilities whenever appropriate. Cross-infection is an important form of iatrogenic disease. Hospital staff when unwell are often highly conscientious and hesitant to take time off work, perhaps because of staff shortages. Respiratory infections in particular can be spread in this way and may be contracted from patients. Occupational health staff can encourage sensible practices such as exclusion of staff from wards while likely to be infectious.

Some particular problems in hospital

Hepatitis B and human immunodeficiency virus (HIV) infection

Precautions with these infections are similar because both are spread primarily by way of blood, blood products, or sexual contact. Shaking hands, sharing bathrooms and eating facilities are unlikely to spread infection unless the infected person is bleeding or has open wounds or sores. Toothbrushes and shavers should not be shared since they may be contaminated with blood. When blood samples are taken or intravenous cannulae used, care must be taken to avoid spillage of blood and accidental trauma to staff (Fig. 17.9). Masks and goggles worn when splashing is anticipated may prevent contamination of mucosal surfaces. Careful disposal of needles and other contaminated materials is essential.

For hepatitis B, active immunization is available for those who may be unavoidably exposed to blood such as casualty staff. When accidental contamination of skin breaks or mucous membranes occurs to those non-immune, high-titre hepatitis B immunoglobulin helps prevent or attenuate infection.

Tuberculosis

Infectivity may be quickly reduced by effective treatment, but until then tuberculosis is highly infectious when organisms are present in sputum. Staff should wear protective masks, for example, when making beds or taking specimens such as sputum or laryngeal swabs. Immunity from BCG administration can decline: repeated tuberculin skin tests and regular chest radiography to detect early infection are important for staff frequently exposed.

Pregnancy

If possible, patients and staff who are pregnant should not be exposed to highly infectious patients. Women who may become pregnant and are likely to come into contact with rubella (for example, doctors or children's ward nurses) should determine their immune status and if necessary be immunized. Chickenpox can

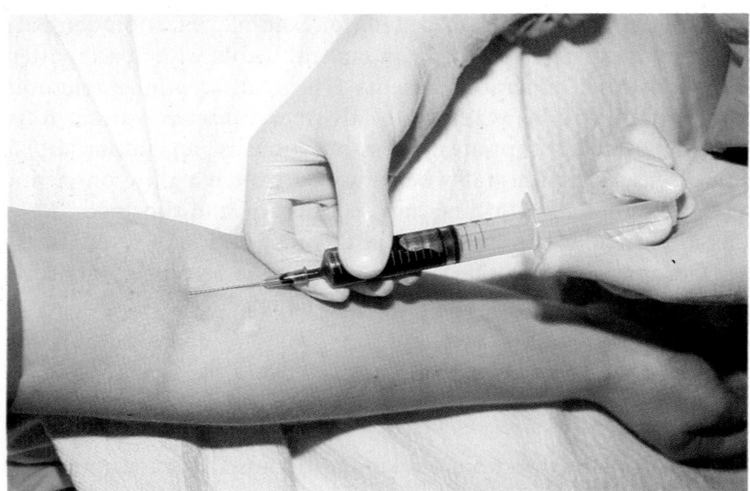

Fig. 17.9 Sample blood with care to avoid contaminating the operator.

be especially serious in pregnancy and non-immune pregnant staff should not care for patients with chickenpox or shingles. If this exposure inadvertantly occurs specific hyperimmune immunoglobulin should be administered. HIV infection itself should not be a special risk but close contact with AIDS patients should be avoided if there is risk of exposure to possible opportunist infections such as tuberculosis or cytomegalovirus.

Debilitated and immunocompromised patients

Debilitated patients are very vulnerable to infection and exposure to infections such as influenza (which may have caused a neighbour's relapse of bronchitis), shingles (varicella-zoster virus in the immunocompromised can cause fulminating varicella pneumonia), or unexplained fever (perhaps shedding pathogens such as staphylococci or measles virus) should be avoided. 'Reverse barrier' nursing may be required for those with severe immunocompromise (for example, transplant patients) to protect them from environmental organisms.

Burns

Morbidity and mortality following severe burns are closely linked to infection through the damaged body surfaces. As with the immunocompromised, these patients may have to be isolated for their *own* protection. Various special facilities are available including laminar flow ventilation units which direct clean-air currents on to and contaminated air away from the patient.

Intensive care units

When establishing these units, attention may be given to design features for positioning equipment and access to the patient at the expense of basic infection control measures such as facilities for washing hands and disposing of contaminated materials, or separation in cubicles. Many such units contain several beds positioned close to one another, with many long intravenous or arterial lines *in situ* and equipment for assisted respiration. Patients often have recent surgical or traumatic wounds and may be on broad spectrum antibiotics. Thus cross-infection of wounds, septicaemias, and respiratory infections with unusual organisms such as pseudomonas with multiple antibiotic resistance are common. Serious problems can arise if no isolation cubicles are available; for example, if a case of diphtheria or poliomyelitis requires intensive care or, more often, a child with severe bronchiolitis due to respiratory syncytial virus may spread this to postoperative or other patients already debilitated.

18

Specific drug treatment of pathogens

Identification of a potentially pathogenic agent in a patient does not automatically mean that specific antimicrobial treatment, if available, should be given. First, the organism may not be causing the illness. Second, even if it is considered to be doing so and if specific treatment is available, this therapy may not be necessary or even helpful: for instance, detection of *Salmonella*, *Shigella*, or *Campylobacter* in mild diarrhoea does not usually warrant antibiotic therapy since clinical resolution is not helped and adverse effects may develop such as prolonged carriage of salmonellas. In some infections the agent may become latent (for instance, herpes simplex) or encyst (for example, toxoplasma), then being unaffected by drugs.

Even when specific therapy is justified, other therapeutic measures must be considered. Surgical intervention may be required—for example, for appendicitis or localized abscess; hyperbaric oxygen therapy should be considered for severe anaerobic tissue infections; antitoxins are required for diphtheria and tetanus and may have an increasing role against endotoxin and other mediators of the inflammatory response in severe Gram-negative infections. Immunomodulating drugs to stimulate the body's own defences are already available and under further development.

The most suitable specific treatments for particular infections are indicated in their own sections of this book. Doses are not stated and those of the *British National Formulary* are accepted except where alternative regimes are specified. Dosages should be checked before prescribing, particularly for infants and children where calculations may be based on age, weight or ideally body-surface area.

The World Health Organization has published a list of 'essential drugs' to aid those countries whose resources are more limited to devise the most effective rational formularies for their needs. Brief details of the listed anti-infective drugs are given in the second section of this chapter.

GENERAL PRINCIPLES

Mode of action

Antimicrobial drugs must have preferential toxicity for the infecting agent rather than the human host. The main mechanisms of attack are:

(i) Against bacterial cell wall synthesis.

(ii) Against protein synthesis, by attaching to ribosomes which are sufficiently distinct from human ribosomes.

(iii) Against nucleic acid synthesis, either by blocking important enzymes or by causing nonfunctional analogues to be formed.

These mechanisms may kill the bacterial cell or cause sufficient damage or stasis to allow the host's defences to handle the infection. The choice between bactericidal and bacteriostatic drugs probably matters only in conditions like infective endocarditis where every organism must be killed because they are less accessible to the body's macrophages. Antimicrobial defence systems are likely to be deficient in immunocompromised patients in whom bactericidal drugs are preferred for severe bacterial sepsis. Some drugs, like gentamicin, have a broad spectrum of activity against many pathogens. Although these are often required because of imprecise initial clinical diagnosis, when the infecting agent is known one should select drugs with the narrowest spectrum to have least effect on commensal flora.

Dosage

A drug must reach its site of action in sufficient quantity—properties such as stability to gastric acid, intestinal absorption, lipid solubility, further metabolism, route and rate of excretion determine the dosage and route of administration. Kidney and liver function are particularly important as they metabolize or excrete most drugs. Toxic levels more readily build up when the function of these organs is impaired; blood levels then should be monitored and smaller or less frequent doses given. Occasionally the converse applies, as in pregnancy, when increased renal excretion and dilution by fluid retention can give ineffective drug concentrations. Levels should be monitored in pregnancy and when treating severe sepsis.

For intravenous administration, repeated single doses (boluses) are preferable to continuous administration. Boluses may not maintain inhibitory levels continuously, but often cause less toxic effects—for instance, the lowest concentration level ('trough') of aminoglycosides measured immediately before another bolus correlates best with toxicity.

Duration

A drug should be given long enough to settle acute symptoms and minimize subsequent relapse. This can vary from single-dose regimes for some urinary tract infections, gonorrhoea, or schistosomiasis, to lifelong treatment for lepromatous leprosy. Guidance is given in the sections in the book about each infection. However, in many infections, such as wound sepsis, medication is stopped somewhat arbitrarily four or five days after fever has settled.

Routes of administration

Severe infections are usually treated by parenteral rather than oral administration to ensure that higher tissue levels of antibiotic are reached rapidly, to avoid possible destruction of drug by gastric acid, or loss by erratic absorption, particularly with vomiting or diarrhoea. Although intramuscular preparations may satisfy these requirements, even 1 ml volumes of non-isotonic preparations may be painful—intravenous administration is then preferable for frequent dosage. However, single daily doses—for example, procaine penicillin for syphilis—are usually given intramuscularly. As a patient improves it is often safe to change to a similar drug orally if there is no gastrointestinal upset.

For less severe infections, drugs may be given orally if suitable preparations are available, perhaps with a single parenteral dose first. For instance, patients with streptococcal tonsillitis could receive one intramuscular dose of a mixture of long-acting penicillins (for example, Triplopen) to give high initial blood levels, followed by a course of oral phenoxymethylpenicillin.

Meningitis due to unusual organisms such as fungi in the immunosuppressed or neonates may sometimes respond only to drugs which do not readily reach the CSF from the blood, thus requiring intrathecal or intraventricular administration. Fortunately, drugs that cross inflamed meninges are available for the common types of bacterial meningitis.

Other methods of administration include inhalation, pessary, suppository, or topical application. Suppositories are particularly being developed for use in children. Agents applied topically to skin are more likely to induce allergy than when taken orally and their use should be limited. Only impetigo is commonly treated by topical antibiotics.

Disinfectants can be used for superficial cleansing; for example, of varicose ulcers.

Anti-infective agents are considered below in families with similar properties.

Toxicity

The safer drugs attack part of the infecting agent with no human counterpart, for example the cell wall of bacteria. Penicillins in general inhibit cell wall synthesis by growing bacterial cells, leading to lysis. They have few side-effects and generally a wide therapeutic ratio. Other effective drugs like aminoglycosides that bind irreversibly to bacterial ribosomes can also to a lesser extent do so to human ribosomes, and thus have more toxic side-effects and a narrow therapeutic ratio. Both the mode of action of a drug and its concentration in tissues (reflecting dosage, solubility, diffusion, and excretion) are important for estimating the likelihood of toxic side-effects. Drugs with narrow therapeutic ratios may need regular monitoring of serum levels and haematological and biochemical indices. Other side-effects may result from idiosyncratic reactions like aplastic anaemia following chloramphenicol, or from allergic reactions. Allergies commonly present as rashes (Figs. 18.1–18.5), less commonly as serum sickness or anaphylaxis. Drugs given to the mother are usually transferred during pregnancy and breast-feeding to the fetus or infant on whose relatively immature system possible effects must be considered. As in all therapeutic decisions, the potential benefits of any treatment must be balanced against its risks.

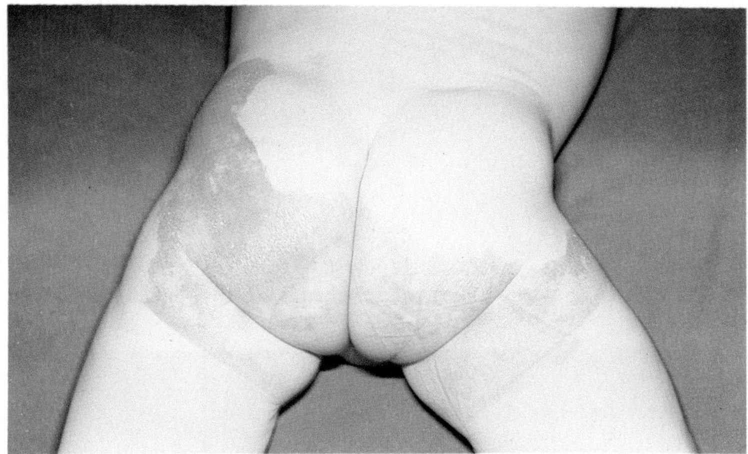

Fig. 18.1 Local distribution of sensitivity rash in napkin area after renal excretion of drug.

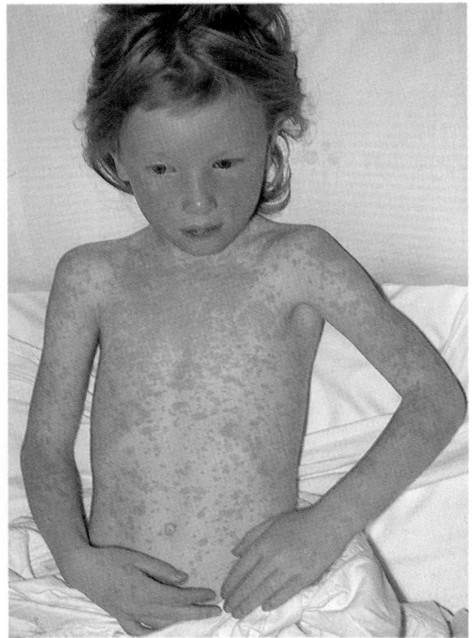

Fig. 18.2 The common ampicillin hypersensitivity rash is often like that of measles, but lack of prodrome and of progression down the body readily distinguish it.

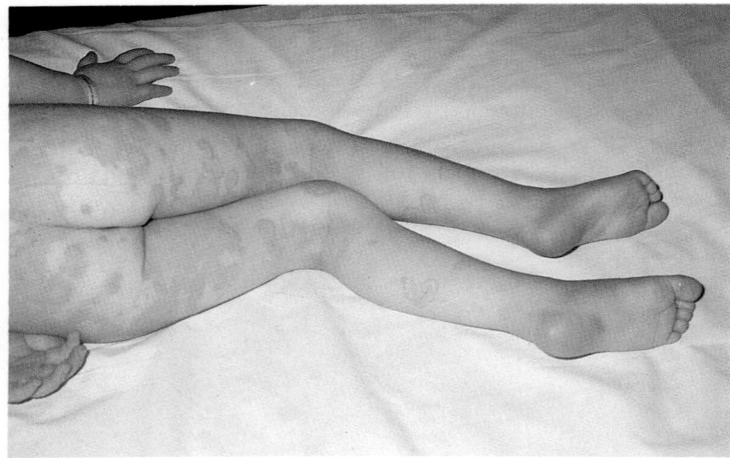

Fig. 18.3 Urticaria after oral phenoxymethyl penicillin.

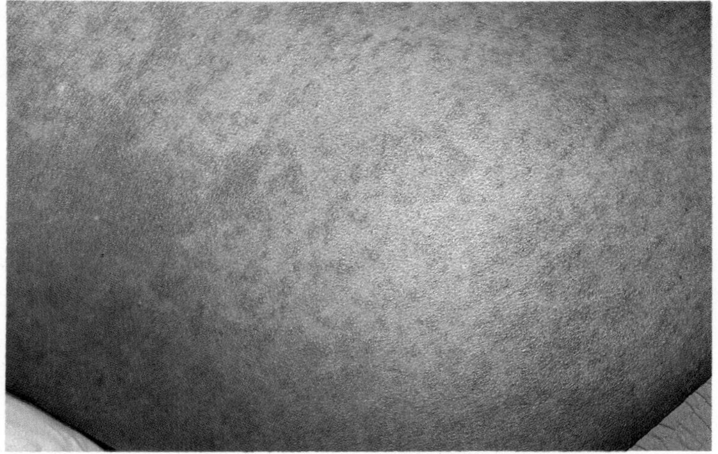

Fig. 18.4 Maculopapular erythematous rash in pigmented skin after erythromycin.

Drug combinations

Combinations of drugs are often used in severe sepsis when the causative agent is unknown and a range of organisms must be covered, or in mixed infection. In treating infections such as leprosy or tuberculosis, combinations reduce the likelihood of resistance developing to any one drug. Some combinations may be synergistic, i.e. the effect of two drugs together may be greater than the sum of the individual components. Synergy usually reflects the action of drugs either at

(a)

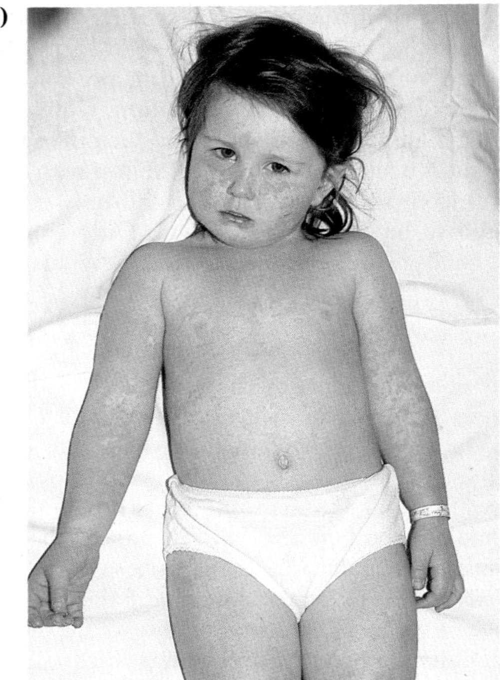

(b)

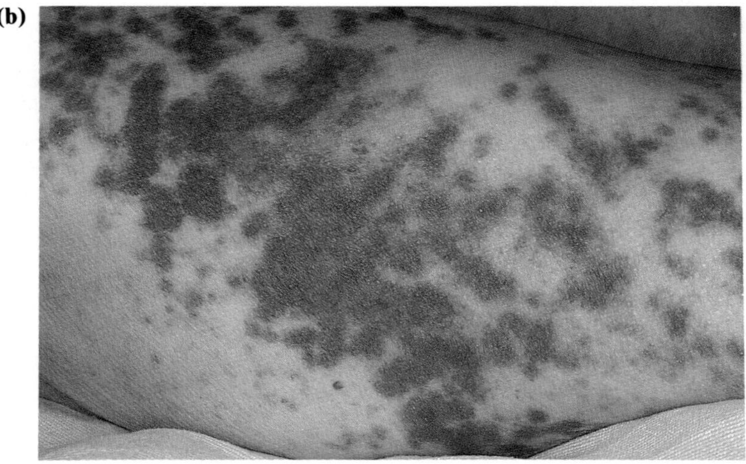

Fig. 18.5 **(a)** Hypersensitivity to co-trimoxazole; drug rashes often have peripheral distribution. **(b)** Less common non-thrombocytopenic purpuric rash after co-trimoxazole.

different sites of the same metabolic pathway—for instance, that of sulphamethoxazole and trimethoprim (co-trimoxazole) on synthesis of folic acid—or there may be enhanced absorption by a pathogen, as when penicillin damage to the cell wall increases aminoglycoside absorption. Conversely, combinations may be antagonistic, as occasionally with fucidin and flucloxacillin in staphylococcal sepsis. A bacteriostatic drug combined with one acting on the cell wall may show antagonism as the latter can kill only growing cells, but the bacteriostatic drug will have inhibited growth. Drug combinations may also lower dosage and hence reduce side-effects from individual drugs.

Resistance

Drugs may have to combat defence mechanisms of the pathogen as they bind to or attempt to enter it to reach their site of action. The commonest such mechanism is production of inactivating enzymes, like penicillinases. Many genes coding for these inactivating or modifying enzymes can be transferred as plasmids (circular strands of extra-chromosomal DNA) between organisms of both the same and different genera. For instance, the plasmid mediating ampicillin resistance in *H. influenzae* is the same as that mediating penicillin resistance in *N. gonorrhoeae*. Other resistance mechanisms are changing receptor sites by the pathogen to prevent uptake or binding of the drug within the cell, or bypassing a metabolic pathway. These often arise through mutations in chromosomal DNA, selection pressure favouring resistant mutants. Resistance to erythromycin and tetracycline can result from development of new membrane transport systems that actively remove the drug from the cell.

Compliance

Studies show that complete compliance by the patient in taking the drug as recommended may be as low as 7 per cent. The many reasons include frequency of administration, duration of treatment, multiple drug therapy, adverse effects, improvement in symptoms, inaccurate dosage, and taste.

Epidemiological feedback

Surveillance systems to record resistance patterns of pathogens at local, national and international levels are increasingly important in refining drug usage as resistant pathogens (usually generated within the hospital environment) become more widespread in the community.

LIST OF ANTIMICROBIALS BASED UPON THE WORLD HEALTH ORGANIZATION'S 'ESSENTIAL DRUG LIST', PLUS ADDITIONAL DRUGS MARKED * MENTIONED IN THIS BOOK

Antibacterials

Penicillins

Mode of action: Modify cell wall synthesis in growing and dividing cells, bactericidal.

Toxic effects: Erythematous maculopapular rashes, diarrhoea, occasionally urticarial rashes and anaphylaxis.

Amoxycillin

Capsule/tablet 250 mg, 500 mg
Powder for oral suspension 125 mg/5 ml

Principal uses: Otitis media, respiratory infections, urinary tract infections, skin infections.

Ampicillin

Powder for injection 500 mg

Principal uses: Parenteral form for severe infection of above or when oral intake impossible.

Benzathine benzylpenicillin

Powder for injection 1.44 g

Principal uses: Long-acting preparation that improves compliance, e.g. against syphilis, prevention of rheumatic fever.

Benzylpenicillin

Powder for injection 600 mg, 3 g

Principal uses: Streptococcal (except faecal streptococci), pneumococcal, anaerobic infections (except *B. fragilis*), meningococcal infection, anthrax, diphtheria, leptospirosis.

Cloxacillin

Capsule 500 mg
Powder for oral solution 125 mg/5 ml
Powder for injection 500 mg

Principal uses: Staphylococcal infections.

Phenoxymethylpenicillin

Tablet 250 mg
Powder for oral suspension 250 mg/5 ml

Principal uses: Similar uses to benzylpenicillin, but less active. Can often substitute for benzylpenicillin when improvement seen. Prophylaxis against rheumatic fever.

Piperacillin

Powder for injection 1 g, 2 g

Principal uses: Increased activity against Gram-negative organisms including pseudomonas, neutropenic patients.

Procaine benzylpenicillin

Powder for injection 1 g, 3 g

Principal uses: Medium duration of action. Improves compliance, e.g. for syphilis.

Other antibacterials

Chloramphenicol

Capsule 250 mg
Oral suspension 150 mg/5 ml
Powder for injection 1 g

Mode of action: Inhibits protein synthesis by binding to ribosomes, bacteriostatic.

Toxic effects: Transient bone marrow suppression in most recipients, rarely aplastic anaemia. Severe neonatal toxicity if used in very high dosage.

Principal uses: Enteric fever, pyogenic meningitis, also for staphylococcal, Gram-negative and anaerobic infections.

Ciprofloxacin

Tablet 250 mg
Infusion (2 mg/ml)

Mode of action: Inhibits DNA gyrase

Toxic effects: Gastrointestinal upset, dizziness, headache, tremor, confusion, rashes, blurred vision, joint pains.

Principal uses: Infections of gastrointestinal tract, bones and joints, enteric fever.

Erythromycin

Tablet/capsule 250 mg
Powder for oral suspension 125 mg
Powder for injection 500 mg

Mode of action: Inhibits protein synthesis, bacteriostatic but slowly bactericidal in higher doses.

Toxic effects: Nausea, vomiting, rashes.

Principal uses: Atypical pneumonias especially mycoplasma and legionellosis; substitute for benzylpenicillin in hypersensitive patients with Gram-positive infection; chlamydiosis, campylobacteriosis, whooping cough, diphtheria carriers.

Gentamicin

Injection 10 mg/2 ml, 40 mg/2 ml

Mode of action: Inhibits protein synthesis by irreversibly binding to ribosomes.

Toxic effects: Nephrotoxicity, ototoxicity; reduced dosage in renal failure; ideally requires monitoring of serum levels as low benefit–risk ratio.

Principal uses: Severe sepsis by gut organisms, including pseudomonas, staphylococcal infection, infective endocarditis (with benzylpenicillin).

Metronidazole

Tablet 200–500 mg
Injection 500 mg/100 ml
Suppository 500 mg, 1 g
Oral suspension 200 mg/5 ml

Mode of action: Interferes with DNA function, bactericidal.

Toxic effects: Gastrointestinal disturbance, rashes, alcohol intolerance.

Principal uses: Pseudomembranous colitis, invasive amoebiasis, giardiasis, trichomoniasis, anaerobic infections.

Spectinomycin

Powder for injection, 2 g

Mode of action: Inhibits protein synthesis, bactericidal.

Toxic effects: Nausea, headache, dizziness, injection site pain.

Principal uses: Gonorrhoea caused by penicillinase-producing gonococci.

Sulphadimidine

Tablet 500 mg
Oral suspension 500 mg/5 ml
Injection 1 g/3 ml

Mode of action: Inhibits folate synthesis: bacteriostatic.

Toxic effects: Nausea, vomiting, rarely blood dyscrasias and crystalluria.

Principal uses: Against sensitive strains of meningococci, and for simple urinary tract infections.

Sulphamethoxazole + trimethoprim (cotrimoxazole)

Tablet
100 mg + 20 mg,
400 mg + 80 mg
Oral suspension
200 mg + 40 mg/5 ml

Mode of action: Sequentially inhibits folate synthesis, bactericidal.

Toxic effects: Rashes, fever, megaloblastic anaemia, crystalluria.

Principal uses: Chronic bronchitis, urinary tract infection, prostatitis, enteric fever, treatment and prophylaxis of pneumocystis.

Tetracycline

Capsule/tablet 250 mg

Mode of action: Inhibits protein, synthesis, bacteriostatic.

Toxic effects: Nausea, diarrhoea, headache, teeth staining up to 8 years of age, can exacerbate renal failure.

Principal uses: Chronic bronchitis, non-specific urethritis, atypical pneumonia (not legionellosis), rickettsial and treponemal infections, brucellosis (+ streptomycin), Lyme disease.

Doxycycline

Capsule/tablet 100 mg
Powder for injection 100 mg

Principal uses: As for tetracycline but daily dose increases compliance.

Nitrofurantoin

Tablet 100 mg

Principal uses: For urinary tract infections only—can cause nausea and peripheral neuritis. Not to be used in renal failure.

Trimethoprim

Tablet 100 mg, 200 mg

Principal uses: As for cotrimoxazole where sulphonamide sensitivity is present.

Antileprosy drugs

Clofazimine

Capsule 50 mg, 100 mg

Toxic effects: Red/brown discoloration of skin, conjuctivae and urine, nausea, abdominal pain, and weight loss.

Principal uses: Leprosy (in combination).

Dapsone — Tablet 50 mg, 100 mg

Toxic effects: Nausea, headache, dizziness, hypersensitivity reaction of fever, rash, eosinophilia, lymphadenopathy, leucopenia, and hepatitis.

Principal uses: Leprosy (in combination), pneumocystis prophylaxis.

Rifampicin — Capsule/tablet 150 mg, 300 mg

See below under antituberculosis drugs.

Antituberculous drugs

All used in combinations to prevent induction of drug resistance.

Ethambutol — Tablet 100–400 mg

Toxic effects: Optic neuritis, red/green colour blindness.

Isoniazid — Tablet 100–300 mg

Toxic effects: Nausea, rashes, hepatitis, peripheral neuritis (give pyridoxine 10 mg daily prophylactically, if malnutrition, alcoholism or diabetes).

Pyrazinamide — Tablet 500 mg

Toxic effects: Nausea, hepatotoxicity, hyperuricaemia.

Rifampicin — Tablet/capsule 150 mg, 300 mg

Toxic effects: Hepatitis, orange-red discoloration of urine, makes oral contraceptive less effective. With intermittent therapy hypersensitivity syndromes, commoner—flu-like, abdominal, respiratory, shock, renal failure, and thrombocytopenic purpura.

Principal uses: Leprosy, tuberculosis and brucellosis (in combination), legionellosis, staphylococcal infections, prophylaxis for meningococcal and *H. influenzae* contacts.

Rifampicin combined with isoniazid — Tablet 150 mg (R) + 100 mg (I), 300 mg (R) + 150 mg (I)

Streptomycin — Powder for injection, 1 g

Toxic effects: Nephrotoxicity, ototoxicity.

Thiacetazone combined with isoniazid — Tablet 50 mg (T) + 100 mg (I), 150 mg (T) + 300 mg (I)

Toxic effects (thiacetazone): Nausea, diarrhoea, vertigo, bone marrow, depression, haemolysis.

Antifungal drugs

Amphotericin B — Powder for injection 50 mg

Toxic effects: Fever, nausea, headache, nephrotoxicity, bone marrow depression.

Principal uses: Broad spectrum activity against systemic fungal infections.

Griseofulvin — Capsule/tablet 125 mg, 250 mg

Toxic effects: Headache, nausea, diarrhoea, tiredness, neurotropenia.

Principal uses: Dermatophyte infections of skin, scalp, hair, and nails.

Ketoconazole — Tablet 200 mg; Oral suspension 100 mg/5 ml

Toxic effect: Hepatoxicity, nausea, rash, pruritis, headache, dizziness.

Principal uses: Broad spectrum of activity but erratically absorbed.

Nystatin — Tablet 500000 i.u.; Pessary 100000 i.u.

Toxic effects: Nausea, diarrhoea.

Principal uses: Gastro-intestinal tract or vaginal candidiasis.

Flucytosine — Capsule 250 mg; Infusion 2.5 g/250 ml

Toxic effects: Nausea, diarrhoea, abdominal distension, bone marrow suppression.

Principal uses: Systemic infections by cryptococci, torulopsis, and candida; often combined with amphotericin B.

Antivirals

Acyclovir*

Tablet 200 mg, 400 mg, 800 mg
Powder for infusion 250 mg, 500 mg
Cream for topical application
Oral suspension 200 mg/5 ml

Mode of action: Antithymidine kinase action.

Toxic effects: Rashes, renal impairment, neurological reactions.

Principal uses: Herpes simplex and varicella-zoster infections.

Zidovudine*

Tablet 100 mg, 250 mg
Syrup 5 mg/ml
Injection 20 mg/ml

Mode of action: Anti-reverse-transcriptase action.

Toxic effects: Anaemia, leucopenia, nausea, headache, abdominal pain, fever, paraesthesia, myalgia, insomnia.

Principal uses: Human immunodeficiency virus infection.

Idoxuridine*

Solution 5%, 40% for topical application

Principal uses: Herpes simplex and zoster skin infections.

Amantidine*

Tablet 100 mg

Toxic effects: Rashes, oedema, visual, nervous system, and gastrointestinal upsets.

Principal uses: Prevention of influenza A in those exposed and at risk of complications.

Anthelminthics

Intestinal anthelminthics

Levamisole

Tablet 50 mg, 150 mg

Mode of action: Paralyses susceptible helminths allowing expulsion.

Toxic effects: Nausea, vomiting, abdominal pain, dizziness, headache. Avoid in advanced liver and kidney disease.

Principal uses: Ascariasis.

Mebendazole

Chewable tablet 100 mg

Mode of action: Irreversibly blocks glucose uptake by susceptible helminths.

Toxic effects: Abdominal pain, diarrhoea.

Principal uses: Ascariasis, enterobiasis, hookworm, trichuriasis.

Niclosamide

Chewable tablet 500 mg

Mode of action: Kills and partially digests worms.

Toxic effects: Nausea, abdominal pain.

Principal uses: Tapeworms, including beef, pork, dwarf, dog, fish.

Piperazine

Tablet 500 mg
Elixir/syrup 500 mg/5 ml

Mode of action: Flaccid muscle paralysis in susceptible worms allowing expulsion.

Toxic effects: Nausea, vomiting, abdominal pain, diarrhoea. Contraindicated in liver or severe kidney disease and epilepsy.

Principal uses: Enterobiasis, ascariasis.

Praziquantel

See under antischistosomals (also used against other flukes, tapeworms).

Pyrantel

Chewable tablet 250 mg
Oral suspension 50 mg/ml

Mode of action: Paralysing susceptible worms allowing expulsion.

Toxic effects: Nausea, anorexia, abdominal pain, diarrhoea.

Principal uses: Ascariasis, enterobiasis, hookworm.

Thiabendazole

Chewable tablet 500 mg
Lotion 500 mg/5 ml

Mode of action: Enzyme inhibitor, interfering with energy cycle.

Toxic effects: Dizziness, drowsiness, anorexia, nausea, vomiting, occasional erythema multiforme.

Principal uses: Visceral and cutaneous larva migrans, guinea worm, strongyloidiasis, invasive stage of trichinosis, adjunct therapy for enterobiasis, ascariasis, trichuriasis, and hookworm.

Albendazole — Chewable tablet 200 mg

Mode of action: Irreversibly blocks glucose uptake by susceptible helminths.

Toxic effects: Abdominal pain, diarrhoea.

Principal uses: Hydatid disease, cysticercosis (also hookworm, ascariasis, trichuriasis, enterobiasis).

Antifilarials

Diethylcarbamazine — Tablet 50 mg

Mode of action: Unknown

Toxic effects: Headache, nausea, precipitates allergic response to released filarial antigens.

Principal uses: Filariasis, including loiasis and onchocerciasis, toxocariasis.

Ivermectin — Tablet 6 mg

Mode of action: Muscular paralysis and subsequent immune-mediated death.

Toxic effects: Allergic reactions less frequent and less severe than with diethylcarbamazine.

Principal uses: Onchocerciasis.

Suramin sodium — Powder for injection 1 g

Mode of action: Unknown

Toxic effects: Immediate nausea, vomiting, shock—give test dose. Later paraesthesiae and nephrotoxicity.

Principal uses: Onchocerciasis, trypanosomiasis

Antischistosomals

Metriphonate — Tablet 100 mg

Mode of action: Anticholinesterase activity.

Toxic effects: Nausea, abdominal pain, diarrhoea, headache, weakness.

Principal uses: Against *Schistosoma haematobium*.

Oxamniquine — Capsule 250 mg; Syrup 250 mg/5 ml

Mode of action: Unknown, male worms more susceptible to death, females prevented from releasing further eggs.

Toxic effects: Dizziness, drowsiness, headache, nausea.

Principal uses: Against *Schistosoma mansoni*.

Praziquantel — Tablet 150 mg, 60 mg

Mode of action: Unknown, but causes strong muscular contraction and vesiculation of skin.

Toxic effects: Headache, dizziness, drowsiness, abdominal discomfort.

Principal uses: Schistosomiasis, other fluke infections, tapeworms.

Antiprotozoal

Diloxanide — Tablet 500 mg

Toxic effects: Flatulence, vomiting, urticaria, pruritis.

Principal uses: Clears amoebic intestinal cyst carriage.

Metronidazole — Tablet 200–500 mg; Injection 500 mg/100 ml; Oral suspension 200 mg/5 ml

Toxic effects: Nausea, rashes, alcohol intolerance.

Principal uses: Invasive amoebiasis, giardiasis, trichomoniasis, anaerobic infections.

Meglumine antimoniate — Injection 30% (=8.5% antimony)

Toxic effects: Anorexia, vomiting, coughing, substernal pain. Infusions must be given slowly and stopped if toxic effects develop.

Principal uses: Visceral, mucocutaneous and extensive or unsightly cutaneous leishmaniasis.

Pentamidine — Powder for injection 200 mg

Toxic effects: Hypotension, hypoglycaemia, nephrotoxicity, hepatotoxicity, leucopenia, thrombocytopenia.

Principal uses: Leishmaniasis, trypanosomiasis, pneumocystis pneumonia and prophylaxis.

Chloroquine — Tablet 150 mg; Syrup 50 mg/5 ml

Toxic effects: Nausea, headache, pruritis, blurred vision, irreversible retinopathy (usually after prolonged high dosage).

Principal uses: Treatment and prophylaxis of vivax, ovale, malariae, and chloroquine-sensitive falciparum malaria.

Primaquine — Tablet 7.5 mg, 15 mg

Toxic effects: Nausea, abdominal pain, haemolysis in G6PD deficiency.

Principal uses: Clearing liver hypnozoites of vivax and ovale malaria.

Quinine — Tablet 300 mg; Injection 300 mg/2 ml

Toxic effects: Tinnitus, nausea, vomiting, vertigo, pruritis, rashes, cardiac conduction defects.

Principal uses: Severe or chloroquine-resistant falciparum malaria.

Mefloquine — Tablet 250 mg

Toxic effects: Nausea, diarrhoea, abdominal pain, dizziness, loss of balance, neuropsychiatric disturbances, bradycardia.

Principal uses: Treatment and prophylaxis of chloroquine-resistant falciparum malaria.

Sulfadoxine + pyrimethamine (Fansidar) — Tablet 500 mg + 25 mg

Toxic effects: Rashes, bone marrow suppression.

Principal uses: After treatment with quinine in severe or chloroquine-resistant falciparum malaria.

Tetracycline — Capsule/tablet 250 mg

Toxic effects: Nausea, diarrhoea, headache, teeth staining up to 8 years of age, exacerbates renal failure.

Principal uses: Treatment with quinine in severe chloroquine-resistant falciparum malaria.

Proguanil — Tablet 100 mg

Toxic effects: Mouth ulcers, nausea, hair loss, skin rashes.

Principal uses: For prophylaxis together with chloroquine in chloroquine-resistant falciparum malaria endemic areas.

Antitrypanosomals

Melarsoprol — Injection 3.6% solution

Toxic effects: Reactive encephalopathy, vomiting, abdominal colic, peripheral neuropathy, albuminuria, diarrhoea, exfoliative dermatitis.

Principal uses: Late-stage African trypanosomiasis.

Pentamidine — Powder for injection 200 mg

Toxic effects: Hypotension, hypoglycaemia, nephrotoxicity, hepatotoxicity, leucopenia, thrombocytopenia.

Principal uses: Early stage *T. b. gambiense* infection, also leishmaniasis, pneumocystis, pneumonia.

Suramin sodium — Powder for injection 1 g

Toxic effects: Immediate nausea, vomiting, shock—give test dose. Later paraesthesia and nephrotoxicity.

Principal uses: Early stage *T. b. rhodesiense* infection also onchocerciasis.

Benznidazole — Tablet 100 mg

Toxic effects: Nausea, abdominal pain, peripheral neuropathy, rashes.

Principal uses: Acute stage *T. cruzi* infection.

Nifurtimox — Tablet 30 mg, 120 mg, 250 mg

Toxic effects: Anorexia, nausea, weight loss, abdominal pain, peripheral neuropathy.

Principal uses: Acute stage *T. cruzi* infection.

19

Infection and the laboratories

INTRODUCTION

Laboratories are used in the diagnosis, management, and screening of patients with diseases of infection. These functions are best performed in close co-operation with clinicians. Ideally, the clinician should consider the points depicted in the flow chart (Fig. 19.1) before requesting a laboratory test. In practice, the workload in laboratories and the number of tests per patient has been increasing, largely because of repeat tests, more than three-quarters of which provide no new, or different, information.

The range of tests and methods used often vary between different hospitals. These variations are usually minor but can affect the type of sample required—thus, one's local laboratory should be visited to establish what is available. Such visits are appreciated by laboratory staff and allow clinicians to be given a better service. Also, if there are queries about interpretation of results there should be no hesitation in contacting the laboratory.

The clinician should also be aware of the normal working hours of the laboratory. Samples sent outside these hours are considerably more expensive to process and, unless the test result affects immediate management, may not be tested until next day. Although time-consuming, specimens must be labelled clearly and request forms completed adequately. If different names, however minor, appear on the specimen and request form, another sample will be required. If no clinical details are recorded on the form, the sample may not be processed or inappropriate tests may be performed. Good service is only obtained when care is taken to ensure that the laboratory has all information necessary for optimal selection of tests and helpful, interpretative reporting.

Samples that may cause serious risk of infection to laboratory staff must be clearly labelled. For instance, where hepatitis B or HIV is suspected, samples should be labelled 'risk of infection' and placed in a separate, sealed polythene bag. The request form should be placed in a separate pocket outside the bag so that it does not become contaminated. In many laboratories, only selected investigations are performed on high-risk specimens. Greater awareness of dangers has reduced laboratory-acquired infection. In hospitals where patients with viral haemorrhagic fevers may be admitted, a more detailed system of grading the risk of infection is appropriate so that, depending on risk, various

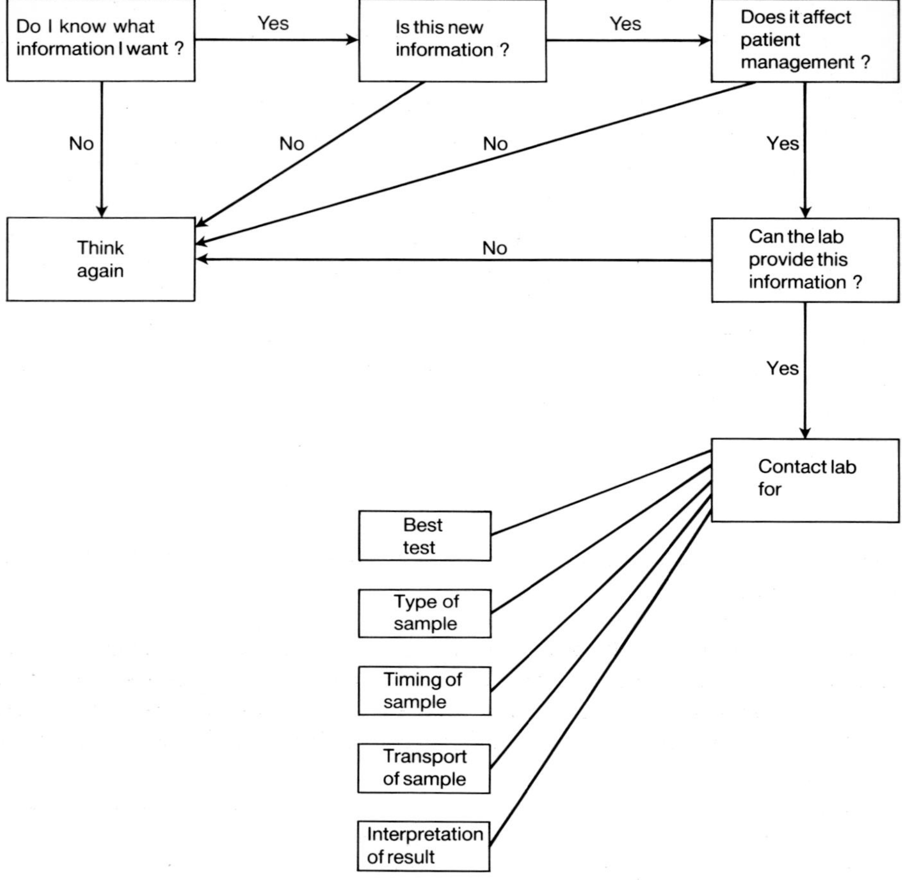

Fig. 19.1 Requesting a laboratory test.

types of investigation may or may not be carried out perhaps under high-security conditions.

This chapter considers the use of the microbiological, virological, and haematological laboratories with regard to infection. Collection of specimens is described. Time is wasted when incorrect samples are sent to the laboratory and it may then be necessary for additional samples to be collected. The routine and special tests offered by each department and their interpretation are discussed.

MICROBIOLOGY

The principal methods of detecting an infective pathogenic agent are as follows:

(i) *Visualize* the organism or its characteristic structures, usually by light microscopy. This method allows rapid diagnosis within a few minutes or hours, whereas growth of *Mycobacterium tuberculosis*, for example, may take six weeks.

(ii) *Detect* its specific components, normally antigen but sometimes part of its nucleic acid. This is especially useful where antibiotics have been given; for example, finding meningococcal or pneumococcal antigen in CSF.

(iii) *Try to culture the agent*. Samples must be taken before starting antibiotics, avoiding contamination, and transported promptly to the laboratory. Results may be available within 24 hours, but much longer if the inoculum is small or the organism slow-growing.

(iv) *Test for specific (immunological) host reactions*, mainly serological (IgM, IgG, rising or high static antibody titres) or as cell-mediated immunity (for example, skin tests).

Samples

Swabs

The various types and uses of swabs are shown in Table 19.1, but there are special considerations for sampling specific sites.

Swabs from lesions. Cotton-wool swabs are often used but only absorb a small quantity of material. For dry lesions, moisten the swab with sterile saline before use. Many laboratories use 'albumin swabs' which enhance survival of bacteria and can also be used for virological specimens. If the swab cannot reach the

Table 19.1. *Types of swabs and their applications*

Swab type	Application
Albumin/plain cotton wool	Lesion Nose Throat Ear Eye Vagina Cervix Urethra Rectum
Calcium alginate on flexible wire	Nasopharynx
Charcoal	Genital or pharyngeal, for *Neisseria*
Small-headed	Urethra Infants

laboratory within four hours, send it in transport medium. Swabs are less satisfactory than collections of pus, scrapings from lesions, or biopsy specimens.

Nasal swabs. Moisten a sterile swab with sterile saline and place it in the anterior nares. Rotate the swab to obtain as much material as possible. Such specimens are better than throat swabs for identifying *Staphylococcus aureus* carriers.

Throat swabs. Swab the pharynx and tonsils, avoiding the buccal mucosa, tongue, and uvula. Throat swabs are useful for isolating streptococci, *Corynebacterium diphtheriae*, *Candida albicans*, and *Neisseria gonorrhoeae* from cases of pharyngitis and from carriers of these organisms or *Neisseria meningitidis*, and enable Vincent's angina organisms to be visualized by microscopy.

Nasopharyngeal swabs. Insert a calcium alginate swab on a flexible wire into the nose at right angles to the face (Fig. 19.2). Keep the tip of the swab in touch with the bottom of the nose cavity and push it gently backwards. When the nasopharynx is reached, rotate the swab and remove it. This is useful in isolating *Bordetella pertussis* and identifying carriers of group A streptococci, *N. meningitidis*, *H. influenzae* type B, and *C. dipththeriae*.

Ear swabs. First clean the auditory meatus to reduce contamination. Sterile saline may be required to remove wax and debris. Take care not to perforate the tympanic membrane.

Eye swabs. With a moistened swab clean the outer lid and eye lashes to remove debris. Evert the eyelid and gently roll a sterile swab along the inside of the eyelid. In eye diseases other than conjunctivitis an ophthalmologist may be required to obtain samples, e.g. scrapings of conjunctiva.

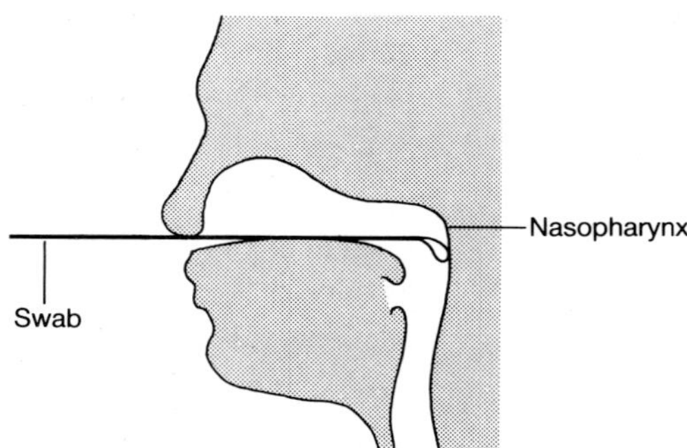

Fig. 19.2 Position of nasopharyngeal swab.

Vaginal and cervical swabs. Since bacterial flora in these two areas may be different, care must be taken to sample the correct site. With vaginal swabs avoid contamination from the labia. Use a bivalve vaginal speculum to expose the cervix. Clear the cervix of mucus before taking samples. Charcoal swabs of cervix, not vaginal swabs, are used to isolate *N. gonorrhoeae.*

Urethral swabs. Slight milking of the penis can produce pus or urethral discharge to be collected on a swab. If not, place a moistened small-headed swab in the urethral canal and rotate it.

Faeces and rectal swabs. Faecal samples placed in a sterile container are superior to rectal swabs which are useless unless faecally stained.

Sputum

The best samples are collected on wakening, after a deep cough or after physiotherapy. Sputum is easily contaminated with oropharyngeal flora making interpretation of culture results difficult. Thus, samples that contain too much saliva (numerous squamous epithelial cells and few leucocytes on microscopy) may be discarded. In pneumonic illness with dry cough, sputum may be induced by nebulized hypertonic saline. Difficulties with contamination sometimes necessitate other methods such as endotracheal suction, transtracheal aspiration, or bronchial lavage. Culture of sputum is important in the investigation of pneumonia and tuberculosis; examination of a Gram-stained smear may identify bacteria or fungi.

Urine

Unless urine is collected properly, the result may be meaningless. Suitable samples may be obtained in several ways:

Midstream specimen of urine. A female patient should first clean the external labia with soap and water. Then, with separated labia, she should void a small amount of urine in the toilet; next, continue micturition into a specimen container which may contain boric acid if delay is likely before examination; finally, any remaining urine is voided. A male should retract the foreskin and clean the glans penis. As with the female, urine is initially voided into the toilet and a midstream sample collected.

Collection bag. For babies and infants, place a sterile self-adhesive plastic bag over the cleansed genitals. Transfer collected urine into a sterile plastic container. Often, during examination, the baby may start to urinate and the opportunity can be taken to obtain a 'clean catch' sample which is preferable to a bag specimen.

Suprapubic aspiration

This procedure is mainly used in babies and infants whose collection bag specimens give equivocal results. The procedure is as follows:

1. Put the patient flat on his back. Palpate and percuss the full bladder. If the bladder is not full, the patient can be given a drink or diuretic.
2. Clean skin in the midline of the anterior surface of the abdomen 2 cm above the pubic symphysis with isopropyl alcohol (Fig. 19.3).
3. Insert a 21G needle on a 20 ml syringe as in Fig. 19.3 and aspirate the urine. Remove the needle, compressing the aspiration site with a swab. Put a plaster over the site. Local anaesthetic is usually unnecessary; the procedure is only slightly more painful then venepuncture.

Ideally, urine should be examined microscopically (see below) before being sent to the laboratory. Urine should reach the laboratory within 30 min of collection for best results. With longer periods bacterial multiplication and lysis of leucocytes may give false results. If a delay of 2 hours or more is likely, refrigerate the sample at 4 °C, use a commercial dip inoculation slide or add a bacterial preservative such as a weak solution of boric acid to a final concentration of 1.8 per cent.

Blood cultures

Three separately collected sets of blood culture bottles in one day are usually sufficient. If the need for treatment is urgent, take three sets at the same time, since

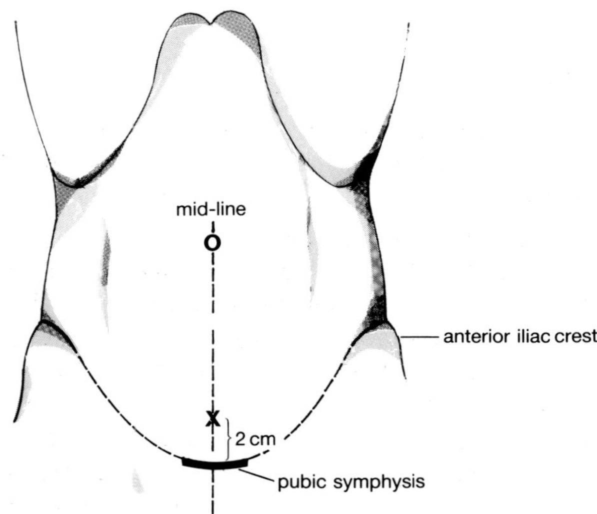

Fig. 19.3 Landmarks for suprapubic aspiration.

the organism is more likely to grow from a greater volume of blood. The procedure for blood culture is as follows:

1. Inspect blood culture bottles for contamination—fluid should be clear. Remove protective covering and clean exposed diaphragm with alcohol.
2. Wash the hands. Gloves should be considered for the inexperienced operator. Gowns or face masks are not necessary unless splashing of blood is possible and infection with hepatitis B, HIV, or viral haemorrhagic fever is suspected.
3. Palpate suitable vein.
4. Cleanse skin with isopropyl alcohol and allow to dry.
5. Perform venepuncture without further palpation of vein. Remove 10–20 ml of blood from an adult (2–5 ml from infants). Discard needle into a 'sharps' container without resheathing.
6. Distribute sample between 2–3 blood culture bottles, using fresh needles for each inoculation. Blood cultures should be inoculated before distributing blood to containers for other investigations.

The laboratory should be told what organisms are suspected to have caused the septicaemia, whether the patient is on antibiotics (as it may be possible to inactivate these drugs) and what antibiotics are likely to be given. Aerobic (one bottle) and anaerobic (one bottle) cultures are usually inoculated for all blood cultures, but special blood culture bottles are required for leptospirosis, brucellosis, and Legionnaires' disease. Semi-automated techniques for detecting bacterial growth are less labour-intensive and allow blood cultures to be more frequently inspected for growth.

Cerebrospinal fluid

Contraindications to lumbar puncture are raised intracranial pressure, bleeding diathesis, and abnormal skin (psoriasis, burns) in the lumbar area. The procedure for lumbar puncture is:

1. Place patient in left lateral position with spine horizontal, lying at the edge of the bed with knees flexed into the abdomen and neck flexed with head supported by pillow. Children are best held in this position by an assistant.
2. Palpate the iliac crests. A line joining these should be vertical (Fig. 19.4) and pass between the third and fourth lumbar vertebrae which is the point of puncture (or one space up or down).
3. Wash hands. Put on gown, gloves, and face mask. Clean skin with chlorhexidine solution or spirit in a wide band between the iliac crests.
4. Infiltrate the skin at proposed point of puncture with local anaesthetic, superficially with 25G needle then with a 21G along the intended track.

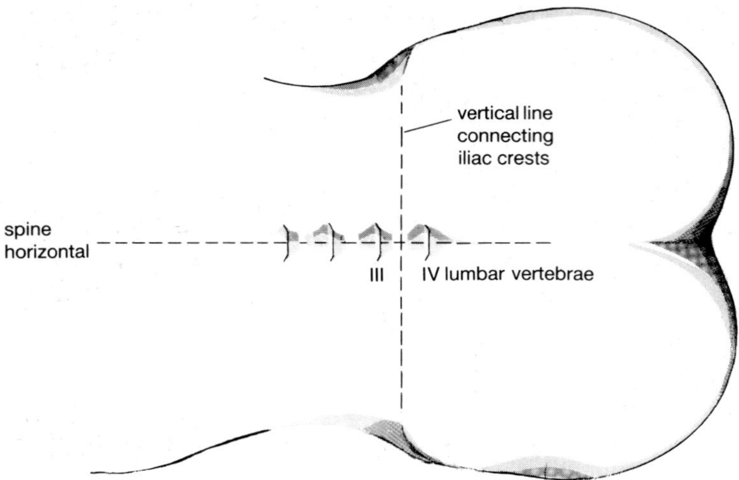

Fig. 19.4 Patient in left lateral position for lumbar puncture.

Anaesthetic is usually omitted in the younger child and infant as its administration is as distressful as the lumbar puncture.

5. Check that the stylet and three-way tap fit into the lumbar puncture needle (Fig. 19.5) and the manometer into the three-way tap (Fig. 19.6). Ensure that sterile containers to collect CSF and for glucose measurement are available.
6. Check that the patient is still in the correct position and the skin anaesthetized. Position sterile lumbar puncture needle and enclosed stylet at the space between third and fourth lumber vertebrae. Press needle steadily inwards, and slightly cranially parallel to the spinous processes. There is a slight 'give' as the needle passes through the dura mater. If inserted too far the needle can pierce the venous plexus on the wall of the vertebral body. This is a common but usually harmless complication with infants. If bone is encountered, withdraw needle to just under the skin, change angle cranially and re-insert. If gap is still not found, try L2/L3.
7. Remove stylet. CSF should come out of the needle; if not, rotate needle. Discard first few drops if blood-stained, to allow fluid to clear.
8. Attach the three-way tap and manometer and measure the pressure. Put fluid from manometer into specimen container. Remove manometer and collect any additional CSF required.
9. Remove three-way tap and replace stylet. Hold skin down around the needle with a sterile gauze and quickly withdraw the needle.
10. Put a plaster over the site of puncture. Lie patient on back for 6–8 hours afterwards to reduce headaches or fainting.

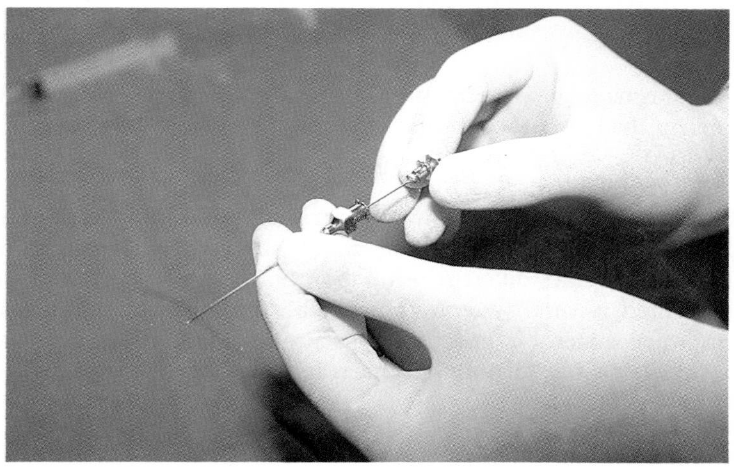

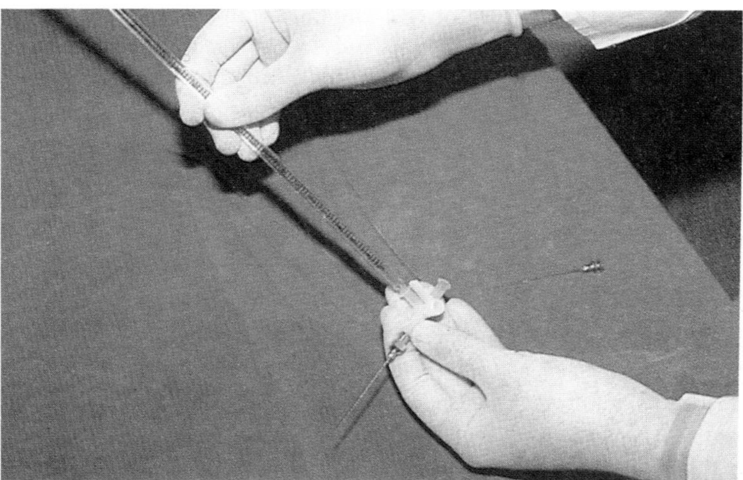

Fig. 19.6 Lumbar puncture needle, without stylet, connected to three-way tap and manometer.

Number sample bottles sequentially if there has been a traumatic tap; later samples are usually clear of blood. Do not use transport medium or refrigerate the specimen. Microscopy, cell count, and culture are performed, usually with simple biochemical tests (see Table 6.3). The chances of a positive result, especially culture, are greatly influenced by the quantity of material. Samples of a few drops may have to be repeated; withdrawal of 2–5 ml does not usually cause the patient greater after-effects.

Special tests

Certain organisms require special procedures, and when they are suspected one should inform the laboratory so that appropriate methods can be used.

Fastidious organisms

The culture media routinely used in a bacteriology laboratory are not ideal for the growth of some organisms. For these, special media should be used (Table 19.2). Sometimes different incubation times, atmosphere, and temperature are required. Several systems are available for collection and transport of specimens when anaerobic infection is suspected. Similarly, a large range of media is available for identifying anaerobes and one example is given in Table 19.2.

Antigen detection

This is particularly useful for rapid diagnosis. The main methods and some uses are: direct fluorescent antibody test (for example, *Treponema pallidum*, *Pneumocystis carinii* in sputum or *Giardia lamblia* in stools); counter-immuno-electrophoresis (for example, to identify polysaccharide antigens such as those of *Strep. pneumoniae*, *N. meningitidis*, and *H. influenzae* type b); latex agglutination

Table 19.2. *Diseases in which special culture media are required*

Disease	Culture media
Yeast and fungus infection	Sabouraud's medium
Gonorrhoea	Thayer–Martin medium: 10 per cent CO_2
Tuberculosis	Löwenstein–Jensen medium and others
Meningococcal infection	Chocolate agar, 10 per cent CO_2
Cholera	Thiosulphate–citrate–bile agar
Campylobacter	*Campylobacter* medium
Brucellosis	Glucose serum agar/broth
Legionnaires' disease	Buffered charcoal yeast extract
Mycoplasma infection	*Mycoplasma* medium
Diphtheria	Tellurite blood agar
Pertussis	Charcoal agar
Leptospirosis	Johnson and Harris's modification of Ellinghausen and McCullough's medium
Non-sporing anaerobes	Wilkins–Chalgren anaerobic agar

(for example, for bacterial or cryptococcal antigens in the CSF); and enzyme-linked immunoassay, for example of urine for *Legionella pneumophila* antigen.

Serology

IgM and IgG antibodies to an increasing range of microbes can be detected. A specimen of clotted blood (5–10 ml) in a sterile container is adequate, but check first with the laboratory what tests are available there or at reference laboratories.

Skin tests

These can detect the immune response but, apart from the tuberculin test, are rarely used.

Microscopy for parasites, ova, and cysts

Stool specimens should be fresh and warm if looking for amoebic trophozoites. In diseases such as amoebiasis and giardiasis, both cysts and trophozoites are produced; water, urine, and cooling destroy trophozoites. Specimens may have to be examined on the ward. The first few drops of urine are useful for seeking *Trichomonas vaginalis*; the last few drops of a midday specimen are best for detecting *Schistosoma haematobium* ova: avoid delay in these specimens reaching the laboratory. Aspiration of abscesses, usually in liver, can be helpful in detecting amoebae, especially if the abscess wall can also be biopsied.

Other tests

The availability of some tests depends on the special interest of the laboratory, for example:

(i) *Limulus* lysate test to detect endotoxin.
(ii) CSF lactic acid levels (usually high in bacterial and low in viral meningitis).
(iii) C-reactive protein, a serum protein increased in non-specific response to inflammatory conditions, especially bacterial.

Surveillance (Chapter 17)

Many microbiological laboratories are involved in regular or occasional surveys of food and water and on-going reporting of infections in the population (Chapter 17). Outbreaks of illness such as Legionnaires' disease in a hospital or food poisoning in the community may also have to be investigated.

Interpretation of results

The clinical microbiologist is responsible for helping to interpret the laboratory's results and can be particularly useful in the following circumstances.

Treatment

Report of the isolation of an organism and of its antibiotic sensitivity is not necessarily an indication for treatment (Chapter 18). That depends on whether the isolated organism is believed to have caused the patient's illness. This belief is strengthened if it is always a pathogen, if the site of infection is normally sterile, if the organism is not a common contaminant, or if there is an unusual preponderance of one of the normal bacterial flora.

Bacterial sensitivity

Susceptibility of bacteria to antibiotics is commonly assessed by a disc-diffusion test, although tests incorporating antibiotic in 50 ml culture medium ('break-point' methods) are coming into use. Bacteria are reported as 'sensitive', 'resistant', or 'moderately sensitive'. Limitations of *in vitro* tests should be borne in mind: most test only bacteriostatic activity, and may not reflect *in vivo* activity. The results of susceptibility tests are not the only guide to choosing an appropriate antibiotic. Other important considerations are whether the drug is bactericidal or bacteriostatic, its toxicity, the integrity of the patient's immune system, the site of infection, and the ability of the antibiotic to penetrate such a site. Drug combinations may be antagonistic or synergistic and in certain conditions, such as bacterial endocarditis, this may have to be determined by *in vitro* tests. Also some hospitals have antibiotic policy recommendations for specific pathogens. Where the disc-diffusion test does not give sufficient information (for example, treatment of infective endocarditis), the minimal bactericidal concentration of antibiotic or combination of antibiotics for the particular pathogen is determined.

Tissue antibiotic levels

Monitoring of some antibiotic levels (such as aminoglycosides or chloramphenicol) in blood and body fluids enables adequate dosage to be given and can prevent drug toxicity. Levels are usually measured at the 'peak' (20 min after intravenous and 45 min after intramuscular administration) and 'trough' (just before next dose). Whilst some antibiotic levels are measured in most laboratories (for example, gentamicin), others such as 5-fluorocytosine must be sent to other laboratories thus delaying results.

VIROLOGY

The general principles of identification of causative viruses are the same as already discussed in the Microbiology section above. Transfer samples to the laboratory as quickly as possible. If this is not possible, hold at 4 °C for not more than 48 hours. Most viruses, other than respiratory syncytial virus, do not deteriorate in this time. Samples for particular clinical situations are shown in Table 19.3. For viral diagnosis there is usually greater emphasis on antigen detection and serology than on growing the virus. Recognition of specific viral

DNA or RNA by hybridization probes is currently under rapid development and may be applied to diagnosis as well as research.

Samples

Swabs

These are collected, as described in the Microbiology section above, from lesions, nose, throat, nasopharynx, sputum, ear, eye, vagina and cervix, urethra, faeces, and rectum. However, they should be broken off in a bottle of virus transport medium. Dry swabs and bacterial transport medium are useless.

Urine

Urine can be collected as previously described, but should be placed in a sterile container and reach the laboratory within a few hours to be acceptable.

Cerebrospinal fluid

Cerebrospinal fluid (1–2 ml) should be put in a sterile container not in virus transport medium. Smaller samples greatly reduce the chance of virus isolation.

Nasopharyngeal aspirate

Nasopharyngeal secretions are required for immunofluorescence tests for viral antigens; for example, respiratory syncytial, parainfluenza, influenza A (Fig. 19.7) or B viruses. Connect a fine nasogastric tube to a secretion trap and a vacuum pump (Fig. 19.8). Introduce the tube into the nasopharynx in the same way as for a nasopharyngeal swab and ask adults to cough. Use virus transport medium to clear the tube of secretions as they are drawn into the secretion trap.

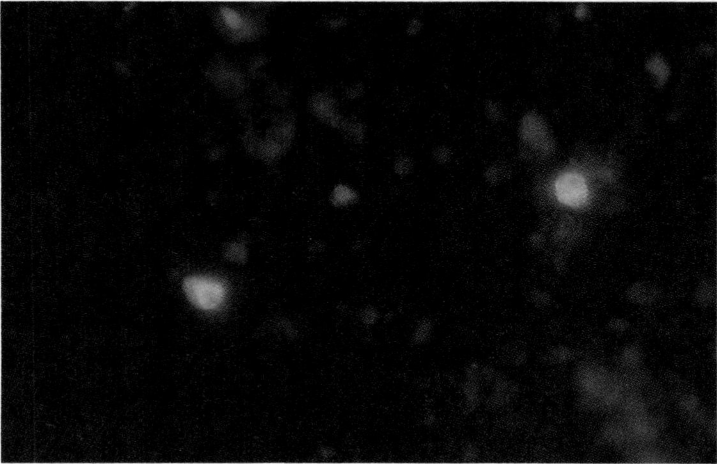

Fig. 19.7 Influenza A identified by immunofluorescence in a nasopharyngeal aspirate.

Table 19.3. *Virological samples for particular clinical conditions*

Clinical condition	Blood*	Faeces†	Throat swab in VTM‡	Other
Aseptic meningitis	Paired	Yes	Yes	CSF†
Carditis				
myo/pericarditis	Paired	Yes	—	Pericardial fluid (if available)
endocarditis	Single blood	—	—	Cardiac tissue (PM or biopsy)
Diarrhoea	—	Yes	—	—
Embryopathy	Single blood	—	Yes	—
Encephalitis				
acute	Paired	Yes	—	Brain biopsy (if available)
SSPE	Single blood	—	—	CSF
Eye lesions	Paired	—	—	Eye swab* or conjunctival scraping in VTM
Glandular	Paired	—	—	—
Hepatitis A or B	Single blood	—	—	—

Myositis	Paired	Yes	—	—
Paralytic disease	Paired	Two specimens	—	CSF†
Pyrexia	Paired	—	—	—
Post-mortem	Single blood	Yes	—	Pathological tissue
Rash/skin lesion	Paired	Yes	—	Lesion sample in VTM, or dry for electron microscopy (contact laboratory)
Respiratory	Paired	—	Yes	Nasopharyngeal aspirate
Rubella contact (pregnancy)	Paired	—	—	Minimum 3 days apart
Rubella susceptibility	Single blood	—	—	—
Stomatitis	Paired	—	Mouth swab	—

**Serological tests:* 5–10 ml clotted blood specimens. Acute and convalescent pairs 10–14 days apart to demonstrate antibody rise. Single sera not normally tested. Exceptions listed above.
†*CSF and faeces* (not rectal swabs): in dry, sterile (or clean, for faeces) containers.
‡*Swabs:* wooden shafted, broken after use into a 2 ml vial of virus transport medium (VTM) from laboratory. Send as soon as possible without freezing.

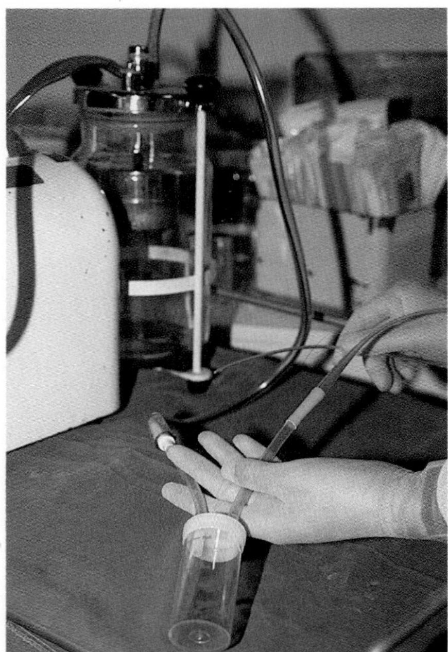

Fig. 19.8 Nasopharyngeal tube connected to secretion trap which is then attached to the vacuum pump.

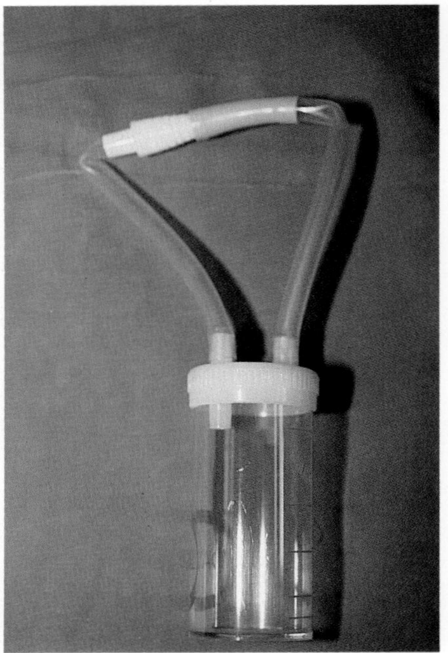

Fig. 19.9 Secretion trap for transport to the laboratory.

Remove one tube and connect the other to the vacant exit of the suction trap (Fig. 19.9). Take secretion trap with contents quickly to the laboratory.

Vesicle fluid

Since vesicles with the most fluid contain the fewest virus particles, choose the smaller vesicles for aspiration using a 1 ml syringe and fine needle. Replace cap carefully on the needle and tape it securely to the syringe. Send capped needle and syringe to the virus laboratory for electron microscopic examination. Although there may not appear to be much material, there is usually enough for examination. The electron microscopic appearances of the herpes group of viruses (e.g. herpes simplex, varicella-zoster, cytomegalovirus) are the same (Fig. 19.10), but differ from poxviruses (for example, orf, molluscum (Fig. 19.11), vaccinia). After aspirating vesicle contents one can use the needle to remove the top of the vesicle and then rotate a dry swab in the exposed base. This swab can then be smeared on a glass slide for immunofluorescence examination or put in virus transport medium for culture.

Others

Place biopsy or necropsy specimens in virus transport medium, *not* in formalin.

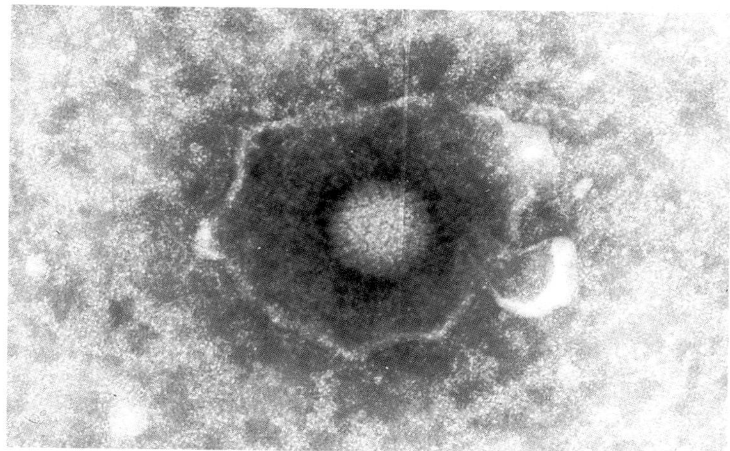

Fig. 19.10 Electron micrograph of herpes group virus.

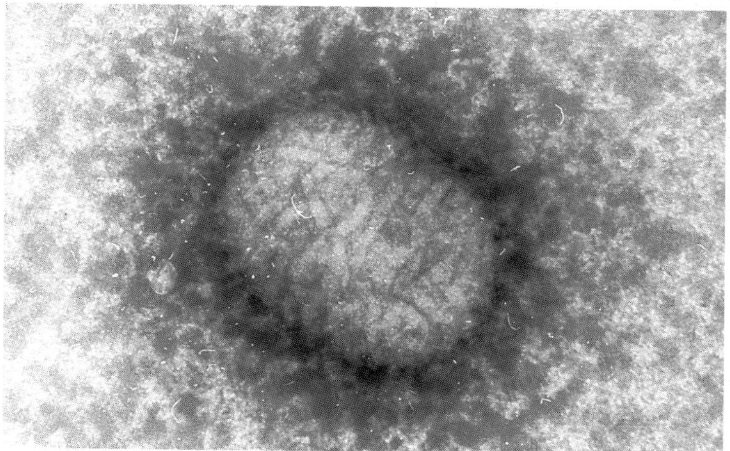

Fig. 19.11 Electron micrograph of molluscum virus.

Antigen detection

The need for rapid viral diagnosis has stimulated development of many new techniques. Monoclonal antibodies are routinely used in most laboratories to detect respiratory viruses and herpes simplex. Enzyme-linked immunoassays (ELISA) can be adapted for antigen detection (for example, chlamydia).

Polymerase chain reaction (PCR)

This extremely sensitive technique in which a specific fragment of viral DNA is amplified up to a million times before recognition by specific probes is being widely applied to virological diagnosis. Diagnostic use of nucleic and hybridization probes is likely to become widespread, but PCR will probably be limited to specialist laboratories.

Serology

Viral antibodies generally start to appear at the end of the first week of illness and reach a peak 2–3 weeks after the start of illness. Where possible, paired samples of blood should be taken, one in the first week and the second 2–3 weeks into the illness in order to reveal any diagnostic change in titre. Sensitive, rapid tests for IgM antibodies can help to diagnose most viral infections before rising titres are detectable. Virus laboratories also usually test blood for antibodies in *Mycoplasma pneumoniae* infection, Q fever, psittacosis, leptospirosis, and toxoplasmosis. Serological tests commonly used are listed in Table 19.4. Many confirmatory tests have been developed using the technique of immunoblotting (Western blotting). By this method specific proteins of an organism are separated by an electric current in a gel and then transferred ('blotted') on to nitrocellulose; when the nitrocellulose is exposed to specific antibodies, an antigen/antibody complex forms which can be stained to produce characteristic bands. This method has been of great value in confirming HIV antibody.

Table 19.4. *Principal serological tests and the type of antibody measured*

Test	Type of antibody
Haemagglutination inhibition	IgG + IgM
Neutralization	IgG + IgM
Complement fixation	IgG (mostly)
Single radial haemolysis	IgG
Radioimmunoassay	IgG/IgM
Immunofluorescence	IgG/IgA/IgM
Enzyme-linked immunosorbent assay (ELISA)	IgG/IgA/IgE/IgM
Agglutination	IgG/IgM
Western blotting	IgG/IgA/IgE/IgM

Special tests

Unusual infections

Tests for antibodies to tropical and unusual virus infections are available at reference centres to which the laboratory can forward specimens.

Investigations of outbreaks

Common outbreaks involve respiratory illness (for example, influenza or respiratory syncytial virus), gastrointestinal illness (Norwalk-like agents, rotavirus), or hepatitis (for example, hepatitis A). It is often easier for laboratory personnel to collect specimens in an outbreak, to ensure that the best samples from the most appropriate patients are taken.

Interpretation of results

Serological tests

A fourfold or greater rising titre of antibody usually signifies current infection. However, consult the virologist if there is doubt about the significance of a single high titre, stationary titres, or about the specificity and sensitivity of a test. The level of antibody which confers immunity is another area in which it may be wise to obtain specialist advice. More than one type of antibody is produced in viral infections and each may have a different significance—for example, IgM generally indicates current infection; IgG alone, past infection; mumps and parainfluenza, also herpes simplex and varicella-zoster cross-react and cannot easily be differentiated.

Virus isolation

The significance of a virus isolate must be interpreted in relation to the clinical circumstances. Viruses are commonly and often asymptomatically present in children. Isolation of a virus does not necessarily mean that the virus has caused the patient's symptoms; for example, herpes simplex from mouth swabs, cytomegalovirus in the urine of an infant, and adenovirus in faeces.

HAEMATOLOGY

Samples

Blood samples

Much routine haematology is done by machines which perform several measurements on one sample. Sequestrenated blood is usually required. For clotting studies, use a citrate anticoagulant. The amount of anticoagulant in the tube corresponds to a particular quantity of blood; if more or less blood than this is added, the result may be affected. Small bottles of anticoagulant are available for clotting studies in children. As clotting factors are labile, analysis must be done promptly.

Blood films

In suspected malaria, blood should be taken before starting treatment and examined promptly. Unfortunately, even overnight storage of blood at 4 °C may make if difficult to identify the parasites. At venepuncture, make thick and thin films unless immediate examination by the laboratory is available. For a thick film, place a large drop of blood in the middle of a glass slide and spread it to cover an area four times its original area; allow to dry thoroughly before sending to the laboratory. For thin films, place a small drop of blood at one end of the slide; with another glass slide which has a smooth edge spread the blood over 70 per cent of the slide (Fig. 19.12).

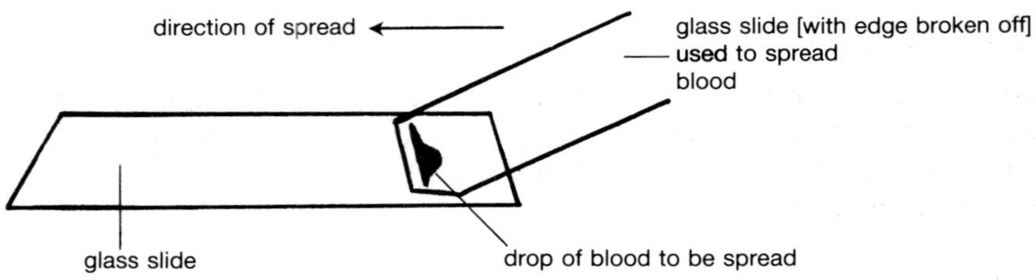

Fig. 19.12 Making a thin blood film.

Capillary samples

In infants, in patients in whom venepuncture is difficult,or in those requiring daily assessment (for example, leukaemics on chemotherapy) most investigations can be done on capillary samples which are usually taken by laboratory staff.

Cold antibodies

Where cold antibodies may be present (mycoplasma infection and infectious mononucleosis), transport the serum sample to the laboratory warm (held in the hand or placed on a beaker of warm water).

Marrow biopsy

This is often performed by the haematologist since, unless the material is properly processed, interpretation may be impossible. Generally, one should not try to aspirate more than 0.3 ml of marrow as thereafter it becomes too diluted with peripheral blood. Smears of marrow must be made quickly before clotting, and the remainder of the biopsy squirted into fixative to separate red cells from particles of marrow. Material can be cultured, as in cases of suspected tuberculosis or typhoid. Organisms such as mycobacteria or leishmania may be seen on staining.

Special tests

Blood film examination for parasites

Some parasites (for example, malaria and trypanosomes) can be seen in an ordinary blood film, but others may require special measures (for example, microfilaria only in films from capillary blood taken at night).

Buffy-coat examination

A buffy-coat preparation is a blood film made from the layer of white cells after centrifuging blood. Atypical, primitive, and abnormal cells are concentrated in this layer. The procedure may also be useful in diagnosing malaria and filariasis since these parasites are concentrated in this layer.

Bleeding problems

Coagulation factor assays, tests for the presence of coagulation inhibitors or assessment of platelet function may be necessary. In severely ill patients, diagnosis and management of disseminated intravascular coagulation can require frequent testing.

Haemolysis

Anaemia and reticulocytosis may result from bleeding as well as from haemolysis. As most haemolytic anaemias are immune-mediated, a direct agglutination (Coombs') test is done. Test for increased urinary urobilinogen, raised serum bilirubin, and reduced serum haptoglobins. Haemosiderinuria is a sign of chronic haemolysis and makes an infectious cause of haemolysis less likely.

Immunology

Blood samples for assessment of cell types (for example, T-helper: T-suppressor cell ratio) or function (for example, defective phagocytosis of neutrophils) must be anticoagulated (usually heparinized) and the cells separated soon after collection. Testing for antibody is done on serum samples, so a clotted specimen in a sterile container is also required.

Interpretation of results

Normal values

There are normal variations in blood indices with age, sex, and time of day. Therapy (such as the use of steroids) or pregnancy can produce neutrophilia. The total white-cell count may be deceptive; for example, 4×10^9/litre in an adult is apparently normal, but if 95 per cent are lymphocytes, the blood picture is abnormal.

Morphology

If blood indices are abnormal, a blood film is usually examined. Neutrophils may show increased granulation or increase in immature cells (left shift), both evidence of infection. Activated lymphocytes suggest viral infection and 'atypical' (Downey) lymphocytes (Fig. 19.13) are present in infectious mononucleosis. Characteristic red-cell disotortion can occur in renal and liver disease, and fragmented red cells are common in DIC.

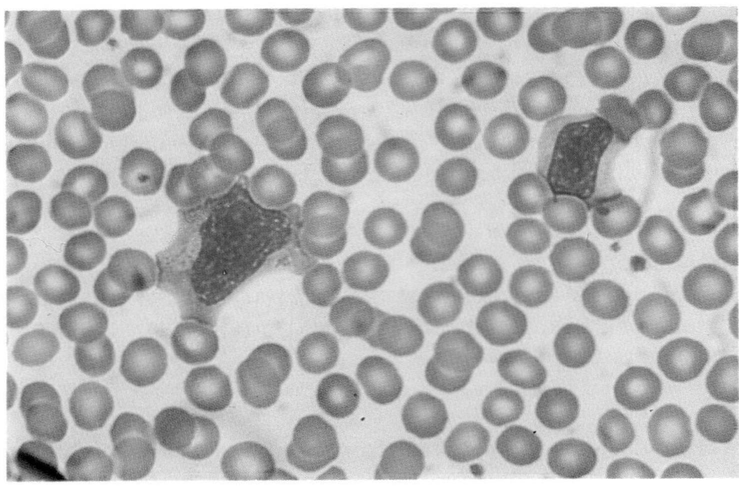

Fig. 19.13 'Atypical' (Downey) lymphocytes.

THE SIDE ROOM

Facilities for side-room tests are available in some hospitals. These tests should be supervised by experienced staff and must be subject to the same quality control and safety standards as the main hospital laboratory. If there is doubt about how to do a test, instruction can usually be arranged with the local laboratory.

Operational procedures

Urine microscopy

It is good practice to examine the urine of all patients. Place a drop of uncentrifuged fresh urine on a microscope slide and cover by a coverslip. Pyuria (over one white cell per high-power field in unspun urine or over five white cells per high-power field in spun urine) should be looked for. Casts may be present in acute or chronic pyelonephritis. Motile bacteria and red cells can also be seen.

Erythrocyte sedimentation rate

Special tubes with citrate anticoagulant are required. During tests the tubes must be vertical. After one hour, the amount erythrocytes have fallen is measured in millimetres. The test should be done within two hours of sampling, or six hours if the sample is refrigerated at 4 °C. An elevated result from this non-specific test of inflammation demands further investigation, but a normal result does not mean that the patient is normal. It can be useful in monitoring therapy in patients with tuberculosis or bacterial endocarditis.

White cell count

In some hospitals the house officer is expected to be able to do a white cell count. Make a 1 in 20 dilution of blood by adding 20 μl of blood to 0.38 ml of diluting fluid (2 per cent acetic acid) in a white cell counting pipette (*not* a mouth pipette). Rotate pipette to mix contents and allow red cells to lyse. Count one hundred cells in as many 1 mm^2 areas (0.1 μl in volume) as required and calculate the average for each 0.1 μl volume (a Neubauer chamber has nine 1 mm^2 areas). If y cells are counted in 0.1 μl, then:

$$\text{White cell count per litre} = y \times 10 \times 20 \text{ (dilution)} \times 10^6.$$

Further reading

General

Christie, A. B. (1987). *Infectious diseases. Epidemiology and clinical practice*, Volumes 1 and 2 (4th edn). Churchill Livingstone, Edinburgh.

Evans, A. S. (ed.) (1989). *Viral infections of humans* (3rd edn). Plenum Press, New York.

Evans, A. S. and Brachman, P. S. (ed.) (1991). *Bacterial infections of humans. Epidemiology and control* (2nd edn). Plenum Press, New York.

Mandell, G. L., Douglas, R. G., and Bennett, J. E. (1990). *Principles and practice of infectious diseases* (3rd edn). Churchill Livingstone, Edinburgh.

Weatherall, D. J., Ledingham, J. G. G., and Warrell, D. A. (ed.) (1987), *Oxford Textbook of Medicine* (2nd edn, Sect. 5). Oxford University Press, Oxford.

Levine, D. P. and Sobel, J. D. (1991). *Infections in intravenous drug abusers*. Oxford University Press, Oxford.

Chapter 1

Mimms, C. A. (1987). *The pathogenesis of infectious disease* (3rd edn). Academic Press, Orlando, Florida.

Tyrrell, D. A. J. (1982). *The abolition of infection. Hope or illusion?* Rock Carling Fellowship Lecture. Nuffield Provincial Hospitals Trust.

Velimirovic, B., Greco, D., Grist, N. R., Mollaret, H., Piergentili, P., and Zampieri. A. (1984). *Infectious diseases in Europe. A fresh look*. World Health Organization Regional Office for Europe, Copenhagen.

Chapter 5

Rains, A. J. H. and Mann, C. V. (1988). *Bailey and Love's Short practice of surgery* (20th edn). H. K. Lewis, London.

Sherlock, S. (1989). *Diseases of the liver and biliary system* (8th edn). Blackwell, Oxford.

Chapter 10

Adler, M. W. (1990). *ABC of sexually transmitted diseases* (2nd edn). British Medical Association.

Chapter 12

Roitt, I., Brostoff, J., and Male, D. K. (1989). *Immunology* (2nd edn). Churchill Livingstone, Edinburgh.

Mindel, A. (1990). *AIDS, a pocket book of diagnosis and management* (1st edn). Edward Arnold, London.

Chapters 14 and 15

Bell, D. R. (1985). *Lecture notes in tropical medicine* (2nd edn). Blackwell, Oxford.

Walker, E. and Williams, G. (1992). *ABC of healthy travel* (4th edn). British Medical Association.
Wilson, M. E. (1991). *A world guide to infections*. Oxford University Press, Oxford.

Chapters 16 and 17

Benenson, A. S. (Ed.) (1990). *Control of communicable diseases in man* (15th edn). American Public Health Association.

Chapter 18

British National Formulary (revised every six months). Joint Formulary Committee of British Medical Association and Pharmaceutical Society of Great Britain.
Lambert, H. P. and O'Grady, F. (1991). *Antibiotics and chemotherapy* (6th edn). Churchill Livingstone, Edinburgh.
WHO Expert Committee (1990). *The use of essential drugs*. World Health Organization Technical Report Series 796.

Chapter 19

Collee, J. G., Duguid, J. P., Fraser, A. G., and Marmion, B. P. (1989). *Practical Medical Microbiology* (13th edn). Churchill Livingstone, Edinburgh.
Shanson, D. C. (1989). *Microbiology in clinical practice* (2nd edn). Butterworths, London.

Index